Lecture Notes in Biomathematics

ctd. on inside back cover

Lecture Notes in Biomathematics

Managing Editor: S. Levin

88

R.H. Bradbury (Ed.)

Acanthaster and the Coral Reef: A Theoretical Perspective

Proceedings of a Workshop held at the Australian Institute of Marine Science, Townsville, Aug. 6—7, 1988

 Springer-Verlag Berlin Heidelberg GmbH

Editor

Roger Bradbury
National Resource Information Centre
GPO Box 858
Canberra ACT 2601, Australia

Mathematics Subject Classification (1980): Primary: 92A15
Secondary: 34C05, 53B40, 62H30, 65C20,
68Q45, 68Q50, 68T05

ISBN 978-3-540-53501-0   ISBN 978-3-642-46726-4 (eBook)
DOI 10.1007/978-3-642-46726-4

2146/3140-543210 – Printed on acid-free paper

PREFACE

In August 1988, the Sixth International Coral Reef Symposium was held in Townsville resulting in an influx of most of the world's coral reef scientists to the city. We seized this opportunity at the Australian Institute of Marine Science to run a small workshop immediately before the symposium on the outbreaks of the crown-of-thorns starfish, *Acanthaster planci*. We invited that small band of mathematicians who had been modelling the phenomenon, (and who may not have normally attended an international meeting so thoroughly dedicated to natural science) to meet with those scientists who had been been actively working on the phenomenon in the field.

John Casti notes in his delightful new book *Alternate Realities* (Wiley, 1989): 'If the natural role of the experimenter is to generate new observables by which we know the processes of Nature, and the natural role of the mathematician is to generate new formal structures by which we can represent these processes, then the system scientist finds his niche by serving as a broker between the two.'

I think our book shows the fruits of that brokerage through the wide range of models explored within its pages, the high level of collaboration and interaction across disciplines evident in the individual papers, and in the emerging synthesis that reflects a far deeper understanding of this complex phenomenon than was possible even a few years ago.

We ran the workshop quite informally, and many of the presentations were very much 'works in progress'. So it has taken some time to finalize them to a standard suitable for publication. For this I must thank the reviewers of each of the papers for the particularly constructive way in which they worked. We strove to make the book accessible to both natural scientists and mathematicians. For most of the papers, I organized two reviewers, one 'mathematically-inclined', the other 'biologically-inclined'. I asked these reviewers to consult with each other and prepare a consensus review. I hope you agree this editorial technique has led to more widely accessible document, with less diminution in rigour, than might be expected given the range of material covered.

I also thank Marcel Dekker, Inc. for permission to republish the paper by Antonelli *et al.* on nonlinear prediction of crown-of-thorns outbreaks, and Oxford University Press for permission to republish the paper by Antonelli *et al.* on the diffusion-reaction-transport model for large-scale outbreak waves. These papers were presented at the workshop and are included here for completeness and because they are central to our current understanding of the phenomenon.

The smooth running of the workshop was a tribute to Rosalie Buck's organizational skills, and the editorial process was made much more enjoyable and efficient with Bruce Macdonald's cheerful assistance.

Roger Bradbury
Townsville, 1990

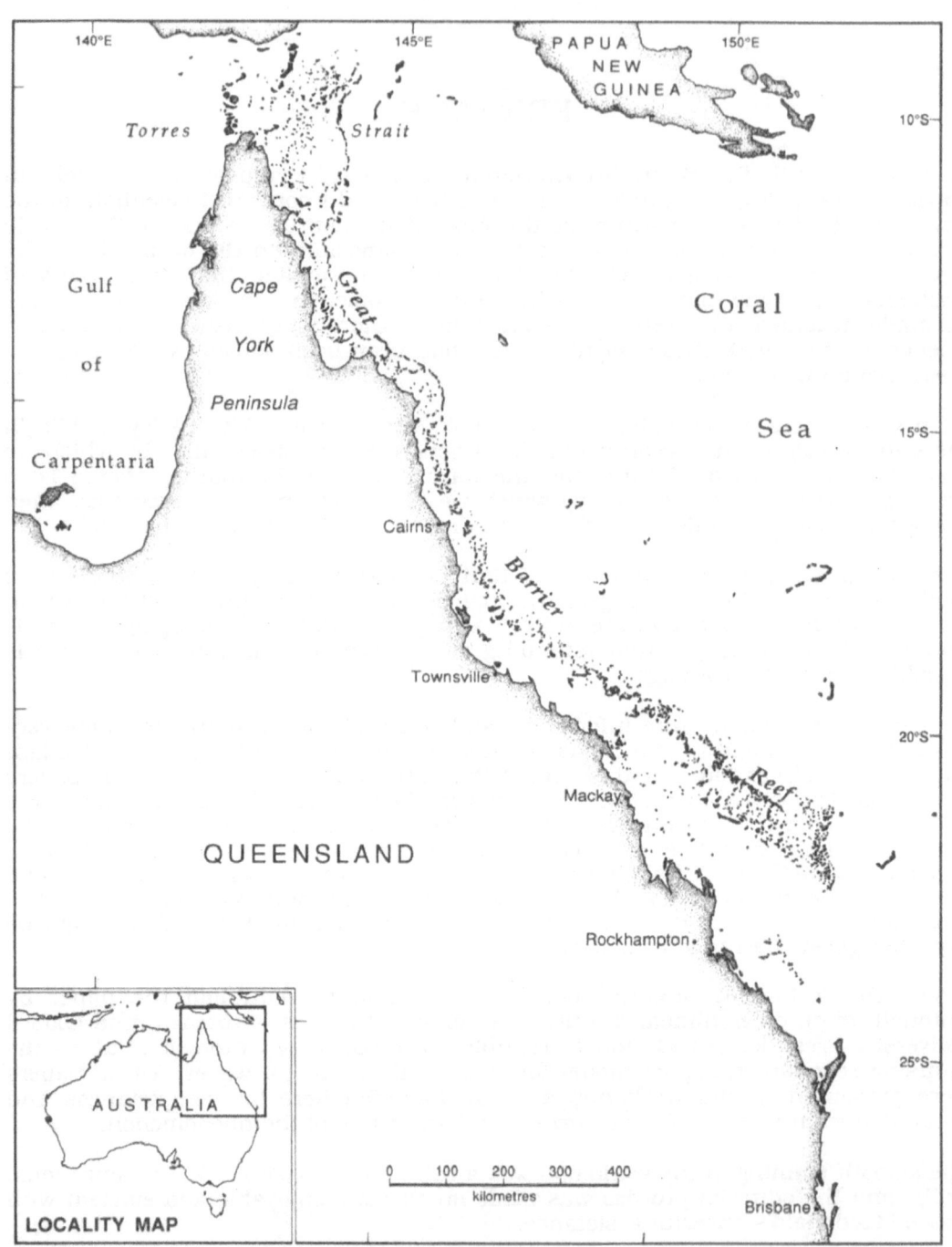

The Great Barrier Reef

TABLE OF CONTENTS

ON MODELLING

JOHN L. FARRANDS

20 The Boulevard, Glen Waverley, Victoria 3150

Under the title of The *Acanthaster* Phenomenon: a Modelling Approach, the organisers of this workshop have invited you to meet to:

. review all current models of the phenomenon, and to identify the results to date and significant gaps in our understanding,

. review some other techniques currently applied to other systems in order to identify those techniques with a high potential of application to the phenomenon,

. to develop an integrated account of the current understanding of the phenomenon of immediate use to ecosystem mangers.

It is good that we have with us so many scientists who are working elsewhere on modelling, that you intend to address ecological modelling in general and the modelling of the *Acanthaster* phenomenon in particular. Of course, I am not wholly sure just what is the *Acanthaster* phenomenon. By it I suppose we mean the unpopular dynamics of the beast which is seen as a "bad thing". Is it different from other outbreaking phenomena?

Many of those present are engaged in modelling work which is far different from this sort of topic, but it is hoped that by transfer of ideas we shall enhance our ability to achieve a socially useful result. During our meeting we shall not doubt be exposed to some very sophisticated models and some equally sophisticated theory on the subject.

With your permission, I intend to use (abuse perhaps?) my position to offer a few thoughts which I think are relevant at least from the point of view of the Chairman of AIMS Council. A sort of emperor's clothes view.

For my part, I have no experience of modelling ecological systems. The closest, I suppose, is in the association with modelling of military systems beginning with simple ones like the Lanchester equations, and ending with the massive force structure studies of the kind conducted by Tony Taylor. In spite of that disclaimer, I put it to you that as an experimental physicist and engineer, I have been modelling all my life.

All mathematical physics, and perhaps all science, is modelling from the plane of reality to some other plane in which the mind can explore the phenomenon, and profit from its understanding. At the risk of seeming trivial, I will take a simple example and draw some lessons from it.

My example is the simple rigid pendulum. The elementary model is a two term second order differential equation which is far removed from reality, yet is one of the most useful results we have. We apply it conceptually in all sorts of fields where there appears to be or there is approximately a linear restoring force. The simple sinusoidal solutions help us to think about vibrating beams, infrared spectroscopy, walking speeds, musical instruments, radio oscillators, the list is endless.

As we become more sophisticated and try to approach reality a little closer, we add in the second moment of inertia of the bar, air resistance, friction at the pivot, temperature and so on, we have become specific to the design of pendulums, or if you will, to clock designing. Our perceptions are no longer general; other applications need new solutions - circular functions become Bessel functions and so on. What we have gained in precision, we have lost in generality.

At the next higher level of sophistication, we shall transform the equations into more general forms with Lagrangian and Hamiltonian formulations. This returns the generality and permits us to apply the original thinking more accurately to things like molecular spectra, and to extend its application to new phenomena such as the libration of the moon. The price we pay for its elegant shorthand is to have eliminated its usefulness to all but the few and possibly to eliminate the insights.

You may be wondering why I am labouring this to a group which includes so many mathematicians. Well I am doing so because I think it is the simplest useful model that we can all understand, and I wish to draw some conclusions from it.

Let us ask ourselves why we want to produce models generally. Initially, I believe, we wish to map the phenomena of the perceived world on to a plane where it can be manipulated intellectually so that our minds find it comfortable in terms of our previous experience. If we can do that, we say we understand the phenomenon. If we cannot, we re-examine our previous experience, and possibly modify our interpretations of that earlier experience. If we still fail then we invoke new explanations, paying due attention to Occam. If we do evoke a new principle or hypothesis then we test it to the limit. (Or someone else does it for us!).

If our model merely explains the observations in terms of our previous knowledge, we have the smug satisfaction of knowing that our previous world view does not yet need modification and that is good. If it leads *necessarily* to a new testable hypothesis, then we are on the road to fame and honour, and that is better. If it yields a new view of the world which is useful, then that is best.

It has been said that nobody believes a theoretical result except the author, and everybody believes an experimental result - except the author. I believe that to be generally true, but in these days perhaps too much trust is put in modelling especially when a computer solution is being offered. The notorious global model of the Club of Rome is evidence of this.

One of the many questions which the community is posing to the marine science community is to provide practical advice on and answers to the following questions:

1. Are *Acanthaster* outbreaks a periodic natural phenomenon?

2. Is *Acanthaster* an essential part of a healthy system in spite of its unattractiveness to tourism?

3. If not "natural" and not essential, are outbreaks the result of human misbehaviour?

4. Are there any means available to us to control them; are they affordable?

Some of us here have chosen to attack this problem from a modelling viewpoint. This is a valid approach so long as we keep our wits about us and recognise the limits we face in experimental data, complexity, and testability.

From preliminary examination of your literature, I believe that the ecological equivalents of the modelling of a simple pendulum are to be found in work of the 1920s. Lotka's work on competition, Volterra's on predator-prey relationships, Elton's observations on swarming, and Lanchester's exchange equations are at the same level of simplistic generality - even simpler, being usually first order - yielding the same sorts of insights as we got with the pendulum. They remain, I suggest, the basis of action of practical men.

As we increase the number of variables, so we generate more and more simultaneous differential or difference equations, and so we lose generality and the chance of obtaining new insights. Let us not forget that we also introduce more and more assumptions. Some of these assumptions are quite untestable *a priori* and must wait for their justification until the model is tested by experimental intervention.

Let us now return to *Acanthaster* and the Reef. It would be fair, would it not, to say that for all practical mathematical purposes, the biological associations of *Acanthaster* or a given coral have to be regarded as almost infinite, and the chemical and physical associations also of an unmanageably high order. Work here has already found that complexity of distributions is not reduced by reductions in geometric scale.

If you accept what I have just said, you would have to agree with me that a complete model of the reef is impossible and so is a complete model of the *Acanthaster* phenomenon. The essential first point is to determine the minimum number of parameters which impact on the gross behaviour of the animal. Now it is interesting to observe that the participants in this workshop are largely mathematicians. Let us hope that we do not stray too far from reality.

I am aware that many of you have the capability to bring to the problem mathematical tools of great power and generality. I would like us to leave here confident that the tools you are going to apply or recommend have been tested in the real world and that data are available or even obtainable to test the application.

We face a dilemma. Von Forerster has made the point that a reductionist approach to problems like ours chops the system into smaller and smaller pieces until they are small enough to understand, at which point one knows a great deal about nothing, and nothing about the interrelationships of the smaller pieces. It lets you know everything about nothing. Holism on the other hand leads to knowing nothing about everything.

I believe that in coming here today, you have involved yourselves in more than an interesting talkfest on the arcana of mathematical modelling, but a serious practical problem. There are several things to be tested - the data base, the mathematical rigour, and the models themselves. An important step is to devise tests by predictions of smaller systems. Perhaps *Acanthaster* is too difficult by its very nature; its cryptic juveniles and unknown predators may be such that our data with which we have to work is wholly inadequate in some areas (*Pace* to Peter Moran). If that were to be so we should test our models on more overt animals. Whether in the open sea or in aquaria is to be decided, but testing at all stages of model development seems to me to be a *sine qua non* of progress or even continuation.

In this plea for continuous testing, I remind you that there are at all stages, two things to test. They are the model itself and the input data.

My penultimate point is to remind you that the end users of any model require that they understand it, can modify it, and believe it. Remember that most decision makers have been educated - *if at all* - in the liberal arts, and are generally not numerate.

It is for you, the experts, to suggest courses of action. I would remind you however, that we have been working on the *Acanthaster* problem for more than six years, and between AIMS and COTSAC we have probably spent in excess of $2,922,000 including $579,729 on modelling. In the commercial world, I would expect a lot of successful R&D for that. Society will not wait much longer, nor in its present mood will it be willing to continue to spend on a topic of academic interest unless it sees a general principle arise which is seminal or of wider application.

We have an obligation now to show that there is a prospect of success in our practical problem-solving by modelling. I personally believe that we can do this only by demonstrations of success, either partial or total, by testing and retesting simple and then increasingly complex models. Indeed one useful addition to our program might well have been a discussion on the validation of models at various stages of their evolution.

I doubt that there is a *science* of modelling because modelling is a human construct, and is not part of the real world. It is, as I said earlier, a tool for the extension of human perception and action. Research into modelling is therefore applied research and must be evaluated in terms of its usefulness in other perceptions of the real world.

DISPERSAL AND CONTROL MODELS
OF *ACANTHASTER PLANCI* POPULATIONS
ON THE GREAT BARRIER REEF

R.E. REICHELT[1]

Australian Institute of Marine Science, Townsville, Queensland 4810

Abstract. This paper summarizes the large scale distribution of *Acanthaster planci* on the Great Barrier Reef (GBR), illustrating that there have been 2 periods in the last 25 years when the number of outbreaks has been high. During each of these 2 periods there was a southward moving wave of outbreaks, presumably being driven by the currents on the continental shelf. These currents move south, on average, during the starfish spawning season. A simple graphical model illustrates this movement of adult populations caused by the advective transport of starfish larvae.

The criteria used for determining when starfish control is feasible have never been clearly defined which has resulted in *ad hoc* management decisions based on limited data. A simulation model of tactical starfish control is presented here to narrow the terms of the debate and provide a starting point for more rigorous assessment of management options. The model illustrates the importance of determining the total size of an outbreak as soon as possible before control is initiated. It is recommended that future control efforts be designed to enhance our knowledge of the critical parameters used in deciding when tactical control is feasible in order that future management options can be defined more rigorously.

INTRODUCTION

The first detailed mathematical modelling study of *Acanthaster planci*, the crown-of-thorns starfish, was an analytical study of the interaction between the starfish and a 2-species coral community (Antonelli & Kazarinoff, 1984). Subsequently there have been modelling studies of the starfish/coral interaction (Bradbury *et al.*, 1985; Reichelt *et al.*, 1990), starfish-predator/starfish interaction (McCallum, 1987), coral population dynamics (Done, 1988), starfish movement around a reef (Reichelt *et al.*, 1988) and starfish dispersal between reefs (Antonelli *et al.*, 1989; James *et al.*, this volume). In addition to the models which include

---

[1]Present address: Bureau of Rural Resources, Canberra, ACT

starfish explicitly, there has been considerable effort on modelling the physical oceanography of the Great Barrier Reef (GBR) which bears directly on the mechanism of dispersal of planktonic starfish larvae (Black, 1988; Williams *et al.*, 1984; Wolanski & Hamner, 1988).

Prior to these studies a number of non-mathematical models, hypotheses concerning outbreak triggers, were described (Potts, 1981; Moran, 1986). All of this modelling work has emphasized gaps in our knowledge of *A. planci* and has been useful for setting directions in experimental work. In contrast to the theories of outbreak triggers, the problem of human control of starfish populations has received very little attention from the modellers. One reason for this lack of interest may stem from the high degree of both political and scientific uncertainty surrounding the control issue. It is not clear whether the political will to undertake control exists and, possibly as a consequence, the scientific effort in this area has been minimal to date (Zann & Weaver, 1988).

The only large scale control effort, made to date by the Japanese in the Ryukyu Islands, was not very successful (Moran, 1986; Yamaguchi, 1986) but the bounty method employed there has the disadvantage of converting the control effort into a fishery rather than achieving the main aim of reducing the starfish population in a local area.

This paper has two aims. The first is to present a brief overview of the available data on the distribution of *A. planci* on the GBR and describe a general model for the progression of starfish outbreaks along the GBR over long time scales. The second aim is to show how a simple model of starfish control can be useful in highlighting the management problems surrounding starfish population control by humans, which is a difficult issue politically, scientifically and practically.

DISTRIBUTION OF *A. PLANCI* ON THE GBR

The available data on historical distributions of *A.planci* on the Great Barrier Reef have been compiled into a single database held at the Australian Institute of Marine Science. This database has been described by Reichelt *et al.* (in press) and contains about 1500 records of highly variable quality. The observations are taken from the published scientific survey data, scientists' field notes, and reports by other users of the GBR such as tourist operators. The latter reports were compiled by the Great Barrier Reef Marine Park Authority (GBRMPA).

The notion of southward moving waves of populations, which move by advective transport of the planktonic larvae rather than adult migration, was proposed by Kenchington (1977). The wave-like movement is apparent in Figure 1. There appears to have been 2 distinct periods of activity, that is, 2 waves of outbreaks, and in both cases the region of maximum activity has been the central GBR region between 17°S and 18°S (Reichelt *et al.*, in press).

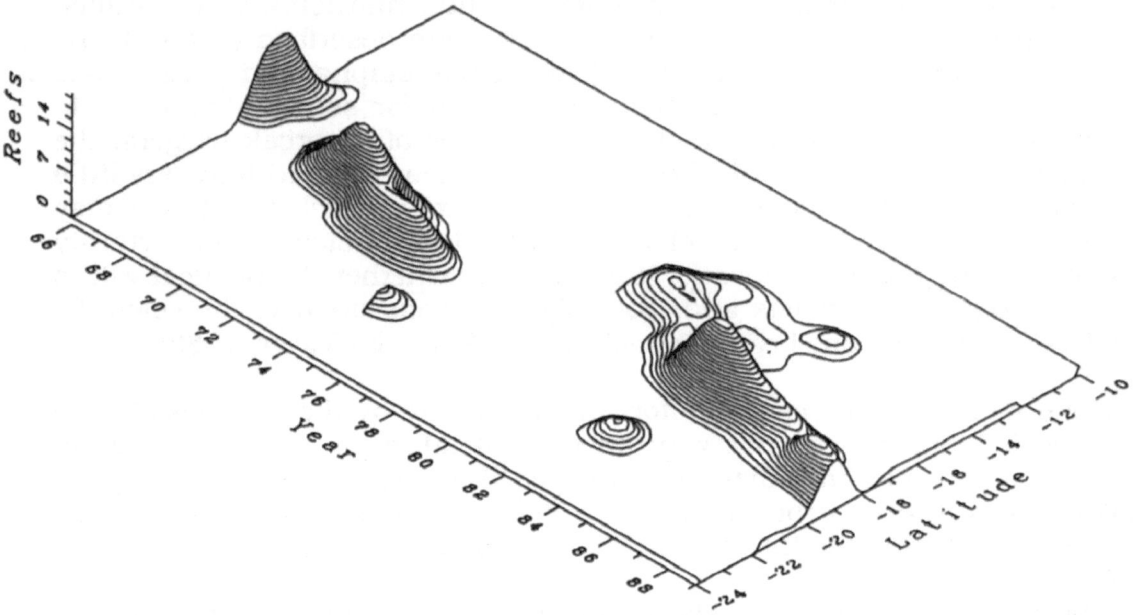

Figure 1. Number of reefs within each latitude band recorded as outbreaking since 1966, shown as a 3 dimensional plot of latitude vs. year vs. number of outbreaking reefs (vertical axis). The GBR is treated as a linear system arranged on an approximately North-South axis.

A simple model of this wave progression is shown diagrammatically in Figure 2. This model is essentially a graphical representation of Kenchington's (1977) hypothesis. However, there is nothing in this model which explains why the outbreaks have started where they seemed to have done. Nor does it explain why the outbreaks should appear as a wave rather than as a more chaotic set of events given the degree of variability in the oscillations of longshore currents that occur in summer in the central GBR (Andrews & Furnas, 1986).

In contrast with the simple model above, the dynamics of the mathematical model presented by Antonelli *et al.* (1989) includes the presence of a wave as a consequence of its structure. Their model is effectively a reaction process (the starfish-coral interaction) with transport and diffusion terms added to represent the larval transport mechanisms.

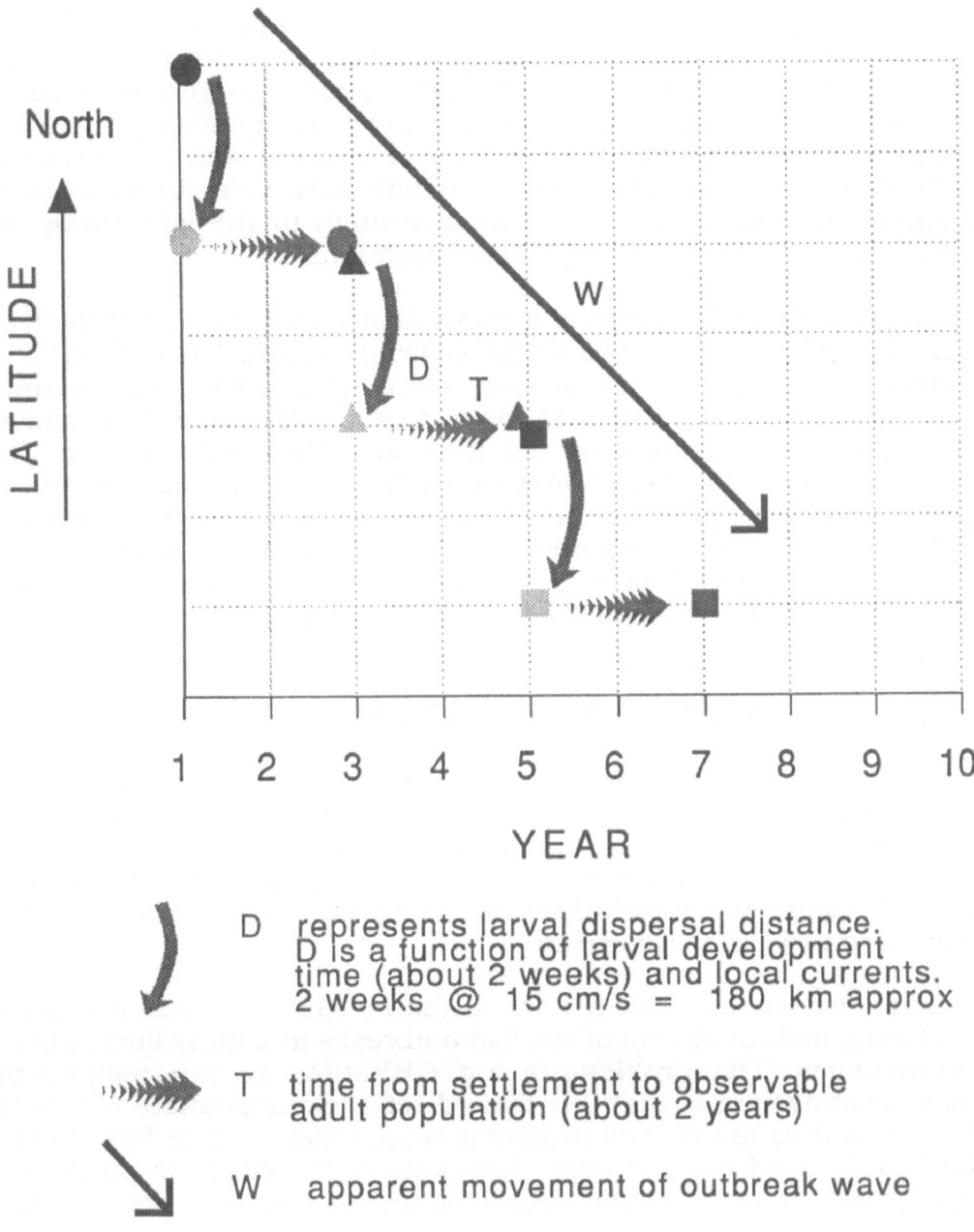

Figure 2. A diagrammatic model of the southward moving wave of starfish outbreaks.

The historical data describing the initiation of outbreaks are relatively poor. Kenchington (1977) suggested that in the late 1950s and early 1960s a region to the north of Cairns served as a 'seed area' where starfish populations reached a kind of critical mass triggering the series of outbreaks to the south. James *et al.* (this workshop) have presented a probabilistic model based on the exchange of larvae between reefs and reef distributions which supports Kenchington's seed idea. Unfortunately, the empirical data for the late 1950s through to the late 1960s are insufficient to test the 'seed' hypothesis adequately.

In spite of the limited amount of starfish distribution data prior to 1979, the similarity between the two waves shown in Figure 1 is striking. It is tempting to suggest that the system is cycling in some stable fashion which will lead to another wave in the mid-1990s. Although the dynamics of the starfish-coral system are much better understood now than they were at the end of the first observed wave in the mid-1970s, it would be highly speculative to propose a third wave in the 1990s. Such a prediction is based mainly on the assumption that the initiation of outbreaks in the central section of the GBR is linked closely to the restoration of coral cover through coral settlement and/or coral growth.

CONTROL OF *A. PLANCI* ON THE GBR

Human intervention in the course of starfish outbreaks has been a controversial issue for many reasons. The issues include the possibility that interference in the course of an outbreak may disrupt the ecology of the coral reef system in some unpredictable way. Another view is that large scale control is not technically feasible and should not be attempted if it is bound to fail in the long run.

The Great Barrier Reef Marine Park Authority (GBRMPA) has a policy of undertaking tactical control of starfish outbreaks in a fairly limited set of circumstances. The problem facing GBRMPA, as the responsible management agency, is that this limited set of circumstances is very ill-defined and thus the control programs undertaken to date have tended to be based on *ad hoc* decisions. In addition, the strategic purpose in adopting this tactic is highly debatable and beyond the scope of this paper.

As a first step in putting bounds on the problem of deciding when control attempts are likely to be feasible, and as a guide to data collection in the future, the following simple model of starfish control is presented.

SIMULATION MODEL OF STARFISH CONTROL

The aim of this simple model is to examine the interaction between the following variables: the initial size of the starfish population; the total amount of control effort, which is equivalent to the total cost of attempting to control an outbreak; the minimum coral cover to which the system is reduced by the outbreak; and the effort expended per time step in attempting to control the outbreak, which is the rate of control.

The starfish-coral interaction model of Reichelt *et al.* (1990) was used as a basis on which some human intervention was overlaid. This non-spatial model has coral growth leading to full recovery of coral cover over 15 to 20 years, and an age structured population of feeding and breeding starfish. The model uses a time step representing 2 months. For the purposes of the control simulation, one large recruitment of starfish was allowed to occur and starfish were detected by humans (and control initiated) when the starfish reached adult stage. No secondary recruitment was allowed to occur.

The starfish are assumed to feed at a rate of $1m^2$ of coral/2 months/starfish. Initial coral cover is 50% of the total area, which is 100 ha. One unit of CONTROL EFFORT is 1 diver working 5 days per 2 months. The control effort is initiated when the number of adult starfish in the system exceeds 10,000 and control is stopped when the number being killed falls below 10 starfish/diver/day. In order to account for the fact that divers can collect a certain maximum number of starfish per day, regardless of total starfish abundance, a maximum harvest rate of 250 starfish/diver/day was reached when there was 10,000 or more starfish. Below 10,000 the rate of collection was reduced linearly in proportion to starfish abundance. If the number killed is K and the number of starfish present is N then,

$$K = 250*N/10000 \text{ for } N < 10,000, \text{ and } K = 250 \text{ for } N > 10,000.$$

A total of 90 separate simulations were run. Two parameters were set as initial conditions for each run - these were the initial size of the outbreak (numbers of starfish) and the effort which would be expended per time step in controlling the outbreak (units of effort).

The initial values are shown on the axes of the graphs in Figures 3 and 4, and the results of the simulations are presented in the body of these plots. Figure 3 indicates the total effort expended on each outbreak from initiation through to cessation of control and Figure 4 shows the minimum coral cover reached. In other words, Figure 3 shows the cost of the control program given the initial conditions indicated on the x and y axes, and Figure 4 shows the impact of the outbreak on the coral cover under the same conditions as those plotted for Figure 3. The duration of each outbreak is not represented here, but varies widely depending on both the size of the outbreak and the control effort expended.

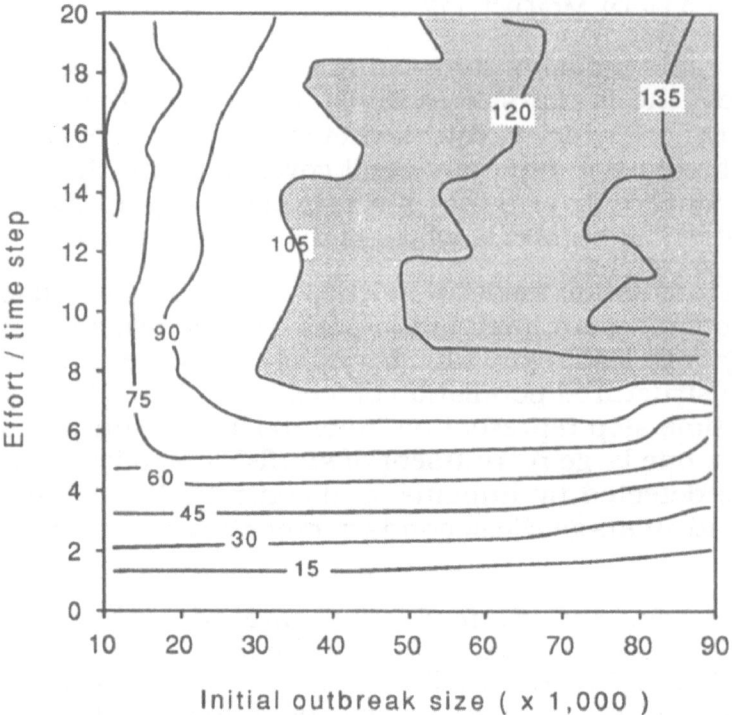

Figure 3. Total effort expended (units of effort) is contoured in the space defined by the x axis, initial outbreak size (thousands of starfish), and the y axis (units of effort/ time step).

The shaded portions of Figures 3 and 4 can be arbitrarily assigned to represent critically important regions. In Figure 3 the shaded area shows where more than 105 units of effort were expended, that is, a large and presumably costly effort. In Figure 4 the shaded area shows where coral cover was reduced from 50% down to 25% or less, in spite of the control attempt, representing a failed control program.

Real managers will need to assign politically and/or logistically realistic values to these parameters. These figures were selected to show that there are areas where these shaded regions overlap if Figures 3 and 4 are overlaid. This area of overlap represents that part of the parameter space where the control effort is considered very costly and where the control effort has failed in spite of the expense - a circumstance which environmental managers would wish to avoid (Yamaguchi, 1986).

Note that the contours in Figure 3 are not arranged on a smooth gradient, as they are in Figure 4. This non-linearity is introduced by the fact that a large effort early in an outbreak shortens the time taken to kill most of the starfish, and lessens the impact on the corals also. Figure 3 also shows that a very low effort results in a low total cost which is independent of the initial outbreak size because most of the coral is eaten reasonably quickly.

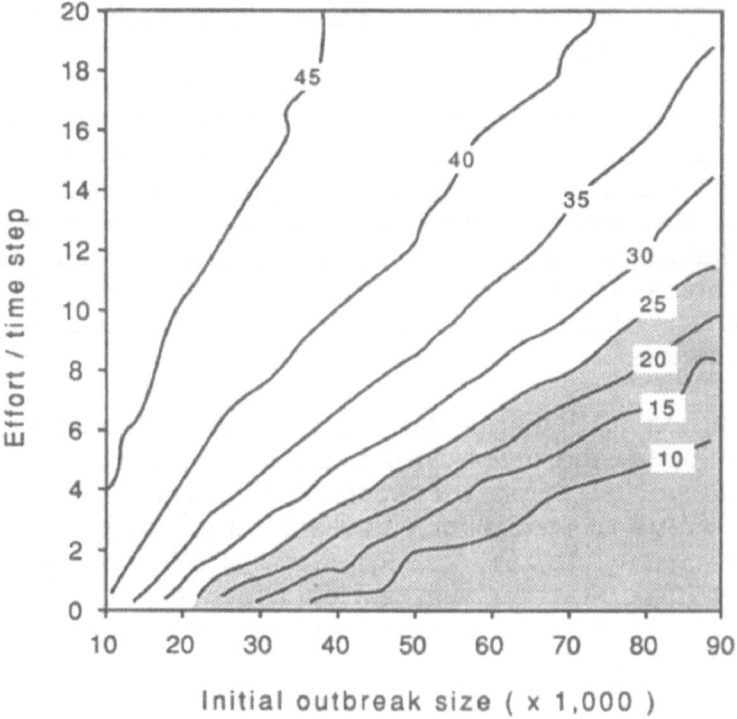

Figure 4. Minimum coral cover reached after the outbreak (percent cover) is contoured in the space defined by the x axis, initial outbreak size (thousands of starfish), and the y axis (units of effort/ time step).

DISCUSSION

The question arises as to whether resource managers could use such a model to define the bounds of their management options a little more clearly than is done at present. It is apparent that the most critical parameter, which must be measured very rapidly, is the initial size of the outbreak. In the model presented here the success of subsequent management tactics depends on this completely. It may be possible to extend the model described here to allow managers to assess the initial state of a reef and its starfish population, then apply some predetermined criteria to decide whether control should be initiated. In other words the manager could look up, on figures similar to those shown here, whether the reef in question is in a parameter region that indicates control is likely to succeed without prohibitive cost.

In terms of improving the reality of the model, it is easy to conceive of some important parameters that may be quite difficult to quantify. Some examples which of great importance to coral reef managers are the aesthetic value, tourism value, and environmental value of a particular reef.

In the real world there are several other parameters that are not represented in the model but would be equally important as initial outbreak size. For example, the spatial extent of the starfish population, not simply the number of starfish but the dispersion pattern, would be critical. In addition, the structural complexity of the reef at geomorphological scales, such as patch reefs size, lagoon structure, and reef slope structure would have a major impact on the ability of divers to search for and kill starfish and on the degree of coral mortality caused by the starfish outbreak. The function describing the kill rate by divers used in the model here is plainly simplistic and would be influenced strongly by these 2 parameters.

If a reliable method of detecting large populations of relatively small starfish is developed, managers will need to consider the option of attempting to kill starfish before they reach spawning age. The benefits of investing large sums of money at this early stage may outweigh the (assumed) high cost when the consequences of allowing spawning to occur in a very large starfish aggregation are taken into account. A probabilistic model of starfish spawning and recruitment may allow this option to be assessed also.

CONCLUSIONS

If the management strategy and decision making process is to be made more rigorous, models such as the one described here should be developed to allow various management options to be assessed. An assessment of the robustness of the model predictions would be part of a sensitivity analysis of the models' parameters.

In implementing a tactical starfish control policy, the future control efforts should be designed in a way that provides data for such models. The control effort should be expended on a tightly restricted set of reef types and then control models may be developed and validated by directing control efforts to appropriate types of outbreaks on such reefs.

Acknowledgments. My colleagues W. Greve (BAH) and R. Bradbury (AIMS) are contributing to the control modelling study and this collaborative work is still in progress; the large scale dispersal work was done with R. Bradbury and P. Moran (AIMS); Financial and logistic support were provided by the Australian government (Australia-Germany Bilateral Science Grants) and the Biologische Anstalt Helgoland in Hamburg.

REFERENCES

Andrews, J.C., Furnas, M.J., 1986. Subsurface intrusions of coral sea water into the central Great Barrier Reef - I. Structures and shelf-scale dynamics. Cont. Shelf Research 6: 491-515.

Antonelli, P.L., Kazarinoff, N.D. 1984. Starfish predation of a growing coral reef community. J. Theor. Biol. 107: 667-684.

Antonelli, P.L., Kazarinoff, N.D., Reichelt, R.E., Bradbury, R.H., Moran, P.J., 1989. A diffusion-reaction-transport model for large-scale waves in crown-of-thorns outbreaks on the Great Barrier Reef. J. Math. Appl. Med. Biol. 6: 81-89.

Bradbury, R.H., Hammond, L.S., Moran, P.J., Reichelt, R.E., 1985. Coral reef communities and the crown-of-thorns starfish: evidence for qualitatively stable cycles. J. Theor. Biol. 113: 69-81.

Black, K.P., 1988. The relationship of reef hydrodynamics to variations in free-borne larval numbers on and around coral reefs. Proc. 6th Int. Coral Reef Symp., Townsville, 1988, Vol. 2, pp. 125-130.

Done, T.J., 1988. Simulation of recovery of pre-disturbance size structure in populations of *Porites* spp. damaged by the crown of thorns starfish *Acanthaster planci.* Marine Biology 100: 51-61.

James, M.K., Dight, I.J., Bode, L. (this volume) Great Barrier Reef hydrodynamics, reef connectivity and *Acanthaster* population dynamics. In: Bradbury, R.H. (ed.) The *Acanthaster* phenomenon: a modelling approach. Springer-Verlag, Berlin.

Kenchington, R.A., 1977. Growth and recruitment of *Acanthaster planci*(L.) on the Great Barrier Reef. Biol. Cons. 11: 103-118.

McCallum, H.I., 1987. Predator regulation of *Acanthaster planci.* J. Theor. Biol. 127: 207-220.

Moran, P.J., 1986. The *Acanthaster* phenomenon. Oceanogr. Mar. Biol. Ann. Rev. 24: 379-480.

Potts, D.C., 1981. Crown-of-thorns starfish - man induced pest or natural phenomenon? In: R.L. Kitching and R.E. Jones (Eds), The Ecology of Pests. CSIRO, Melbourne, Vic., pp. 55-86.

Reichelt, R.E., Bradbury, R.H., Moran, P.J., in press. Distribution of *Acanthaster planci* outbreaks on the Great Barrier Reef between 1966 and 1989. Coral Reefs.

Reichelt, R.E., Green, D.G., Bainbridge, S.J., 1988. A simulation study of crown of thorns starfish outbreaks on the Great Barrier Reef. Math. Comp. Simul. 30: 145-150.

Reichelt, R.E., Greve, W., Bradbury, R.H., Moran, P.J., 1990. *Acanthaster planci* on the Great Barrier Reef: a starfish-coral site model. Ecol. Modelling 49: 153-177.

Williams, D.McB., Wolanski, E., Andrews, J.C., 1984. Transport mechanisms and the potential movement of planktonic larvae in the Central Region of the Great Barrier Reef. Coral Reefs 3: 229-236.

Wolanski, E., Hamner, W.M., 1988. Topographically controlled fronts in the ocean and their biological influence. Science 241: 177-181.

Yamaguchi, M., 1986. *Acanthaster planci* infestations of reefs and coral assemblages in Japan: a retrospective analysis of control efforts. Coral Reefs 5: 277-288.

Zann, L., Weaver, K., 1988. An evaluation of crown of thorns starfish control programs undertaken on the Great Barrier Reef. Proc. 6th Int. Coral Reef Symp., Townsville, 1988, Vol. 2, pp. 183-188.

GREAT BARRIER REEF HYDRODYNAMICS, REEF CONNECTIVITY AND *ACANTHASTER* POPULATION DYNAMICS

M.K. JAMES, I.J. DIGHT AND L. BODE

Department of Civil and Systems Engineering,
James Cook University of North Queensland, Townsville, Queensland 4811

Abstract. Models capable of simulating large-scale hydrodynamics and *Acanthaster* larval dispersal are presented for the Cairns Section of the Great Barrier Reef Marine Park. The models identify asymmetries in the patterns of larval dispersal which can explain many observed features of the recent episodes of *Acanthaster* activity. The northern part of the Cairns Section, between the ribbon reefs and the mainland, is seen to be hydrodynamically distinctive, with the potential to be self-seeding, and therefore the source of *Acanthaster* activity. Implications of the emerging pattern of stochastic connectivity among reefs are explored using a simple population model. This model is based on a hypothetical assemblage of reefs, with patterns of connectivity derived from the results of the larval dispersal simulation. Experiments with the model demonstrate the potential for the long-term maintenance of populations in the northern region. By simulating the propagation of population outbreaks through the system, it is shown how such a model could be used to test certain hypotheses concerning the cause of outbreaks and to identify the most likely primary outbreak areas.

1. INTRODUCTION

This paper is motivated by research results which indicate that the process of water movement, with resulting larval transport between reefs, may be one of the principal determinants of large-scale *Acanthaster planci* population dynamics. To gain an understanding of the *Acanthaster* phenomenon one must consider the patterns of reef connectivity generated by the hydrodynamics, together with the life-history characteristics of *A. planci*.

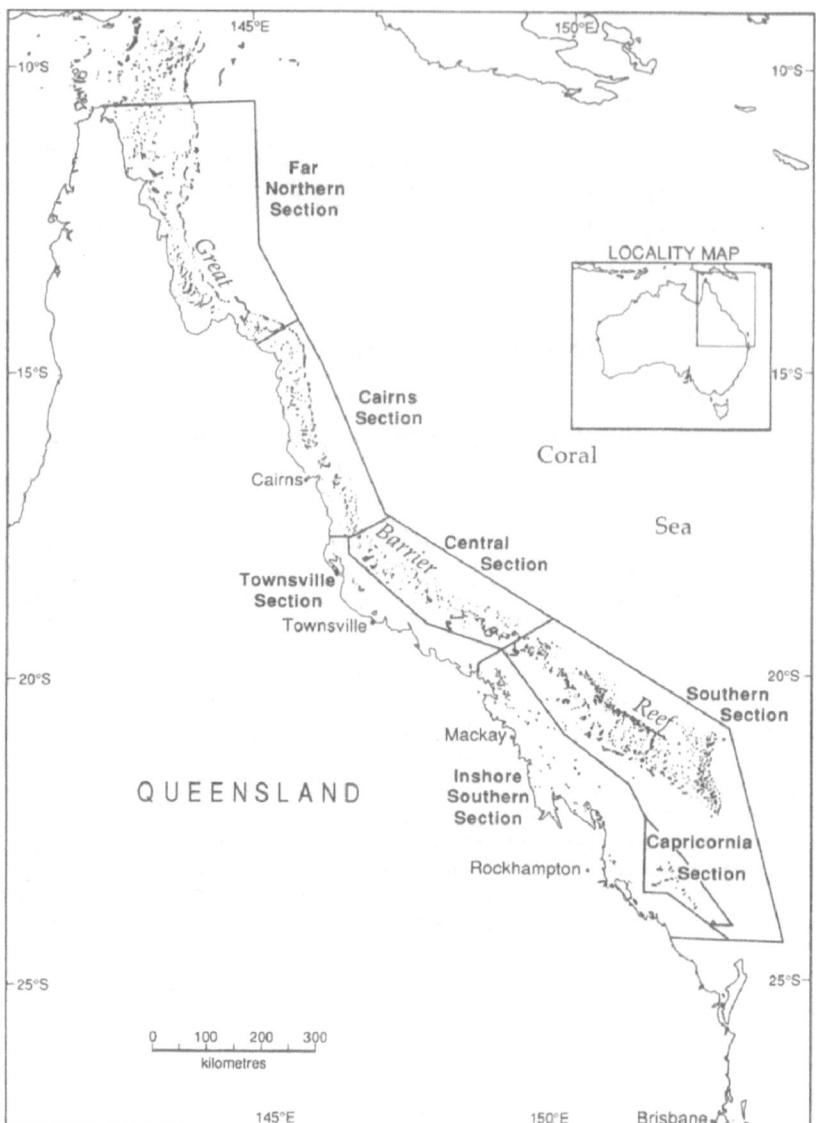

Figure 1.1 Great Barrier Reef province showing Cairns Section.

More generally, the dispersal process is believed to play an important role in recruitment of juveniles to coral and reef fish populations. For example, there is growing evidence (Sale 1980) that fish community species composition is determined by the pool of larvae in the plankton that is carried to isolated reefs by water currents.

The research reported here is based on the use of large scale numerical models which can simulate the hydrodynamics under different modes of forcing. The hydrodynamic models are used in conjunction with a particle tracking model to simulate and analyse larval dispersal trajectories from selected reefs in the Cairns Section of the Great Barrier Reef (GBR) Marine Park (Figure 1.1). This approach leads to a more comprehensive and robust picture of water movement and larval dispersal than do progressive vector plots (e.g. Williams et al. 1984; Frith et al. 1986) which are based on data from a limited number of moorings.

The dispersal simulations can ultimately be used to estimate the probabilities that spawning events on selected source reefs will result in competent larvae becoming available for settlement on associated sink reefs. The computation of these probabilities is the first step in describing the GBR system as a stochastic network (i.e. a graph with non-deterministic arcs) whose dynamics may be rich enough to provide adequate explanations of many observed large-scale features of *A. planci* population fluctuations.

In modelling phenomena in a system as complex as the GBR, there is a strong need to balance the apparent realism and precision of complex models of biological interactions, against the extreme difficulty of obtaining the information needed to develop and test such models. Many of the interesting large scale features of behaviour can often be captured by relatively simple models based on a network representation of the system (see Roberts (1976) for several examples). It is proposed here that the natural basis for a useful model of *Acanthaster* population dynamics is provided by the stochastic network which the larval transport patterns define.

The numerical modelling approach to hydrodynamics and larval dispersal is described briefly in Section 2. For more details of methods and results, see Dight, James & Bode (1988; 1990) and Dight, Bode & James (1990). Results are discussed in Section 3, where it is shown that significant features of the two recorded episodes of *Acanthaster* activity in the GBR can be explained qualitatively in terms of the simulated patterns of larval dispersal.

Section 4 investigates a simple model of *A. planci* population dynamics based on a hypothetical stochastic network whose structure embodies the results of the dispersal simulations. The objective here is to demonstrate the potentially dominant role of reef connectivity in determining the major large scale features of those dynamics.

2. NUMERICAL MODELS: HYDRODYNAMICS
AND LARVAL DISPERSAL

2.1 The Hydrodynamic Model

The simulation of larval dispersal is based on the numerical hydrodynamic model, SURGE, of Sobey et al (1977). SURGE was originally developed to predict storm surges induced by tropical cyclones, but has since been further developed and modified to study tidal and wind-driven flows within the GBR (Bode et al 1988). The model has been tested and validated within the GBR, and can reproduce both the magnitudes and directions of such currents (Frith & Mason 1986; Griffin et al. 1987; Andrews & Bode 1988; Middleton & Bode 1990).

The model employs the Navier-Stokes equations for a homogeneous, incompressible fluid, reduced to two spatial dimensions by integrating through the water depth. Three different forcings are applied separately:

(i)   tide;
(ii)   wind;
(iii)   East Australian Current.

Net water transport at any point is found by linear superposition of the three components.

The seven major tidal constituents (M2, S2, N2, K2, K1, O1, P1) were modelled separately. The results agree well with published data of amplitude and phase for all seven constituents. To enable the simulation of water movement under a time-varying wind field, the hydrodynamic model was used to develop data files characterising the steady-state transport patterns for various combinations of wind speed and direction. In all, 88 different combinations were modelled. A similar technique using only 8 wind directions and monthly mean velocities was implemented by Prandle (1978) to simulate the response to wind forcing in the southern North Sea. The East Australian Current (EAC) was modelled by application of steady-state mass transports across the deep water open boundary. The resulting flows were then calibrated to match published data on current velocity from Church & Boland (1983) and Church (1987).

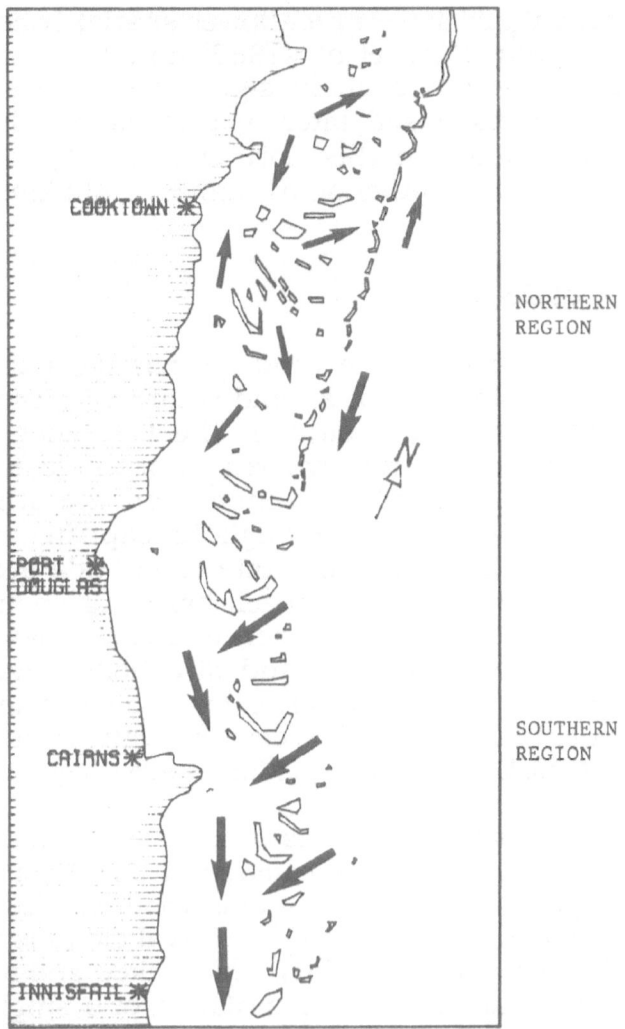

Figure 2.1. Dominant flow patterns from December to March, when *A. planci* larvae disperse.

Overall, the model results correspond well with our understanding of the large-scale features and dynamics of water circulation on the continental shelf within the Cairns Section of the GBR Marine Park. The results also indicate a markedly different hydrodynamic regime in the northern part of the Cairns Section behind the ribbon reefs, compared with the southern part. Notably, in the north, the dense reef matrix extends across most of the shelf and effectively restricts the development of appreciable wind driven circulation to narrow passages. The differences are illustrated in Figure 2.1. During the summer months when the larvae of *A. planci* disperse, flows in the southern part of the Cairns Section are dominated by on-shore and long-shore (polewards) movement, whilst currents in the northern part have much lower spatial coherence, with little or no defined pattern. Frith et al. (1986) note that such regional differences in circulation may result in significant differences in the movement of larvae. As discussed later in this paper, the emerging picture of particle movements in this northern area appears to have significance for the long-term population dynamics of *A. planci.*

2.2 Simulation of Larval Dispersal

Nineteen years of twice-daily surface wind records from the Low Isles weather station drive the simulation. A wind record with more frequent sampling would clearly be more desirable. The Low Isles record was selected because it came from within the region of interest, it was less affected by the land-sea breeze phenomenon than the other stations available and was considered long enough to capture most of the variability experienced in this area. The state of the tide over a 28 day cycle and the wind record at which particle tracking commences are randomly selected to represent the occurrence of each spawning event. The ensuing sequence of wind speeds and directions determines the variation in wind-driven current patterns for the duration of larval transport. Only segments of the wind record relevant to the spawning and dispersal of the starfish (December to March) are used.

Acanthaster larvae are assumed to behave as passive particles whose advection in one-hourly time steps is calculated directly from the net water current at each point. The movement of larvae following a spawning event is simulated for a period of 28 days. The larval cloud is represented by seven particles, the movements of which are tracked and intersections with reefs recorded. An example of the simulation of a single spawning event is presented in Figure 2.2.

Larvae are assumed to be precompetent and unable to settle during the initial two weeks. Only intersections with reefs following the two week precompetent period are therefore considered. For the present purpose, once an intersection occurs during the competent period, the remaining track and any further intersections are ignored.

Source reefs with some record of *Acanthaster* activity were selected from which to simulate spawning events and the subsequent movement of larvae. Thirty reefs were selected in the Cairns Section: fifteen in the northern region between the ribbon reefs and the mainland, and fifteen in the southern region. Three cross-shelf zones were equally represented: the outer edge of the reef matrix; central within the matrix; the inner edge of the reef matrix, bordering the main lagoon.

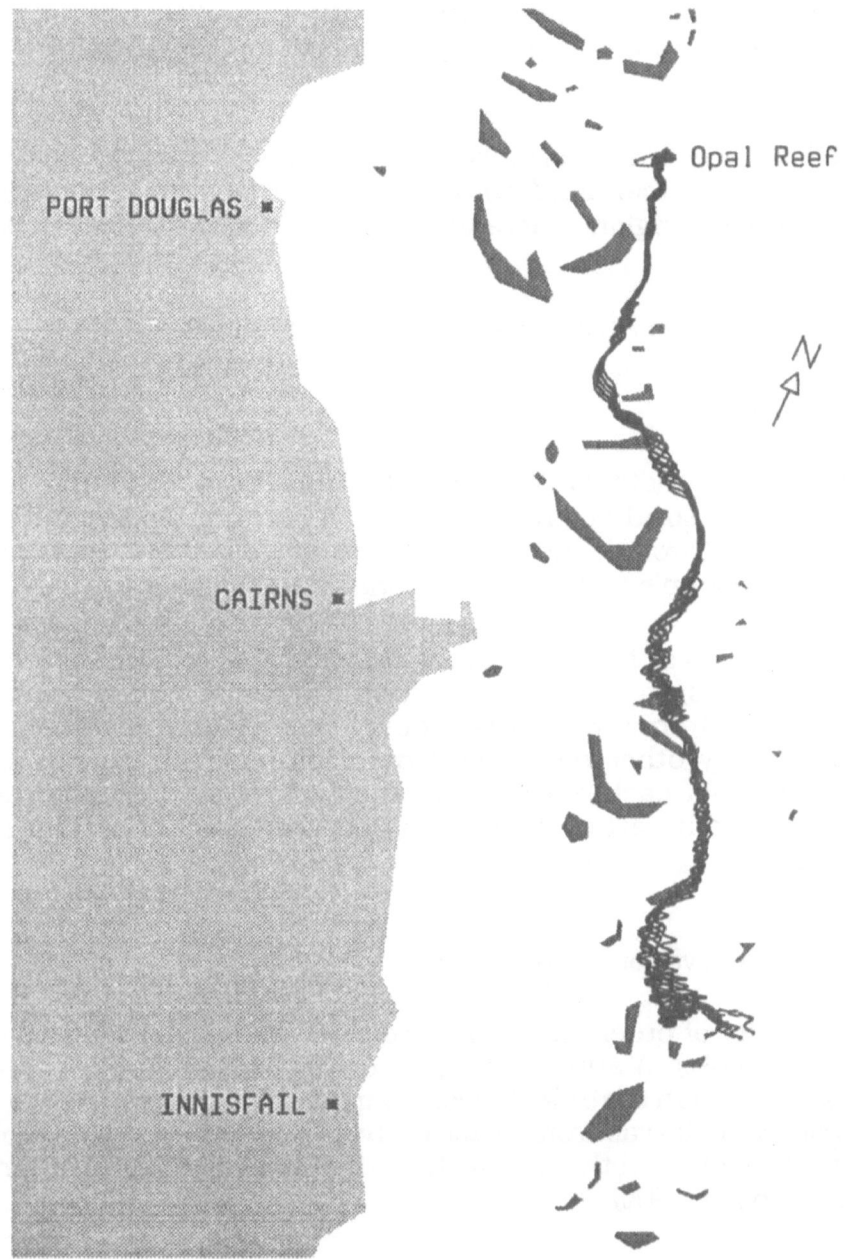

Figure 2.2. Representative trajectory of a single dispersal event from Opal Reef. Tracks of the seven particles representing the larval cloud are plotted for the full simulated period of 28 days.

Forty spawning events were simulated and analysed for each reef, using randomly selected tide and wind histories. The reefs with which particles intersected were recorded and classified with respect to cross-shelf zone, and with respect to long-shelf location (north or south) relative to the source reef.

In what follows, the term "northern region" refers to the area between the ribbon reefs and the mainland, in the northern half of the Cairns Section. The term "southern region" refers to the remainder of the Cairns Section.

3. LARVAL DISPERSAL PATTERNS

3.1 Southern Region

Although there is considerable variability in the data, examination of the trajectories in the southern region reveals a distinct tendency for larvae to move onshore through the reef matrix and long-shore in a southerly direction. Some sample trajectories are shown in Figure 3.1.

Numerical analysis of the trajectories enables reef connectivity (source-sink relationships) to be expressed in terms of the proportion of competent larvae from a particular source reef which are transported to the vicinity of any other reef. The general pattern can be extracted by pooling the data for each of the cross-shelf zones, as presented in Table 3.1 for thè southern part of the Cairns Section (based on a total of 4200 particle tracks).

The following observations can be made:

(1)   connectivity of outer matrix source reefs was primarily with central matrix reefs (45.0%). A substantial proportion of larvae still reached the vicinity of inner matrix reefs (8.8%), marginally more so than to outer reefs (7.2%). Many larvae from outer matrix reefs were not transported to the near vicinity of any other reef within the time constraints of larval life that were imposed (39%).

(2)   connectivity of central matrix source reefs was fairly evenly distributed between both central (37.6%) and inner matrix reefs (39.9%). Only relatively rare events resulted in larvae being transported offshore to the outer matrix reefs. In general, most dispersal events resulted in larvae being transported to the near vicinity of other reefs during their competent period of development (80.4%).

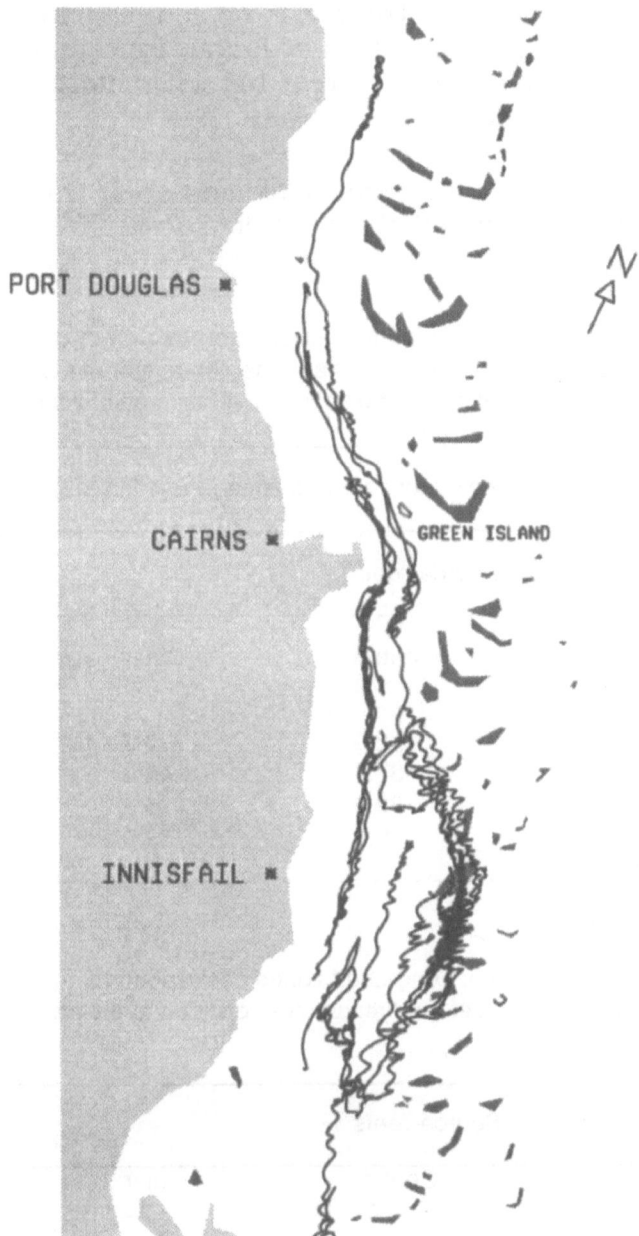

Figure 3.1. Particle tracks representing ten dispersal events from Green Island. Each track corresponds to the centroid of the larval cloud during the two week competent period of larval development and is based on a random selection of tide and wind records.

(3)   connectivity of inner matrix source reefs was almost exclusively with other reefs bordering or within the lagoon (55.5%). Again, only relatively rare events resulted in larvae being transported offshore, this

time to the central matrix reefs. There was no connectivity indicated with outer matrix reefs. A high proportion of larvae from inner matrix reefs was not transported near another reef (41.5%), more than from the central matrix reefs in particular.

Source-sink relationships with respect to long-shelf transport are also presented in Table 3.1. Connectivity is directed strongly southwards.

Table 3.1. Source-sink relationships for the southern region of the Cairns Section. Connectivity is expressed as a percentage of the total number of particles from all simulated dispersal events in each zone.

(a) Across the continental shelf. (zones are defined in Section 2.2).

Sink reefs	Source reefs		
	Inner %	Central %	Outer %
Outer	0.0	2.9	7.2
Central	3.0	37.6	45.0
Inner	55.5	39.9	8.8
None	41.5	19.6	39.0

(b) Along the continental shelf. Long-shelf zones correspond to north and south relative to the source reef, and recruitment back on to the parent reef (source).

Sink reefs	Source reefs		
	Inner %	Central %	Outer %
North	9.1	2.7	4.3
South	46.9	77.2	56.5
Source	2.5	0.4	0.2
None	41.5	9.7	39.0

3.2 Northern Region

Obvious differences are observed when larval trajectories in the northern region are examined. An example is presented in Figure 3.2. There is a tendency for larvae to move to reefs located offshore from, or to the north of the source. Table 3.2 shows that, while the southerly trend is still dominant, there is no cross-shelf asymmetry in this region, and the connectivity pattern is considerably more random than in the southern region.

3.3 Discussion

Results indicate that:

(1)   within the southern region there is a strong tendency for larvae to be transported southwards;

(2)   particular reefs are most likely to receive larvae; and

(3)   certain regions of the GBR are hydrodynamically distinctive, with different patterns of dispersal and recruitment.

The first observation is in agreement with the apparent southward spread of *A. planci* populations, which was conceptually modelled by Kenchington (1977). The distances over which the modelled larvae disperse are consistent with observations and Kenchington's model.

The second observation relates to the susceptibility of particular reefs to 'outbreaks' of *A. planci*. This arises through large-scale features of the hydrodynamic regime which lead to strong on-shore movement of larvae across the reef matrix, and long-shore movement within the main lagoon of the coastal/reef complex. As a result, reefs located within or along the edge of the lagoon are more likely to receive larval recruitment. Reefs such as those off Townsville (John Brewer), Innisfail (Gibson, Howie, Peart and Feather) and in the vicinity of Cairns (Green Island, Upolu and Arlington) are clear examples. These reefs have a history of repeated outbreaks, with Green Island, Feather and John Brewer Reefs suffering extensive coral mortality during the two recorded episodes of *Acanthaster* activity.

These results indicate that it is not necessary to invoke hypotheses based on human interference with the environment to explain the high incidence of secondary outbreaks on these inner matrix reefs.

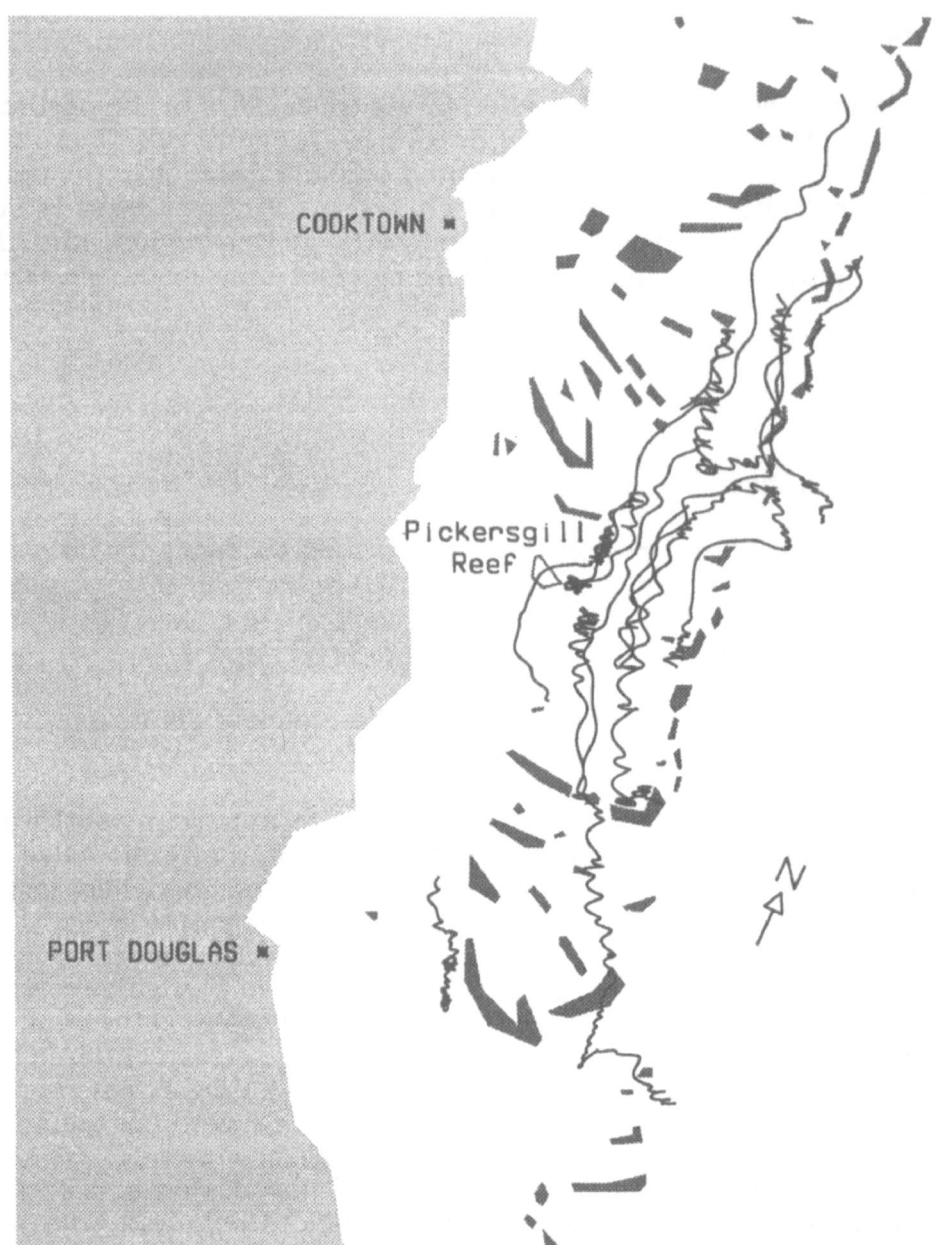

Figure 3.2. Particle tracks representing ten dispersal events from Pickersgill Reef. see caption, Figure 3.1, for details.

The third observation relates particularly to the area behind the ribbon reefs. In contrast to the region further south, there is no cross-shelf asymmetry in larval transport. Furthermore, while southward transport

Table 3.2. Source-sink relationships for the northern region of the Cairns Section. Connectivity is expressed as the percentage of the total number of particles from all simulated dispersal events in the zone.

(a) Across the continental shelf. (Zones are defined in Section 2.2).

Sink reefs	Source reefs		
	Inner %	Central %	Outer %
Outer	2.0	11.8	34.0
Central	25.6	43.2	18.4
Inner	38.9	15.2	1.0
None	33.5	29.8	46.6

(b) Along the continental shelf. Long-shelf zones correspond to north and south relative to the source reef, and recruitment back on to the parent reef source.

Sink reefs	Source reefs		
	Inner %	Central %	Outer %
North	15.7	11.7	5.9
South	48.4	56.8	39.7
Source	2.4	1.7	7.8
None	33.5	29.8	46.6

is still dominant, the probability that larvae will be carried to reefs to the north of their source is greater than in the southern region. Such behaviour points to the possibility that there may always be significant (but not necessarily outbreaking) populations of A. *planci* within this region, which from time to time may propagate southwards. In the face of strongly directed onshore and southward transport of larvae in most of the region south of Cairns, this helps explain the long term persistence of populations of A. *planci* in the GBR system.

Model results strongly suggest that primary outbreaks in the area off Cairns and further to the south would be unlikely to result in the observed distribution of A. *planci* populations to the north and further off-shore. Rather, the indications are that primary outbreaks have occurred in the northern region, with subsequent propagation of large populations to the south.

This is in agreement with Kenchington's (1977) hypothesis that the source of *Acanthaster* populations in the region of Cairns lay further to the north.

4. POPULATION MODEL

The majority of reef organisms are sedentary in the sense that, once settled on a particular reef, they do not migrate to other reefs. This is particularly so for slow-moving (or attached) benthic organisms, and is probably true for *A. planci*. On the other hand, during the planktonic larval phase of the life cycle of most coral reef species, dispersal away from the natal reef occurs (Sale 1980; Williams et al. 1984). In these circumstances, it is appropriate, when considering the population dynamics of such organisms, to model the reef system as a network in which each reef is a node, and the process of larval dispersal generates the links or arcs. The directional nature of larval transport means that directed arcs are necessary, transforming the network into a directed graph, or digraph.

For many reef organisms, such as *A. planci*, spawning takes place during well-defined seasons which are short compared with a full year, or in comparison with the developmental period leading up to breeding status. Thus major changes occur at essentially discrete times, and it may be most appropriate to employ mathematical models which focus on discrete events, rather than deal explicitly with continuous changes. The 'continuous' processes (e.g. predation, growth, feeding, etc) which take place in the intervals between the discrete events may be parameterised so that the integrated effects of those changes are modelled, rather than the changes themselves.

4.1 Digraphs and Weighted Digraphs

A digraph model of the reef system which is appropriate for studying the *Acanthaster* phenomenon is readily defined by the pattern of reef connectivity developed from the larval dispersal simulations. In this model, a directed arc from reef **i** to reef **j** implies that larval transport can occur from **i** to **j** (Figure 4.1(a)). Digraphs may be characterised in a number of ways (see Roberts (1976) for a good discussion). Some of these features may have relevance to the ecological structure of the GBR.

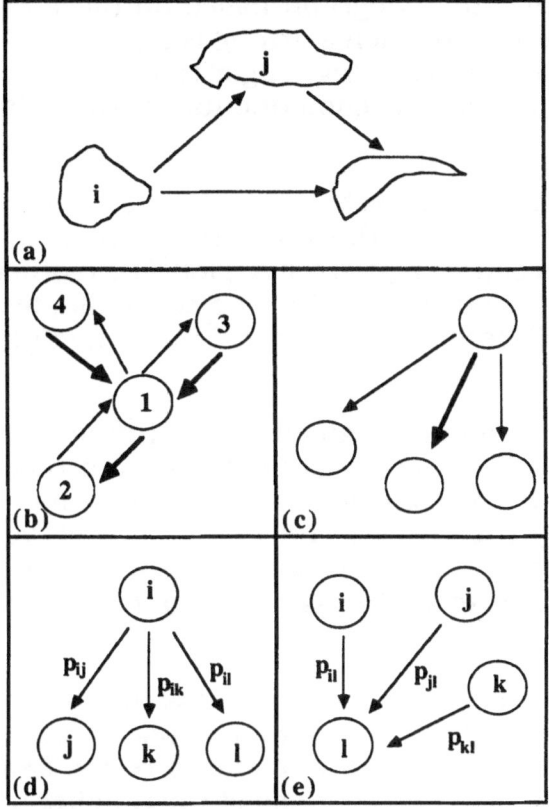

Figure 4.1. Digraphs and weighted digraphs.
(a) Reefs represented by nodes; directed arcs represent potential larval transport.
(b) Typical pattern in northern region. The thickness of an arrow is a relative measure of the probability assigned to it.
(c) Typical pattern in southern region.
(d) Pulse process from source to multiple sinks.
(e) Pulse process to sink from multiple sources.

Reachability: A reef **v** is reachable from reef **u** if a path, made up of one or more arcs, exists from **u** to **v**. In such a case, spawning by a population on reef **u** may eventually result in the establishment of a population on reef **v**.

Strong connectedness: The reef system would be strongly connected if for each pair of reefs **u** and **v**, **v** is reachable from **u**, and **u** is reachable from **v**. For example, if there existed a closed path (a loop, or cycle) through all the reefs, the system would be strongly connected.

Strong components: Given the apparent asymmetries in larval transport identified by the dispersal simulations, it appears unlikely that the reef system is strongly connected. However it may be possible to identify subgraphs which are strongly connected. The results discussed in Section 3 indicate that connectivity among reefs located in the northern region may be such that the subgraph based on these reefs is strongly connected. The significance of such a strong component is that it may be regarded as potentially self-seeding. Algorithms are available for the identification of all the strong components of any digraph (Roberts 1976).

Vertex basis: This is defined as a set of reefs, B, such that every reef not in B is reachable from some reef in B. In the present context, a vertex basis could play the role of a source region for *A. planci* (and other) populations. Such key reefs, once identified, should become candidates for protective zoning. Roberts (1976) describes an algorithm for finding all the vertex bases of any digraph.

A weighted digraph model is obtained if weights are assigned to the directed arcs. The weight assigned to an arc should reflect the relative importance of the arc in the context of the process being modelled. Here, the appropriate weight would be the probability of larval transport between the reefs, given that a spawning event has occurred on the source reef. The resulting graph is referred to in this paper as a stochastic network.

The concepts of reachability and connectedness are essentially the same as for the (unweighted) digraph. Additional useful concepts relate to integrated measures along paths from reef **u** to reef **v**. Examples are the "length" (measured, for example, in terms of the number of time steps) of the shortest path from **u** to **v**; and the probability of the most probable path from **u** to **v**.

4.2 Network Dynamics

Simulated larval transport data for the GBR are still incomplete. It is therefore not possible at this time to compute the topological characteristics referred to above. However, as discussed in Section 3, sufficient simulations have been performed to enable broad patterns of larval dispersal to be discerned. Those patterns are closely correlated with observations of the *Acanthaster* phenomenon. They also make it possible to deduce the structure of an appropriate digraph model.

To help justify the interpretations made in Section 3, and also to investigate further the important role of reef connectivity generated by the hydrodynamics, a simple model of *A. planci* population dynamics was developed. A weighted digraph based on a hypothetical reef assemblage was used. The hypothetical system (Figure 4.2) embodies the major qualitative features of the Cairns Section, with a distinctive northern region having a dense line of ribbon reefs on the outer edge. The density and arrangement of reefs are intended to reflect properties of the Cairns Section, although the number of reefs in the model (68 in all) is much smaller than the number actually present

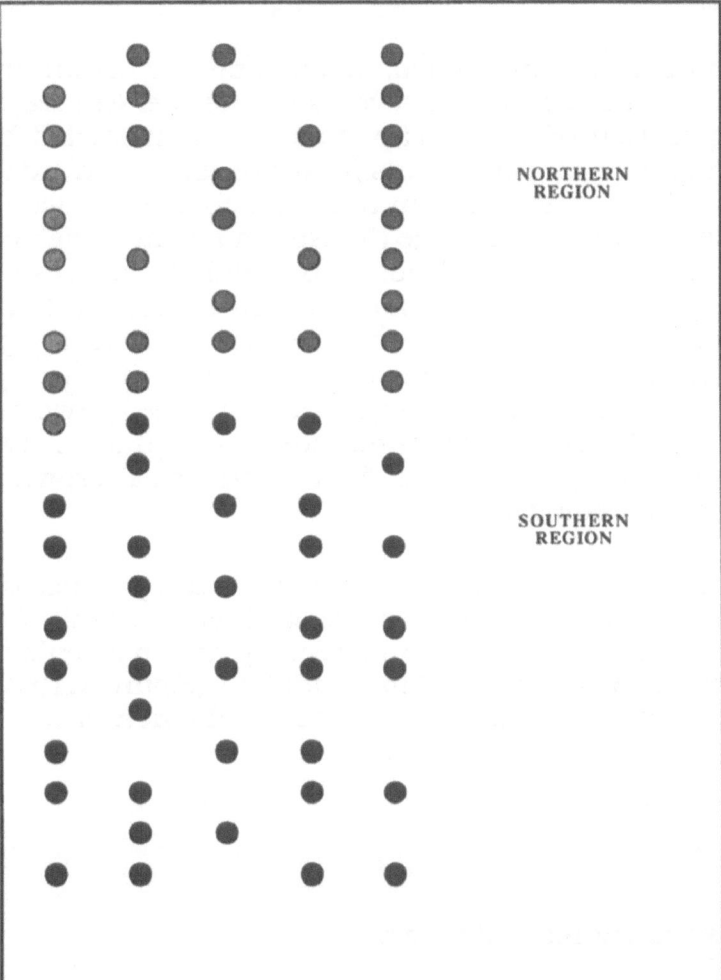

Figure 4.2. Hypothetical reef system

A stochastic network was generated by defining arcs and assigning probabilities as discussed above. The network represents a generalization of the qualitative and quantitative results of the dispersal simulations. Depending on its location, each reef is assigned one of several possible stereotyped connectivity patterns. For example, reefs in the northern region are typically linked to their neighbours in the manner illustrated in Figure 4.1(b), reflecting the potential for larval transport in all directions, but with a dominant southerly trend. The thickness of an arrow is a relative measure of the probability assigned to it. Figure 4.1(c) shows how typical reefs in the southern region are connected, with strong tendencies for movement on-shore and polewards. The assigned probabilities are based on the results of numerical analysis of the simulated data.

In order to model the dynamics, some specific assumptions are necessary to define the effects that events at one reef can have on other reefs. The propagation of *A. planci* populations is modelled by a pulse process. Deterministic models of this kind are discussed extensively in Roberts (1976). At a discrete point in time, t, a change ("pulse") in a state variable at node i produces a change ("pulse") in a state variable at node j, one time step later. The pulse at node j is equal to the pulse at node i, amplified by the weighting factor a_{ij} assigned to the arc from i to j. A stochastic version of the process is employed here: A breeding population on reef i, represented by a unit pulse, has probability p_{ij} of producing another unit pulse, or breeding population, on reef j two years later (Figure 4.1(d), (e)). Thus the time step is equal to two years, corresponding to the time required for an individual crown of thorns starfish to achieve breeding status.

In a single breeding season, a unit pulse on reef i can produce unit pulses on each of the reefs with which it is linked (Figure 4.2(d)). Conversely, individual unit pulses arriving simultaneously at a particular reef from more than one source are assumed to be additive (Figure 4.1(e)). Thus the pulse sizes can grow as they propagate through the network:

$$x_j(t+1) = \Sigma_i \ a_{ij} \ x_i(t);$$

where

$x_i(t)$ = pulse size at reef i at time t;

a_{ij} = 1 if connection i -> j exists at time t;

a_{ij} = 0 otherwise.

For simplicity, a population is assumed to be active for only one spawning season, and is then no longer considered. This implies strong regulation of the population (for example, by disease). Since very few, if any, outbreaking populations have been observed to occupy any reef for more than a few years, this may not be particularly unrealistic. However the model could be readily modified to deal with greater longevity.

4.2.1 Deterministic case

It is instructive first to consider the deterministic process. This corresponds to the assignment of a probability value of unity to each arc, so that $a_{ij} = 1$ on all arcs at all times. The stability of the process is of particular interest. Because of the assumption that pulses arriving simultaneously at a reef are simply added together, the possibility exists for the pulse magnitudes to grow rapidly as they propagate through the network. In certain configurations, unbounded growth (i.e. instability) can occur. The stability properties are determined entirely by the pattern of connectivity among the reefs. For example, pulse processes on reefs connected as in Figure 4.1(b) are unstable. This is readily demonstrated by the following time history of the simple pulse process which commences with a unit pulse on reef 1:

Time	Pulse Magnitude			
	Reef 1	Reef 2	Reef 3	Reef 4
0	1	0	0	0
1	0	1	1	1
2	3	0	0	0
3	0	3	3	3
4	9	0	0	0
5	0	9	9	9
6	27	0	0	0

The pattern of Figure 4.1(b) is a common feature of the present model. Therefore, deterministic, simple pulse processes in the hypothetical reef system are expected to be unstable. The dynamics of such a model are readily studied via the adjacency matrix $A = [a_{ij}]$ (Roberts 1976). In particular, if A has eigenvalues with magnitudes greater than unity, there will be at least one reef from which a pulse process will be unstable. Instability of the hypothetical system was confirmed by computing the eigenvalues of the adjacency matrix.

4.2.2 Stochastic case

The deterministic process is useful for demonstrating the fundamental importance of the pattern of reef connectivity. However the dispersal simulations display considerable variability (e.g. Figure 3.2). Reef connectivity is clearly non-deterministic, so real insight into *A. planci* dynamics can be gained only by considering the stochastic network in which

$$\Pr \{a_{ij} = 1\} = p_{ij}$$

Thus both the spatial pattern of connections and the numerical weights (probabilities) are taken into account. An analytical investigation does not appear to be feasible. However a number of issues were addressed by Monte Carlo simulations: (1) the possibility of 'unstable' growth of the population magnitudes as they propagate through the reef system; (2) the question of the long-term persistence of low-density populations; and (3) the occurrence of primary outbreaks in the northern region, with subsequent propagation of large populations into the southern region. Each of these issues has important implications for the understanding (and, possibly, management) of the *Acanthaster* phenomenon.

4.3 Simulations of Population Dynamics

4.3.1 Stability

Behaviour of the deterministic process indicates that the potential for unstable population growth exists in the GBR purely because of the pattern of reef connectivity in the northern region. However a number of other issues are important in determining if that potential is realised. One of these concerns the inherent variability in the larval transport process. In the stochastic system, it is not possible to state that a process is either stable or unstable. It is more appropriate to estimate the probability that the process will result in population levels exceeding some (large) value. As a partial (and preliminary) investigation of this possibility, individual processes were simulated and tested to see if pulse magnitudes larger than 10 units were achieved.

Fifty replicate pulse processes, starting with a unit pulse, were simulated independently for each reef. In every case, the pulse magnitudes failed to reach or exceed 10 units before the process died away. Evidently the probabilities assigned to each arc are too low to sustain the unstable growth characteristic of the deterministic processes.

The assigned probabilities may be unrealistically low, since they are based directly on the results of the dispersal simulations of individual spawning events. That is, no account is taken of the (strong) likelihood that many spawning events may occur, at random times, on any single populated reef during any breeding season. On the other hand, the probabilities may be unrealistically high, because the larval settlement process has been treated as deterministic: all larval trajectories intersecting a reef are assumed to result in recruitment to the population on that reef. In fact, the ultimate fate of larvae reaching the vicinity of a reef will be, in part, determined by the details of the fine-scale circulation around the reef, which the hydrodynamic model used here is unable to resolve. Larvae approaching a potential sink reef may not all become entrained within its circulation. Rather, many may be transported further afield. Thus, recruitment from a single spawning event is likely to be spread over more than a single reef.

Density-dependent effects may also be important, but have not yet been incorporated into the model. For example, the number and magnitude of spawning events during a breeding season may reasonably be assumed to increase as the spawning population increases. Thus, for greater pulse magnitudes, the probability of connection to other reefs will be higher. This could be the point at which distinctive features of A. planci become important: for example, its very high fecundity and the apparent tendency of some populations to aggregate into high-density clumps.

Finally, the possibility of negative feedback affecting population growth should be considered. An obvious example would be the effect of significant coral depletion as populations of A. planci become larger. This may or may not be important, depending on which features of the network structure are responsible for growth of the pulse magnitudes. It should be clear from the earlier consideration of Figure 4.1(b) that features like closed loops or cycles which have some reefs in common can be responsible for unstable growth. If the loops involved are tight, then reefs in the loops will be occupied by pulses at frequencies so high that coral recovery between pulses may not be possible, and a strong negative feedback could develop. On the other hand, if large structural features with relatively long loops are involved, the frequency of occupation of reefs could remain low enough for coral recovery to take place, so that no significant negative feedback can occur.

These and other issues, such as the possibility of self-seeding of individual reefs (Black 1988), are the subjects of continuing work with the model.

4.3.2 Persistence of low-density populations

As pointed out in Section 4.1, reefs in the northern region may constitute a "strong component" of the GBR network. The possibility that, as a result, the region may act as a self-seeding "nursery" for *A. planci* populations has already been mentioned. This requires that populations persist indefinitely in the region. Since all recruitment to any reef is assumed to come from sources external to the reef, this in turn requires sufficiently high levels and strengths of connectivity among reefs (as defined in Dight, James & Bode 1990). Simulation experiments were performed to investigate this possibility. If only low-density populations are considered, then negative feedback effects due to coral depletion can be ignored. It is then reasonable to assume that each unit pulse represents a population which is large enough to achieve significant spawning success, but not large enough to cause significant reduction in coral cover.

The persistence of such populations in the northern region was investigated by simulations in which initially a given proportion, Q, of reefs in the region carried populations. These populated reefs were arbitrarily distributed. For each value of Q, 50 replicate simulations were performed, and the probability of extinction within 100 generations was estimated. The results are shown in Figure 4.3, and it is clear that the potential exists for long-term persistence of the population, provided a significant proportion of reefs in the region become populated.

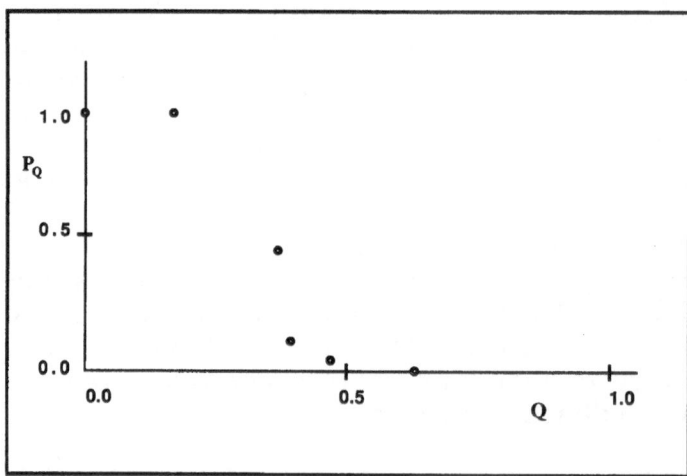

Figure 4.3. Persistence of low density populations. Q is the proportion of northern reefs initially populated. P_Q is the probability of extinction within 100 generations (200 years).

In this case, each simulation represents the simultaneous realisation of a set of simple pulse processes, each originating on one of the selected starter reefs. The processes are not independent of each other, since pulses from different processes frequently occupy the same reefs at the same time. They are then subject to the same probabilistic fate at the next time step. The interactions among processes would be strengthened, and persistence would be enhanced, if the density-dependent effect discussed in Section 4.3.1 were incorporated.

While low-density populations may persist indefinitely in the northern region, the strong southward drift in the southern region means that populations can be maintained in this region only by recruitment from the north. That is, from time to time, populations must propagate southward from the northern region. The question of the frequency of southward propagation was briefly studied under conditions of sustained low-density population in the northern region. During the simulations described above, a selected reef on the north-western edge of the southern region was monitored and found to be occupied (by low-density populations) at highly random intervals, with a mean frequency of once every 23 years.

4.3.3 Propagation of outbreaks

As discussed in Section 3, in light of the dispersal simulations, the observed distribution of outbreaking populations in the GBR is consistent only with the hypothesis that primary outbreaks occur in the northern region. Subsequent propagation of populations into the southern region results in the observed secondary outbreaks. Simulation experiments were run to illustrate this process and to demonstrate the consistency between the resulting pattern of secondary outbreaks and the observed distribution. In this case, a unit pulse represents a large 'outbreak' population, and each pulse process represents the occurrence of a primary outbreak on some reef in the northern region, and the resulting spread of the outbreak through the network.

A 'propagation matrix' was computed from the results of 1000 simulated pulse processes originating and developing independently from each reef in the northern region. An element P_{ij} of this matrix is an estimate of the probability of a primary outbreak on reef i eventually propagating to reef j in the south. In terms of the stochastic network, P_{ij} is simply a measure of the reachability of reef j from reef i. The matrix can be used to compute a measure of the long-term relative frequency of secondary outbreaks on any reef j in the southern region, given the hypothesis that primary outbreaks occur only on some subset **i** of reefs in the northern region. If all the reefs in **I** are equally likely to suffer primary outbreaks, the measure is

$$F_j = \Sigma_{i\epsilon I}P_{ij}$$

Figure 4.4 depicts the relative frequency distribution for the case in which I contains all of the reefs in the Northern Region. Values are normalised with respect to the reef with the highest relative frequency. The relatively high susceptibility of mid-shelf reefs, and especially those in the north-west sector of the southern region, is clearly evident, and is in agreement with the observed features of *Acanthaster* activity. Also in agreement is the diminishing frequency to the south, with the clear possibility that outbreaks would rarely be observed south of some cut-off point.

This modelling approach could be used to investigate different hypotheses concerning the origin and location of primary outbreaks. Many hypotheses (e.g. human interference; terrestrial runoff) imply that primary outbreaks will occur in definable areas. By considering all primary outbreaks to be restricted to such an area, the propagation matrix could be used to deduce the long-term relative frequency distribution of secondary outbreaks which is consistent with the underlying hypothesis. For example, if primary outbreaks are restricted to the northern ribbon reefs, the distribution in Figure 4.5 results. On the other hand, primary outbreaks restricted to the southern ribbon reefs generate the relative frequencies depicted in Figure 4.6.

The differences in patterns suggest that it might be possible both to invalidate certain hypotheses and to identify the most likely primary outbreak areas.

5. CONCLUSIONS

Larval dispersal simulations in the southern part of the Cairns Section have identified asymmetries in reef connectivity related to:

(i)   the long-shelf transport of larvae,
(ii)   the cross-shelf transport of larvae,
(iii)   source-sink relationships.

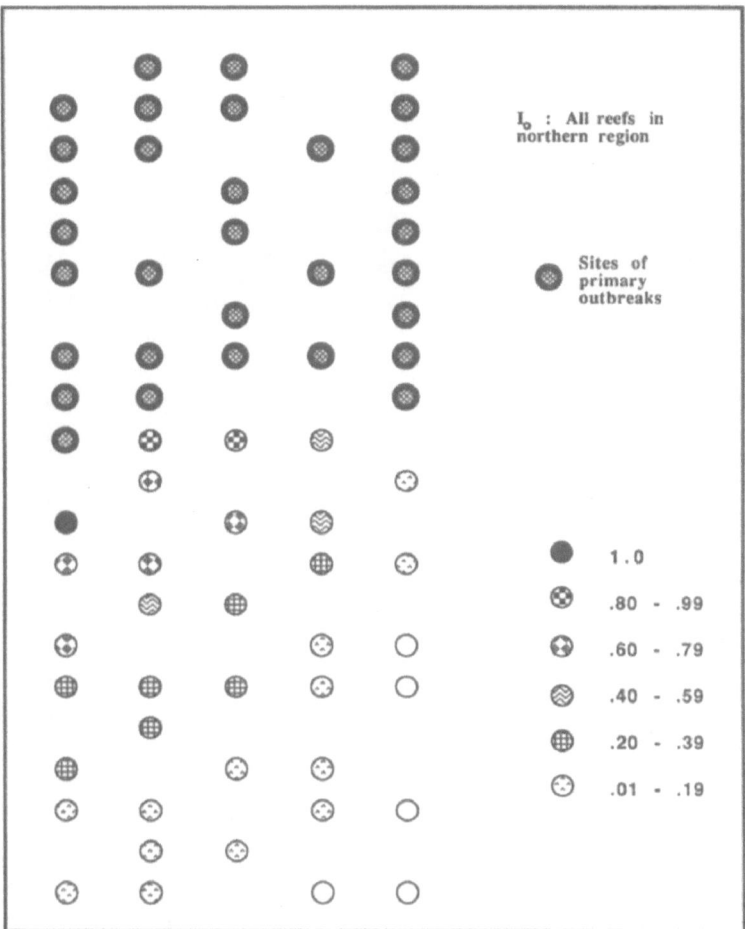

Figure 4.4. Long-term relative frequencies of secondary outbreaks on southern reefs, if all reefs in the northern region are equally likely sites of primary outbreaks.

These asymmetries can account for the apparent southward spread of *Acanthaster* populations from the region of Green Island, the high incidence of outbreaks on mid-shelf reefs, and the relatively high susceptibility of reefs such as Green Island, Feather and John Brewer. It is apparently not necessary to invoke human interference in the ecology of these more accessible reefs in order to explain their high levels of *Acanthaster* activity.

A simple population model based on the findings of the dispersal simulations has demonstrated the potential of the hydrodynamic approach to investigate and explain large-scale features of observed

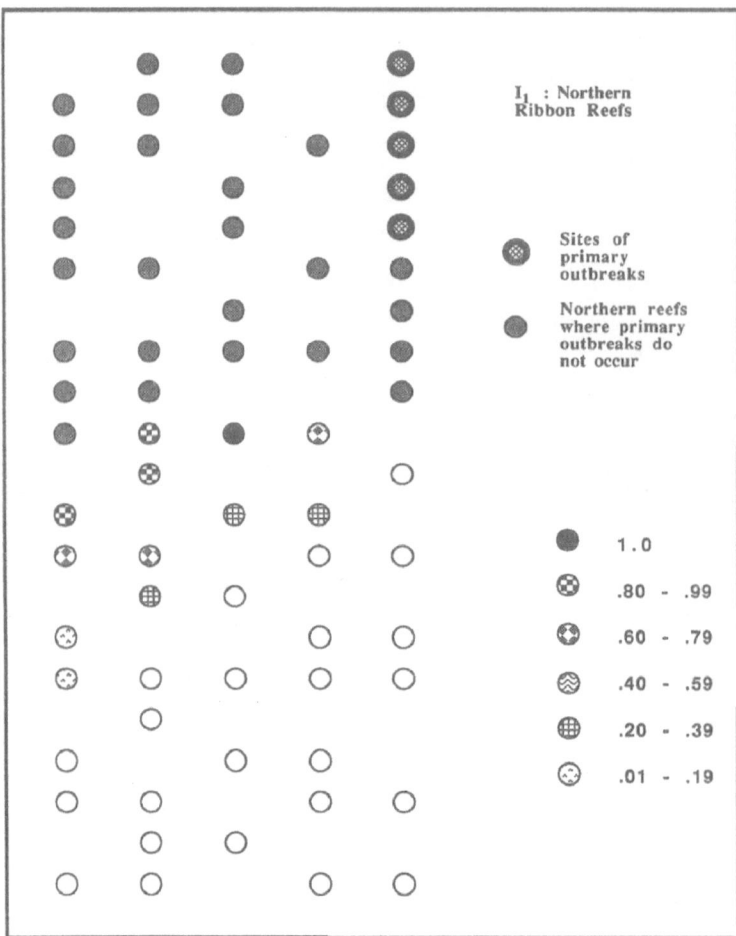

Figure 4.5. Long-term relative frequencies of secondary outbreaks on southern reefs, if primary outbreaks are confined to the northern ribbon reefs.

A. planci distributions. While the model could be modified in a number of ways to improve its biological realism, it is felt to be unlikely that the general patterns of behaviour would change markedly.

The origin of primary outbreaks and the possibility (and likely frequency) of repetitive outbreak episodes are still open questions. However it would seem reasonable to consider the hypothesis that natural temporal and spatial variability in spawning, fertilization success, larval development and reef connectivity are the dominant factors in answering those questions, without the need to invoke external triggers like human interference and terrestrial runoff.

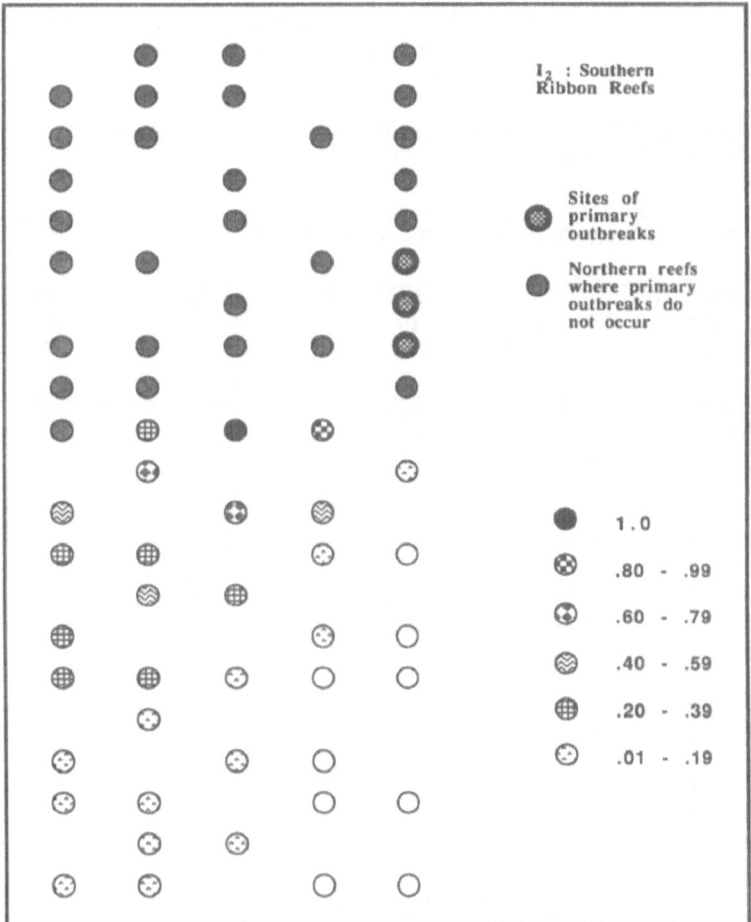

Figure 4.6. Long-term relative frequencies of secondary outbreaks on southern reefs, if primary outbreaks are confined to the southern ribbon reefs.

Acknowledgements. This research was partially supported by a COTSAC grant to the Department of Civil and Systems Engineering, James Cook University of North Queensland. The authors gratefully acknowledge the support and encouragement of Professor R.E. Jones and the late Professor K.P. Stark, and the valuable programming assistance of Mr. Luciano Mason and Mr. Chris Gossett.

REFERENCES

Andrews, J.C., Bode, L. (1988). The tides of the Central Great Barrier Reef, Cont Shelf Res 8: (in press).

Bode, L., Mason, L.B., Middleton, J.H. (1988). The response of a shelf sea/barrier reef system to tidal oscillations: Analytical and numerical methodologies. J. Phys Oceanogr (submitted).

Black, K.P. (1988). The relationship of reef hydrodynamics to variations in the free-borne larval numbers on and around coral reefs. Proc. 6th. Int. Coral Reef Symp., Townsville 2:125-130.

Church, J.A. (1987). The East Australian Current adjacent to the Great Barrier Reef. Aust J Mar Freshwater Res 38 : 671-683.

Church, J.A., Boland, F.M. (1983). A permanent undercurrent adjacent to the Great Barrier Reef. J Phys Oceanogr 13 : 1747-1749.

Dight, I.J., James, M.K., Bode, L. (1988). Models of larval dispersal within the central Great Barrier Reef: Patterns of connectivity and their implications for species distributions. Proc. 6th Int. Coral Reef Symp., Townsville 3:217-224.

Dight, I.J., Bode, L., James, M.K. (1990). Modelling the larval dispersal of *Acanthaster planci*: I. Large scale hydrodynamics, Cairns Section, Great Barrier Reef Marine Park (in press).

Dight, I.J., James, M.K., Bode, L. (1990). Modelling the larval dispersal of *Acanthaster planci*: II. Patterns of reef connectivity (in press).

Frith, C.A., Leis, J.M., Goldman, B. (1986). Currents in the Lizard Island region of the Great Barrier Reef lagoon and their relevance to potential movements of larvae. Coral Reefs 5:81-92.

Frith, C.A., Mason, L.B. (1986). Modelling wind driven circulation One Tree Reef, Southern Great Barrier Reef. Coral Reefs 4:201-211.

Griffin, D.A., Middleton, J.H., Bode, L. (1987). The tidal and longer-period circulation of Capricornia, Southern Great Barrier Reef. Aust J Mar Freshwater Res 38:461-474.

Kenchington, R.A. (1977). Growth and recruitment of *Acanthaster planci* (L.) on the Great Barrier Reef. Biol Conserv 11:103-118.

Middleton, J.H., Bode, L. (1990). Modelling Great Barrier Reef tides: northern section. J Phys Oceanogr (submitted).

Prandle, D. (1978). Residual flows and elevations in the southern North Sea. Proc R Soc Lond A. 359:189-228.

Roberts, F.S. (1976). Discrete Mathematical Models. Prentice-Hall, New Jersey.

Sale, P.F. (1980). The ecology of fishes on coral reefs. Oceanogr Mar Biol Ann Rev 18:367-431.

Sobey, R.J. Harper, B.A., Stark, K.P. (1977). Numerical simulation of tropical cyclone storm surge. Research Bulletin CS14, Dept Civil and Systems Engineering, James Cook Univ Nth Qld.

Williams, D.McB., Wolanski, E., Andrews, J.C. (1984). Transport mechanisms and the potential movement of planktonic larvae in the central region of the Great Barrier Reef. Coral Reefs 3:229-236.

IMA Journal of Mathematics Applied in Medicine & Biology (1989) **6**, 81–89

A Diffusion–Reaction–Transport Model for Large–Scale Waves in Crown-of-Thorns Starfish Outbreaks on the Great Barrier Reef

P. L. ANTONELLI

Mathematics Department, University of Alberta, Edmonton, Canada T6G 2G1

N. D. KAZARINOFF

Mathematics Department, S.U.N.Y. at Buffalo, Buffalo, New York, 14214 USA

R. E. REICHELT, R. H. BRADBURY, AND P. J. MORAN

Australian Institute of Marine Science, Cape Ferguson, Queensland 4810, Australia

[Received 22 April 1988 and in revised form 24 November 1988]

Effects of diffusion and transport of larvae through space are included with previous limit-cycle dynamics of the mesoscale of the Great Barrier Reef (GBR). For this nonlinear diffusion–reaction–transport process, Neumann boundary conditions are employed for an essentially elliptical model GBR. It is demonstrated that the newly discovered wave of outbreaks at the macroscale (the Reichelt wave) should have period in agreement with the mesoscale limit-cycle period. This conclusion is corroborated by the data. It is suggested that the lack of large-scale synchronous outbreaks extending to the southern end of the GBR may be due to weakened or erratic properties of the larvae transport process, at least in part. Results in this paper extend previous work of the authors on modelling starfish outbreaks on the GBR.

Keywords: nonlinear diffusion–reaction–transport; limit cycles; lumped-parameters model; plane-wave solutions; Neumann boundary conditions; starfish outbreaks; Great Barrier Reef.

1. Introduction

Acanthaster planci (the crown-of-thorns starfish) is a large predatory starfish, which has shown a propensity to outbreak in a 'boom then bust' fashion on many coral reefs of the Indo-Pacific region (Moran, 1986). Recent analysis of the distribution of this reef-building coral predator over the Great Barrier Reef (GBR) of Australia by R. E. Reichelt and colleagues has revealed a very-large-scale wave pattern (the Reichelt wave) with two southward progressions of (adult) starfish population outbreaks and a peak of outbreak abundance within each wave (see Bradbury & Munday (1989) for further data and interpretation of this wave). In this paper, we relate this wave pattern (Fig. 1) to the previously observed cyclic dynamical behaviour of *A. planci* and scleractinian corals upon which it preys, observed at the mesoscale of individual reefs (Bradbury *et al.* 1985).

82   P. L. ANTONELLI *et al.*

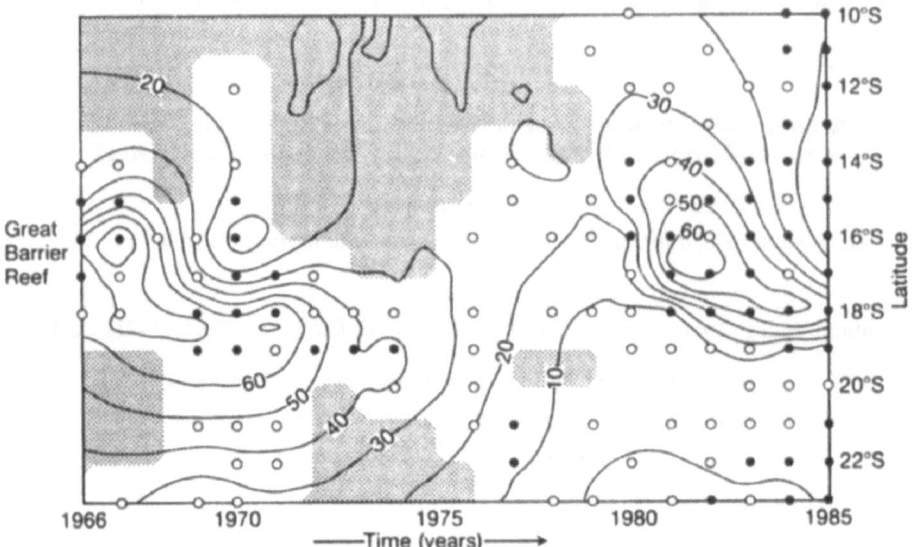

Fig. 1. Contour plot of percentage of reefs with high starfish abundance in samples from latitude–year sectors of the GBR. Solid dots indicate sectors where more than 5% of the total number N of reefs in that sector were sampled (where $78 < N < 599$). Shaded areas indicate no samples taken. Note that of the two peaks apparent, the second is much better documented than the first. (After Reichelt *et al.*, 1988).

The mesoscale dynamics of secondary outbreaks on the GBR has been modelled by Antonelli & Kazarinoff (1984, 1986, 1988) and Antonelli *et al.* (1987), using ordinary differential equations involving a crucial starfish 'cooperative' term. Identification of parameters and stochastic filtering was studied in Antonelli & Skowronski (1988) and Antonelli *et al.* (1987). In the original 2 corals/1 starfish model, the cooperative term was taken as γF^2, where the positive parameter γ was interpreted as *level of aggregation* in the population F of starfish on a reef. More recently, γF^P $(1 < P \leqslant 2)$ has been used to include density-dependent effects (Antonelli & Kazarinoff, 1988). Although fertilization rates of *A. planci* are unknown, the proximity of spawning males is likely to be a significant factor in larval development (Pennington, 1984) and thus lends itself to the interpretation we give to γF^P $(1 \leqslant P \leqslant 2)$. It has been shown numerically that, for $\gamma > \gamma_c > 0$, the system $(P = 2)$ admits a large-amplitude stable limit cycle, with γ_c the computed critical Hopf value for bifurcation to a cyclic dynamics (Antonelli *et al.*, 1987). Stability is guaranteed by 'food preference asymmetry', that is, for the interaction terms $-\delta_1 C_1 F$ and $-\delta_2 C_2 F$ in the two coral equations, $\delta_2/\delta_1 < 1$. As yet unpublished mathematical proofs (X. Lin's Ph.D. thesis, University of Alberta) for the case of 1 coral/1 starfish indicate stability for $P < 2$, and no asymmetry whatever is involved. In the present paper, we consider only the original 2 coral/1 starfish model of the mesoscale secondary outbreak dynamics (Antonelli & Kazarinoff, 1984).

In trying to relate the Reichelt outbreak wave of Fig. 1 to the observed cyclic pattern of adults at the mesoscale of the GBR, it is natural to add diffusion and transport through space to the three equations mentioned above as well as no-flux boundary conditions. The resulting diffusion–reaction system is (2.1) below. This nonlinear system of parabolic equations is a diffusion–reaction–transport system. But how does one interpret diffusion and transport for sessile organisms like corals? Also, there is no evidence to suggest that adult *A. planci* migrate *en masse* more than a few kilometres, and then only on individual reefs rather than between reefs (Moran, 1986). However, during the two-week long planktonic period for larvae, besides diffusion, there is a range of hydrodynamic forces including high-frequency oscillatory weather-band currents and tidal actions which strongly influence larval mobility and numbers (Andrews & Furnas, 1986; Wolanski & Pickard, 1985). A measure of the *average drift or advection* of the southerly (and slightly east) directed East Australian current over the central GBR has been determined by oceanographers (Andrews & Furnas, 1986) during the summer months when spawning occurs. They calculated a residual southward drift of $15 \, \text{cm} \, \text{s}^{-1}$. Thus, in a two-week period, the larvae can be transported as far as 180 km before settlement. But, these newly settled larvae cannot contribute to an outbreak for another two to three years, during which time the typical larva grows from a few millimetres to an average of 30 cm in diameter. Thus, at the macroscale, we may expect adult starfish to be 'transported' at a third to half the rate of their larvae. Moreover, the same scenario is true for coral larvae and other planktonic species that spend two to four weeks in the plankton and then reach maturity after several years. (Reichelt *et al.* 1988). Corals can increase their abundance asexually by 'budding', as well as sexually. However, the relative contribution to coral cover increases on individual reefs due to sexual and asexual recruitment and to growth is not considered here.

From the above reasoning, it follows that the phase velocity of the outbreak wave is approximately 60 km per year. Presently, we show that (2.1) admits a wave solution whose spatial average over the whole model GBR (the macroscale) closely approximates the cyclic dynamics of the mesoscale. It follows that the 10–15 year coral recovery period after an outbreak (see Moran, 1986, for evidence) should be close to the empirically derived period of the Reichelt wave of Fig. 1. For example, see latitude 16°S in Fig. 1, where the period between wave peaks is about 15 years. That this is so, strongly corroborates our model. However, while (2.1) predicts the period for the Reichelt wave (Fig. 1), the true value F of the starfish numbers, and hence the true amplitude of the empirical wave, cannot be accurately estimated because of the sheer impossibility of counting, or otherwise accurately estimating, such large numbers (Moran, 1986). Instead, we plot isotherms of percentages of reefs in the outbreaking state, in a large number of regions over the GBR. Thus, from the contours in Fig. 1, we conclude that the true amplitude is large. This is corroborated by (2.1) via the CHS (Conway, Hoff, & Smoller, 1978) theorem below and the results on the lumped-parameters model reported by Antonelli *et al.* (1987). For example, for coefficient values $\bar{\alpha}_1 = \bar{\alpha}_2 = 0.05$, $\beta = 3.50$, $\delta_1 = 0.50$, $\delta_2 = 0.45$, $\mu_1 = \mu_2 = 2.00$, $\mu_3 = 1.00$, and $\gamma = 0.145 \times 10^{-3}$, a period of 12.38 years and an amplitude of

10.34 is obtained (consult the above-cited paper for more numerical results; see also Antonelli & Kazarinoff (1984, 1986)).

In Section 2, we present the diffusion–reaction–transport equations and discuss briefly, and in our context, the basics of the CHS theory of weakly coupled parabolic systems of nonlinear partial differential equations (with Neumann boundary conditions) which we employ. Use is also made of the work on plane waves of Kopell & Howard (1973), which yields a useful approximation applicable in the mid-portion of the GBR. In the final section, we discuss biological consequences. A key assumption for the mathematical theory is that the diffusion part of the process must be strong compared with the strength M of the mesoscale dynamics and with the rate of the (larvae) transport process (see 3.1). Also, the no-flux boundary condition is important in linking the Reichelt wave with the limit-cycle dynamics of the mesoscale.

2. The reaction–diffusion–transport model

Letting C_1, C_2 and F denote the coral cover percentages and starfish population numbers, respectively, D_1, D_2, D_3, the positive diffusion constants, and T_i^j, the (continuous) advection functions, we may write the diffusion–reaction system as

$$
\left.
\begin{aligned}
\frac{\partial C_1}{\partial t} = D_1 \triangle C_1 + T_1^1 \frac{\partial C_1}{\partial x} + T_2^1 \frac{\partial C_1}{\partial y} \\
- 2\bar{\alpha}_2 C_1 C_2 + \bar{\alpha}_1 (C_2^2 - C_1^2) - \delta_1 F C_1 + \mu_1 C_1, \\
\frac{\partial C_2}{\partial t} = D_2 \triangle C_2 + T_1^2 \frac{\partial C_2}{\partial x} + T_2^2 \frac{\partial C_2}{\partial y} \\
- 2\bar{\alpha}_1 C_1 C_2 - \bar{\alpha}_2 (C_2^2 - C_1^2) - \delta_2 F C_2 + \mu_2 C_2, \\
\frac{\partial F}{\partial t} = D_3 \triangle F + T_1^3 \frac{\partial F}{\partial x} + T_2^3 \frac{\partial F}{\partial y} + \beta F(C_1 + C_2) + \gamma F^2 - \mu_3 F.
\end{aligned}
\right\} \quad (2.1)
$$

Here, $\triangle = \partial^2 / \partial x^2 + \partial^2 / \partial y^2$ is the Euclidean Laplacian, with x and y denoting latitude and longitude in a flat ellipse Ω about 20 times as long (2000 km) as wide and oriented with its major axis in the approximate north–south direction. Letting n denote the outward-pointing normal vector to the boundary of Ω, we assume no-flux boundary conditions, reflecting the fact that starfish and coral adult populations remain within the confines of the GBR; thus, we have the following assumption.

Assumption A (*no-flux*):

$$
\frac{\partial F}{\partial n} = \frac{\partial C_1}{\partial n} = \frac{\partial C_2}{\partial n} = 0 \quad \text{on } \partial\Omega \times \mathbb{R}^+,
$$

(i.e. on the boundary of Ω, for all positive time).

The advection functions T_j^i determine three gradients, one for each species density. The so-called 'reaction terms' are just the terms of the 2 corals/1 starfish model of Antonelli & Kazarinoff (1984, 1986). The most important terms are

$-\delta_1 FC_1$, $-\delta_2 FC_2$, and γF^2, which reflect the food preference of the starfish adults and their aggregation. Note that the original ordinary differential equations model may here be viewed as resulting from (2.1) above by application of the well-known lumped-parameter assumption. We shall use this terminology throughout for the system without diffusion or convection. The coefficients $\bar{\alpha}_1$, $\bar{\alpha}_2$, δ_1, δ_2, μ_1, μ_2, β are all positive constants.

The model and the consequences derived below do not actually depend on the precise elliptical shape of the model GBR we have chosen. Indeed, any figure of that general shape would do as well. However, the figure should have a smooth boundary and be approximately convex. A large concavity could produce a space lag in the generated wave, causing reflection at the southern end, something which seems biologically absurd.

The following assumption seems reasonable in view of the results of Andrews & Furnas (1986).

Assumption B (*constant southward drift*):
Average transport in x = const. directions vanishes. That is, $T_2^i \equiv 0$ $(i = 1,2,3)$, so that only diffusional effects operate through fixed latitudes. Also, $T_1^i = l > 0$ $(i = 1,2,3)$, with l constant. This means that average transport in y = const. directions is southward (approximately) and constant. (The direction in increasing x is southward).

As usual, one transforms the equilibrium state of variables to the origin in (2.1) by setting

$$\bar{C}_i = C_i - C_{0i} \quad (i = 1,2) \qquad \text{and} \qquad \bar{F} = F - F_0,$$

where (C_{01}, C_{02}, F_0) is the unique equilibrium in the positive orthant for the lumped-parameters model (Antonelli & Kazarinoff, 1984). Using Assumptions A and B, we can rewrite (2.1) as

$$\left. \begin{array}{l} \partial_t \bar{C}_i = D_i \triangle \bar{C}_i + l \partial_x \bar{C}_i + f_i(\bar{C}_1, \bar{C}_2, \bar{F}) \quad (i = 1,2), \\ \partial_t \bar{F} = D_3 \triangle F + l \partial_x \bar{F} + f_3(\bar{C}_1, \bar{C}_2, \bar{F}). \end{array} \right\} \tag{2.2}$$

The precise computed forms of $F = (f_1, f_2, f_3)$ are complicated and, since we do not require them, we refer the reader to the original paper for more details (Antonelli & Kazarinoff, 1984).

In order to discuss the plane-wave solutions to (2.2), we now introduce the phase variable

$$z = ct - \alpha_1 x - \alpha_2 y,$$

and set

$$\bar{C}_i = U_i(z) \quad (i = 1,2) \qquad \text{and} \qquad \bar{F} = U_3(z).$$

We substitute these into (2.2) and further require U_i to be 2π-periodic in z. Such solutions, when they exist, are called plane waves (or wavetrains). They do not satisfy Assumption A, however. The phase velocity is the angular frequency C divided by the wavenumber $\alpha_1^2 + \alpha_2^2 \equiv |\alpha|^2$. The vector α indicates the direction of the wavefront. Equations (2.2) become the vector equation

$$\bar{C}\partial_z U = F(U) + |\alpha|^2 D\partial_z\partial_z U, \tag{2.3}$$

where $\bar{C} = C + \alpha_1 l$ is the *phase-shifted angular frequency, and D* is the 3×3 diagonal matrix of diffusion constants. Because of the constancy of the advection current coefficient l, this phase shift is well defined. This transformation reduces the equations to the no transport case. The phase-shifted wave has phase velocity $\bar{C}/|\alpha|^2$ and direction α. Here, we take α_2 small compared with α_1, so that the wave is directed approximately southward. By a known theorem (see e.g. Kopell & Howard, 1973), $(\bar{C}_1, \bar{C}_2, \bar{F})$ is a plane-wave solution of (2.2) if and only if $U(z)$ satisfies (2.3). These authors proved that, for the no transport case, there is a large family of plane-wave solutions parameterized by the wavelength $2\pi/|\alpha| = \bar{\lambda}$ and that large $\bar{\lambda}$ gives good approximations to the limit cycle of the lumped-parameters model. As usual, the frequency is given by $v \equiv \bar{C}/2\pi$, so that $\bar{\lambda}v$ is the phase velocity. In particular, these plane waves can have large amplitude and phase velocity.

We wish now to use the CHS (Conway, Hoff, & Smoller, 1978) theory to study the relationship of the above given family of wavetrain solutions to the limit cycle of the lumped parameter model.

There are several factors of importance to the CHS theory.

(a) The smallest eigenvalue of $-\triangle$, denoted λ. For the ellipse modelling the GBR, λ is inversely proportional to the square of the length of the major axis.

(b) The (Euclidean) norm M of the Jacobian matrix of F maximized over the (compact) *invariant set* $\Sigma \subseteq \mathbb{R}^3$ defined for (2.1). Recall that, for such a compact set, if the values of a solution of (2.1) or (2.2) are contained in Σ for one value of t, the same is true for all later times. The existence of Σ is assured by the existence of compact invariant sets containing an attractor for the lumped-parameters system. Such a set is rectangular in phase space. Solutions satisfying Assumption A exist and are unique for specified initial time functions U_0.

(c) The minimum of the diffusion constants D_i $(i = 1,2,3)$. This is denoted d. This must be relatively large in the ensuing theory.

(d) The maximum size (Euclidean norm) of the advection vectors $T^i = (T^i_1, T^i_2, T^i_3)$ $(i = 1,2,3)$ over the entire GBR, denoted a. In our case, $a = l$.

These four components serve to define a critical rate constant

$$\sigma = d\lambda - M - a(3\lambda)^{\frac{1}{2}}, \tag{2.4}$$

which measures the rate of convergence of any wave solution of (2.1) or (2.2), whose initial values lie in Σ, to the limit-cycle solution of (2.1). (which is independent of space variables). In fact, this spatially homogeneous solution is a (Liapunov) stable solution of the lumped-parameters system and is also stable as a wave solution of (2.1) or (2.2), if $\sigma > 0$. The CHS theory makes the following definition of wave stability.

A bounded solution $\Gamma(t)$ of the lumped-parameters model is a stable solution of (2.1) or (2.2) if, for every $\varepsilon > 0$, there is a $\delta > 0$ and $t_\varepsilon > 0$ such that, if

$$\text{dist}(U_0, \Gamma) + \|\text{grad } U_0\|_{L^\infty} < \delta,$$

then $\|U(\cdot, t) - \Gamma(t)\|_{L^{\infty}} \leq \varepsilon$ for all $t \geq t_{\varepsilon}$. Here, L^{∞} indicates the uniform sup norm for functions almost everywhere on Ω (and L^2 denotes the usual norm of square integrable functions on Ω). U_0 is a vector solution at time $t = 0$.

THEOREM (CHS)   *Let $\sigma > 0$ and let U be any solution of (2.1) and Assumption A, with U_0 in the invariant set Σ determined by the stable limit cycle of the lumped-parameters system. Then, there are constants $C_i > 0$ $(i = 1, 2, 3)$ such that*

$$\|\text{grad } U(\cdot, t)\|_{L^2} \leq C_1 e^{-\sigma t},$$

$$\|U(\cdot, t) - \bar{U}(t)\|_{L^{\infty}} \leq C_2 e^{-\sigma t},$$

where $\bar{U}(t) = (1/|\Omega|) \int_{\Omega} U(x, t) dx$, with $t \geq 0$, is the spatial average of $U(\cdot, t)$ over Ω, whose area is $|\Omega|$. Furthermore, $U(t)$ satisfies

$$\frac{d\bar{U}}{dt} = F(\bar{U}) + g(t),$$

where $|g(t)| \leq C_3 e^{-\alpha t}$.

This important theorem can be paraphrased as saying that any no-flux solution starting out at $t = 0$ in the invariant set Σ of the lumped-parameters model converges, at rate $\sigma > 0$, to its averaged value over the whole Ω and that, for large times, this average value is a good (uniform) approximation to the stable limit cycle. In particular, the large time behaviour of the wave solution is determined completely by the limit cycle, and convergence is relatively rapid. Therefore, for large times, the plane-wave solution will well approximate this wave away from the boundary of Ω.

3. Consequences of the wave model

Because of the relatively slow and more variable increase in coral abundance compared with the boom in starfish abundance (Moran, 1986), we may take D_3 greater than D_2 and D_1 in (2.1), so that $d = \min\{D_1, D_2\}$ in (2.4) and our formula for the rate of convergence of the wave onto the limit cycle is

$$\sigma = \frac{db}{L^2} - M - \frac{l}{L}(3b)^{\frac{1}{2}} > 0, \tag{3.1}$$

where L is the length of the major axis of the ellipse and b is some positive constant. Positivity of σ means that coral diffusivity d must be large relative to L, and relative to the rates of reaction M and relative advection l/L. This must be assumed before any consequences of the CHS theory can be brought to bear on the model (2.1). These statements apply to the plane-wave solution, which for us models the central portion of the GBR, at large times.

HYPOTHESIS   *The plane wave's phase velocity $\bar{C}$ is completely determined by the advection current in the central GBR.*

Recalling that $\bar{C} = C + \alpha_1 l$, we must let $C \equiv 0$, for in this instance, and only in this instance, is the hypothesis fulfilled. Then (2.3) becomes

$$F(U) + |\alpha|^2 D \partial_z \partial_z U = 0. \tag{3.2}$$

For the corresponding wave equation of form (2.2) with $l = 0$ and Assumption A, we have necessarily, for large times,

$$\partial_t \bar{C}_i \approx 0 \approx \partial_t \bar{F} \quad (i = 1, 2), \tag{3.3}$$

by the CHS theorem. Assuming $\partial_t \bar{C}_i = 0 = \partial_t \bar{F}$ ($i = 1, 2$) exactly, it then follows from CHS (Conway *et al.*, 1978: p. 10) that $\sigma > 0$ implies the no-flux boundary problem has only constant solutions (i.e. solutions which do not depend on either space or time variables). However, $F(U) = 0$ must give all solutions since all are constant. But there is only one biologically meaningful solution to this ($U = 0$), because there is only one equilibrium $C_i = C_{0i}$ ($i = 1, 2$) and $F = F_0$ in the positive orthant, and this is a specific case of $U = 0$, owing to transformation to the origin in (2.2).

The above argument shows that large-scale synchronous outbreaks in the central GBR could not occur if there were no net advection current. Although outbreaks on individual reefs may still occur, depending on local conditions, these would not be propagated at all; rather they would be smeared out by diffusive or high-frequency random effects. However, even if outbreaks were occurring throughout the GBR, the Kopell & Howard (1973) theorem would imply the existence of a plane wave with direction α at our disposal to choose as we please, so that there would be many α from local short-time wavefronts, but wholly unpredictable at the larger scale. This discussion, then, lends weight to the argument that isolated local outbreaks occurring in the north are propagated southward by the East Australian current and produce a large-scale wave of outbreaks every 15 years or so, as observed in Fig. 1.

Moreover, the model suggests that, if the advection current dies out or diminishes significantly in strength (l) at the lower end of the GBR, then the Reichelt wave will not be observed there. Also, fluctuations in α_1 can cause negative interference in the wave, causing, for example, a decrease in amplitude. Ocean conditions may not be favourable for starfish larvae or for high recruitment success at the southern end of the GBR and again the wave would not be observed. This is because the adult starfish population would not be large enough to support sufficiently the high levels of aggregation necessary to produce a limit cycle.

The general result described above shows the dependence of the outbreak wave on the long-timescale seasonal advection current and on the mesoscale dynamics. It suggests a mechanism by which the wave could be disrupted or even modulated. Long-term changes in advection current behaviour, over intervals of two to ten years, could radically influence characteristics of the wave. A gross change in the East Australian current or its interaction with the continental shelf break could disrupt the Reichelt wave completely. Alternatively, if control efforts on a number of individual reefs were effective in reducing the abundance of adult starfish, then the large-scale outbreak wave would be affected, in turn. Whereas we are only able to observe, rather than cause, disruption of the East Australian current, we may be able to alter mesoscale dynamics by serious efforts to control local outbreaks.

Acknowledgements

The authors would like to thank NSERC A-7667 for partial support of this paper. Thanks are also extended to Vivian Spak of the University of Alberta Mathematics Department for her excellent AMS-Tex typesetting. This is contribution number 444 from the Australian Institute of Marine Science.

References

ANDREWS, J. C., & FURNAS, M. 1986 Subsurface intrusions of coral sea water into the Central Great Barrier Reef—I. Structures and shelf-scale dynamics. *Continental Shelf Res.* **6**, 491–514.

ANTONELLI, P. L., ELLIOTT, R. J., & SEYMOUR, R. M. 1987 Nonlinear filtering and Riemannian scale curvature. *R. Adv. Appl. Math.* **8**, 237–53.

ANTONELLI, P. L., FULLER, K. D., & KAZARINOFF, N. D. 1987 A study of large-amplitude periodic solutions in a model of starfish predation on coral. *IMA J. Math. Appl. Med. Biol.* **4**, 207–14.

ANTONELLI, P. L., & KAZARINOFF, N. D. 1984 Starfish predation of a growing coral reef community. *J. Theor. Biol.* **107**, 667–84.

ANTONELLI, P. L., & KAZARINOFF, N. D. 1986 Comments on starfish/coral cycles after R. Bradbury *et al. J. Theor. Biol.* **119**, 501–2.

ANTONELLI, P. L., & KAZARINOFF, N. D. 1988 Modelling density-dependent aggregation and reproduction in certain terrestrial and marine ecosystems: A comparitive study. *Ecol. Mod.*, **41**, 219–27.

ANTONELLI, P. L., & SKOWRONSKI, J. M. 1988 Identification of parameters in a model of starfish predation or corals. *Math. Comput. Mod.* **10**, 17–25.

BRADBURY, R. H., HAMMOND, L. S., MORAN, P. J., & REICHELT, R. E. 1985 Coral reef communities and the crown-of-thorns starfish: Evidence for qualitatively stable cycles. *J. Theor. Biol.* **113**, 69–81.

BRADBURY, R. H., & MUNDY, C. N. 1989 Large scale shifts in biomass of the Great Barrier Reef ecosystem. In *Biomass and Geography of Large Marine Ecosystems* (K. Sherman & L. Alexander, Eds.) Boulder, CO, Westview.

CONWAY, E., HOFF, D., SMOLLER, J. 1978 Large time behaviour of solutions of systems of nonlinear reaction–diffusion equations. *SIAM J. Appl. Math.* **1**, 1–16.

KOPELL, N. & HOWARD, L. 1973 Plane wave solutions to reaction–diffusion equations. *Studies Appl. Math.*, (4), 291–324.

MORAN, P. J. 1986 The *Acanthaster* phenomenon. *Oceanogr. Marine Biol.*, *Ann. Rev.* **24**, 379–480.

PENNINGTON, J. T. 1984 The ecology of fertilization of echinoid eggs; the consequences of sperm dilution, adult aggregation, and synchronous spawning. *Biol. Bull.* **169**, 417–30.

WOLANSKI, E., & PICKARD, G. 1985 Long-term observations of currents on the central Great Barrier Reef continental shelf. *Coral Reefs* **4**, 47–57.

REEF SYNTAX: AN EXPLORATORY DATA ANALYSIS OF THE *ACANTHASTER* PHENOMENON USING STRINGS AND GRAMMARS

M.B. DALE

CSIRO, Cunningham Laboratory, St. Lucia, Queensland 4067

R.H. BRADBURY[1] AND R. E. REICHELT[2]

Australian Institute of Marine Science, Townsville, Queensland 4810

Abstract. This paper is concerned with the analysis of historical records of abundance of coral and crown-of-thorns starfish (*Acanthaster planci*) on the Great Barrier Reef. Analysis of the literature permitted the identification of 88 reefs for which coral and starfish abundances had been estimated on at least 3 occasions. Unfortunately observations were both very sparse and very patchy until the establishment of regular surveys in 1980, so that standard methods of pattern analysis could not be employed. This paper examines two relatively new exploratory methods.

The first technique used a Levenshtein similarity measure as a basis for numerical classification and ordination of the time series, after coding the observational data to cope with variations in observer precision. This similarity measure permits the comparison of entire sequences and does not require major adjustment if gaps exist in the sequences. Following the classification analysis, it was determined using the Friedman-Rafsky test, that some of the groups identified were **regularly** distributed along the reef from north to south. The precise nature of the groups requires further investigation. The ordination analyses gave suggestions of various forms of cyclic structure but were insufficiently clear for useful interpretation.

To attempt to further identify these tentative cyclic structures, the historical data were analysed using techniques of grammar inference and the results displayed using the derivation sets. This permitted the identification of several classes which reflected the processes of an outbreak of starfish. While in some cases recovery of coral after attack seems possible, at present there is no evidence of an overall fall in starfish numbers which would presumably be a significant indicator of cessation of the outbreaks.

---

[1] Present address: National Resource Information Centre, Canberra, ACT.
[2] Present address: Bureau of Rural Resources, Canberra, ACT.

INTRODUCTION

The Great Barrier Reef is the world's largest coral reef tract (Hopley, 1982). It occupies about 230,000km^2 of the Queensland continental shelf. It stretches down the Queensland coast for about 2300km from New Guinea to the Tropic of Capricorn, and spans about 15° of latitude.

The crown-of-thorns starfish, *Acanthaster planci*, is a coral eating starfish which has a tendency to outbreak (Moran, 1986). When it does, it causes massive mortality of the hard coral cover, one of the important architectural components of the reef ecosystem. During the last few years, outbreaks have been observed on many reefs in the system prompting widespread public (Raymond, 1986) and scientific (Endean & Cameron, 1985) concern. Widespread outbreaks were observed in the system in the late 1960s and early 1970s. In this period large scale mortality of hard corals was reported in many areas (Pearson, 1981).

The dynamics of the *Acanthaster* phenomenon at the scale of the entire Great Barrier Reef are poorly understood. There are some empirical descriptions (Moran, Bradbury & Reichelt, 1988, Bradbury & Mundy, 1989) and some analytical models (Antonelli, Bradbury & Reichelt, in press, Antonelli *et al.*, 1989). However there remains a need for some exploratory data analysis and pattern analysis, to determine the nature of the patterns present in historical data. It is the application of such methods which forms the basis for this paper, in particular classification methods and associated inference of grammatical models.

There are many questions concerning the *Acanthaster* problem which might be approached through the analysis of historical data. For example, questions concerning the existence of cycles of outbreaks, and of dynamically stable chronic infestations, can only be answered by historical evidence, although the possibility of such events can be deduced from simulation models of various kinds (Bradbury *et al.*, 1985; Bradbury & Antonelli, this volume). The major problem with such analyses is the lack of consistent recording of the abundance of coral and starfish over a range of reefs for a sufficient period of time. However with appropriate analytic techniques some questions can be addressed even with the limited data available

DATA

The data set derives from the Great Barrier Reef Marine Park Authority, an Australian federal agency responsible for managing a large part of the

Great Barrier Reef. They have collected an "all-source" data set of records describing the distribution and abundance of starfish and corals on individual reefs of the system. These data are derived from reports of fishermen, divers and other reef users as well as being culled from the scientific literature. They are of mixed quality but, nonetheless, represent a unique and valuable resource. This original data set contains more than 1000 records involving records for about 500 reefs over the period 1962-1985.

Our colleagues Scott Bainbridge, Johnston Davidson, Craig Mundy and Peter Speare examined and assessed each of these records and extracted reliable records of starfish and coral abundances observed on the reef during a calendar year. From this set we extracted the records of reefs for which both starfish and corals had been observed in each year for at least three different, but not necessarily successive, years. This provided records for 88 reefs geographically spread throughout the system but concentrated in the more heavily surveyed central third (Cairns-Townsville). Dates ranged from the mid-1960s to 1985, with most records from the 1980s.

Because it was derived from many observers, the data set contained a mixture of data types. We transformed this mixture into 3 categories (low, medium, high) for both starfish and corals using the rules shown in Table 1. We then assigned each record a category in a 9-state variable formed by combining the codes for starfish and coral as shown in Table 2. Thus each reef, at each date at which it was observed, provides a **single** code value indicating the abundances of starfish and corals. These codes form, for each reef, a sequence indicating the temporal changes so far as they are recorded; Table 3 shows some of the final data. In addition to the starfish and coral, the position of each reef is also known[3].

METHODS

We undertook two investigations in our attempt to elucidate patterns in the sequence data. In the first we are concerned with whether the reefs show similarities in the order in which the code symbols appear in our records. If they do then we may be able to associate common patterns in the sequences with common geographic position, or common history. To do this requires the definition of a measure of dissimilarity between pairs of reefs with subsequent investigation of patterns in the dissimilarities. However, the commoner measures of dissimilarity, such as the

---

[3] The identification number assigned to each reef largely indicates its position on a Northwest-Southeast latitudinal axis although the reefs are not regularly spaced along this line.

Table 1. Coding rules for starfish and corals

---

1. Starfish coding (Kenchington & Morton, 1976)

If numbers of starfish were counted, then

Code	Number
low	0-3
medium	4-15
high	>15

If starfish counts had already been categorised on a 4-state scale. then

Code	State
low	0-1
medium	2
high	3

---

2. Coral coding

If live coral cover is scored out of 100,
not as percentage of surface area covered

Code	% of total
low	0-30
medium	31-70
high	71-100

If expressed as percentage of coral cover in low cover category
(Kenchington & Morton,1976)

Code	% low coral
low	61-100
medium	41-60
high	0-40

If actual percentage of live coral was measured then

Code	% live coral
low	0-5
medium	6-30
high	31-100

Marczewski-Steinhaus (1958; also known as the Bray-Curtis or Jaccard coefficient) measure, Gower's (1971) mixed data measure, Williams, Lambert & Lance's (1966) information measures or Goodall's (1966) probabilistic measure, all require that the items to be compared are described by a set of single-valued attribute values; for example that they are of this length, that weight and such a colour. This clearly does not include the string descriptions which are the primary descriptors of the reefs in the present study, since each string can be of differing length. Some alternative is clearly required.

The second investigation involves the inference of a context-free grammar[4]. This seeks to describe the patterns of symbols in the reef strings in a succinct manner by the identification of common substructures in the sequences. For example we might determine that all sequences show a common pattern of the form **beginning, middle, end**, although there may still be variation within each of these categories. While grammars have been widely applied as a means of describing essentially infinite variety with a finite set of rules, in most cases reliance has been placed on human induction of the necessary set of rules. Automated inference procedures have received less attention. Unlike the previous investigation, a grammar describes patterns by seeking structure within the sequences as well as between sequences. The combined structure is represented as a set of hierarchically arranged rules which can be used to generate all the sequences.

Table 2. Coding rules for reefs combining starfish and coral codes.

Reef Code	Starfish Code	Coral code
1	low	low
2	low	medium
3	low	high
4	medium	low
5	medium	medium
6	medium	high
7	high	low
8	high	medium
9	high	high

The objective of most inference procedures is the simple grammar which generates all and only the observed sequences. One possible measure of simplicity is just the number of rules, another the number of nonterminal symbols which are introduced. Both of these are still dependent on the type of grammar we seek; we might expect a regular

---

[4] A brief account of grammars is given in the Appendix, together with an account of the Backus-Naur format for their description.

grammar to require more symbols than a context-free grammar simply because the rules of the former are more constrained. Since regular grammars are a restricted subset of context-free grammars, we are presumably entitled to regard them as simpler than a general context-free grammar. Our inference procedure can, and with other data has, produced a result which was a regular grammar, that is the result lacked any rules with 2 nonterminals on the right hand side. In such a case we have a very simple result.

The restriction to generation of **all and only** the observed sequences is also something of a nuisance and we would probably prefer to relax it a little. There are almost certainly other sequences which have occurred but which we have not observed. The problem is that, in the absence of any sequences which are **known** to be impossible, any relaxation of the constraint produces grammars which can generate **many** other sequences. This means that a unique inference procedure is impossible, except for a very restricted class of grammars. One solution is to seek the simplest grammar which, in some sense, most closely approximates the observed strings, with any potential grammar being assessed both by its complexity and by its adequacy. In any case the inference of a unique context-free grammar is "in the limit", and can require an infinite number of sequences. The procedure we use is a sequential one and the result dependent on the order of presentation.

The dissimilarity measure chosen was a Levenshtein distance defined as *the minimum number of changes necessary to convert one string of code symbols into another*. Dale (1989a, b, c) provides a more detailed discussion of such measures, and here it is only necessary to indicate that the changes we permitted were **insertion** and **deletion** of code symbols only, with equal weights for all changes independent of position and symbol. As such our measure is certainly a rather crude means of reflecting the processes which actually occur to change the abundance of starfish and coral. With more knowledge of these processes, it should be possible to improve the measure. For example some changes are certainly easier to accomplish than others and in such cases it would be possible to differentially weight them. It would also be possible to differentially weight changes according to position in the strings, or to allow for variations in the rates of change resulting from different processes by using "time-warping" as a permissible change[5]. Given the lack of information on reef processes such embellishments are hardly warranted.

---

[5] In time-warping, a single symbol in one sequence is mapped to several instances of the same symbol in the other sequence.

The dissimilarities can be used in various ways to organise the resemblances between reefs. In the present study we classified them using a standard agglomerative classification method the Lance-Williams (1967) flexible sorting strategy using $\beta=-0.25$, to yield 9 groups, as indicated by Mojena's (1977) test, or 4 groups, as indicated by the possibly more powerful Ratkowsky-Lance (1978) test. We investigated the relationships between these groups by first calculating the minimal spanning tree[6] (Prim, 1957) and also a two-neighbour connection graph[7] following Williams (1980, see also Abel & Williams, 1981) TWONET procedure. This latter has proved an effective method in previous applications, for example Williams & Tracey (1984), although in the present case we did not distinguish the strength of connections (Williams *et al.*, 1980).

Using the location data, we can further analyse the data using statistical tests to examine if the groups are non-randomly distributed in space. To do this we first generate a spanning tree using the location data for the reefs and then use the Friedman-Rafsky (1979) test to determine if the distribution of sequence group labels on this tree is random[8]. For this analysis we used the 4 groups identified by the Ratkowsky-Lance test, as otherwise some groups had too few members for effective testing.

In contrast to these almost traditional procedures of exploratory data analysis, the grammar inference procedure represents a relatively new and untried method of identifying patterns in the data. Thus the results presented here represent only the first attempts to use what promises to be a powerful technique of analysis. We used the Chirathamjaree & Ackroyd (1980) procedure for inferring a non-recursive context-free grammar[9] for the strings and from this determined the derivation sets, which represent the various possible paths for describing the strings. In a previous study (Dale & Barson 1989) these sets provided a suitable means of displaying information contained in the grammar. It is not suggested that we have obtained an optimal grammar; the inference procedure is chosen as much for its computational tractability as its

---

[6] The minimal spanning tree for a set of items connects the items so that the sum of distances between the items connected by the tree is minimal. It may not be unique if some distances are equal in value.

[7] A minimal spanning tree connects all nearest neighbours and adds those extra linkages required to connect all items in two-neighbour analysis we connect each item to its nearest and second nearest neighbours. The resulting graph may be disconnected.

[8] We wish to test if the sequence groups we have derived are clustered in their locations as well as their biological history. We therefore examine the links on the location spanning tree and determine the group labels attached to the items at each end of the link. If the reefs are clustered in location then there will be few connections between groups with different labels, whereas if they are regularly spaced then there will be too many such connections. To make the test we need to determine the expected number of such connections if the groups are truly randomly distributed.

[9] See Appendix for a definition.

efficacy. It would no doubt be desirable to introduce stochastic elements, to permit parallel operation of rules, to permit recursive structures or to make use of programmed grammars where each rule determines its potential successors. Any of these additional devices might improve the fit to our data, reduce the overall complexity of the grammar necessary to describe the data with the same degree of fit, or even to improve biological interpretability. Although procedures for inference of all such grammars have been proposed, in the absence of effective programs we have no real choice. In any case there is something to be said for understanding a simple procedure before introducing complications in the interpretation and the simplicity of inference at least permits a relatively large data set to be examined.

RESULTS

In Table 3 we show some of the clusters of reefs formed at the 9-group level of the Lance-Williams hierarchical classification. Ideally each group would be representable by a common subsequence representing those parts of the reef strings which were held in common. It is at once apparent that these groups are difficult to define in any such simple manner. While some common substrings can be identified for groups of 2 or 3 reefs, for the larger groups shown the longest subsequences are at most 2 or 3 symbols. This difficulty of definition is, of course, a common characteristic of polythetic agglomerative methods applied to sparse data and we have noted this difficulty in analyses of other data (Dale &.Barson, 1989). In the present study the difficulty is compounded since it is the common characteristics of sets of strings which must be identified. We cannot here turn to monothetic or oligothetic methods in an attempt to determine groups which **do** have a simple definition since such methods are not presently defined for string data. Thus the poor overall quality of our data preclude the traditional reification of the clusters formed.

Although it would certainly be useful to be able to identify specific features of the strings associated with each cluster, this does not mean that the groups are without usefulness. We are interested in the spatial patterning of the groups as well as their semantics. In Figure 1 we show the minimum spanning tree and the two-neighbour connections. The results for the tree suggest four major components exist in the system, groups [A-C], [B-E-G-F], [D] and [H-I] , with [B-E-G-F] forming a central core and the others peripheral. The two-neighbour graph could also be

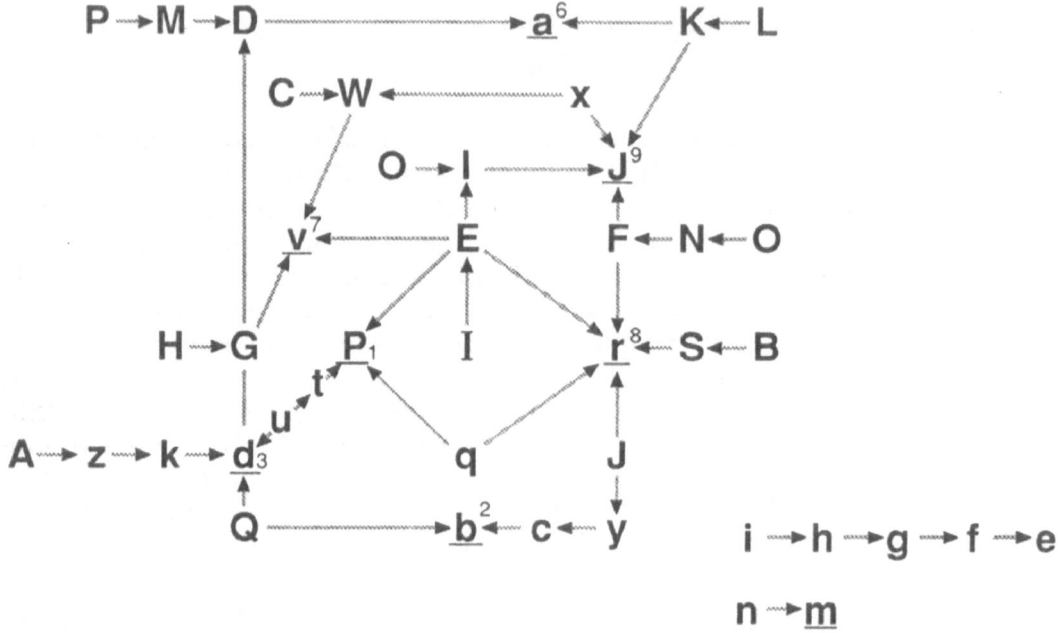

Figure 1. Minimal spanning tree and two-neighbour graph for Levenshtein measures.

KEY

→ Derivation Path
Arrow indicates direction from left to right side of rule

A Nonterminal symbol

r Terminal symbol

1 Combined Starfish / Coral code

Figure 2. Derivation sequences for context-free grammar.

Table 3. Examples of groups formed in the Lance-Williams classification. Numbers in body of table are reef codes.

Reef	Year of Record.																			
	66	67	68	69	70	71	72	73	74	75	76	77	78	79	80	81	82	83	84	85
49	.	.	.	9	5	.	.	.	.	.	.	.	.	.	.	3	.	8	7	9
55	.	.	.	7	4	.	.	.	.	.	.	.	.	.	.	6	6	7	7	7
57	.	.	.	7	7	.	.	.	.	.	.	.	.	6	.	.	3	6	8	9
59	.	.	.	7	7	.	.	.	.	.	.	.	2	.	.	.	3	7	8	7
63	.	.	.	2	3	.	.	.	.	.	.	.	.	.	.	3	3	6	8	8
64	.	.	.	8	8	7	.	.	.	.	.	.	.	.	6	2	6	.	6	5
16	.	.	.	.	1	.	.	.	.	.	.	.	.	8	.	9	1	.	.	1
20	.	7	.	.	1	.	.	.	.	.	.	.	.	7	.	9	.	.	.	1
34	7	.	.	1	.	.	.	.	.	.	.	.	.	.	.	9	.	5	.	1
39	.	.	.	7	1	.	.	.	.	.	.	.	.	.	.	.	9	7	1	.
42	.	.	.	7	.	.	.	.	.	.	.	.	.	.	.	.	9	5	.	.
44	.	.	.	.	7	.	.	.	.	.	.	.	.	.	.	.	5	8	7	9
45	.	.	.	.	.	.	.	.	.	.	.	.	.	.	.	6	9	8	8	.
46	.	.	.	3	7	.	.	.	.	.	.	.	.	.	.	6	3	8	.	1
50	.	.	.	.	1	.	.	.	.	.	.	.	.	.	.	.	2	7	9	7
56	.	.	.	.	.	8	.	.	.	.	.	.	.	.	.	.	.	7	9	8
62	.	.	.	.	9	8	.	.	.	.	.	.	.	.	.	.	.	6	2	5
65	.	.	.	.	7	7	.	.	.	.	.	.	.	.	.	3	2	.	8	.
69	.	.	.	.	1	6	.	8	5	.	.	.	.	.	.	.	.	.	.	4
70	.	.	.	.	2	6	.	8	8	.	.	.	.	.	.	.	.	.	.	1
85	.	.	.	.	.	.	.	.	.	.	.	.	.	.	.	9	6	2	8	6
5	4	4	.	.	.	.	.	.	.	.	.	.	.	.	.	8	7	5	7	4
23	7	.	.	.	3	.	.	.	.	.	.	2	.	.	.	8	.	.	3	3
35	8	.	.	1	1	1	.	.	.	.	.	.	.	.	.	9	9	.	.	1
41	.	.	.	7	7	7	.	.	.	.	.	.	.	.	.	.	5	.	7	.
58	.	.	.	.	7	.	.	.	.	.	.	.	.	.	.	2	7	7	7	7
19	.	.	.	.	1	.	.	.	.	.	.	.	.	.	.	2	.	5	3	2
22	1	.	.	.	2	.	.	.	.	.	.	.	.	.	.	.	.	.	3	9
47	.	.	.	.	3	.	.	.	.	.	.	.	.	.	.	6	.	2	2	2
53	.	.	.	.	.	.	.	.	.	.	.	.	.	.	.	2	.	2	2	3
67	.	.	.	.	.	.	.	3	2	.	.	.	.	.	.	.	.	.	3	2
74	.	.	.	.	.	.	.	2	5	.	.	.	.	.	.	.	.	.	3	2

taken to indicate some cyclic pattern in groups D to I, with the A, B and C groups peripherally placed. While not unequivocal, there is a possibility that the reefs are undergoing a cyclic process of change under the impact of the starfish. The peripheral groups A,B, and C are in fact all associated with high coral and low starfish conditions and probably represent uninfested reefs which have been free of starfish for some considerable period of time. The cyclic pattern would then represent reefs which are presently infested or have recently been so infested and now recovering.

The groups prove more useful when coupled with the location spanning tree. Table 4 shows the results of the Friedman-Rafsky test for spatial pattern. The surprising result is that 3 of the 4 groups used are **regularly** distributed along the reef; that is they form a pattern such as ABCABCABC repeating at regular periods. If there are many repetitions, then the obvious interpretation is that the starfish are moving in waves along the reef and the regular structure results from the processes of invasion, destruction and recovery. Partial recovery is necessary if the starfish can re-invade the same reefs at a later date. Mechanisms have been proposed for the establishment and maintenance of wave structures e.g. Antonelli *et al* (1989), Reichelt (this volume), but this is, so far as we are aware, the first direct observational evidence that regular patterns exist in the reef responses

Table 4. Results of Friedman-Rafsky tests using location spanning tree and 4 groups formed by classification of reef strings.

Group being compared	Observed Adjacencies	Expected Adjacencies	Standard Deviation	Normal Deviate	Pattern
A-B	3	8.18	1.31	-8.54	**Regular**
A-C	11	16.36	2.12	-2.53	**Regular**
A-D	19	18.00	2.27	0.44	Random
B-C	8	4.55	0.91	-5.02	**Regular**
B-D	4	5.00	0.96	-1.04	Random
C-D	11	10.00	1.50	0.67	Random

If we plot the locations of the groups along the reef, we can discover something of the periodicity although this is made difficult by the unequal sampling densities. However it seems that only a few repetitions occur so an alternative explanation might also be acceptable. If the origin of the starfish invasion was initially centrally positioned on the Barrier

Reef and then diverged north and south, we would again obtain some regularity of distribution. This does not preclude the possibility that further waves may follow or that reflection of the present wavefronts of infestation may not occur when the ends of the reef are approached

The possibility of cyclic processes is also indicated by the context-free grammar. The 88 reefs are described by a set of 29 rules, approximately 1/3 of the number of reefs. This indicates that there is indeed common structure to be found within the sequences, and that this structure is shared between the sequences. Table 5 shows these rules, in Backus-Naur form, with what we shall call *interesting* rules specially marked. Interesting rules we take to be rules where the right hand side contains 2 **non-terminal** symbols; this indicates that the string is partially composed of two (or more) separate components each of which appears in other strings in isolation[10].

Detailed examination of the rules poses problems because of the intermittent nature of the records. For example rule 7 shows that it is possible to change from [high coral, high starfish] to any of 7 other states! The interesting rules show some interesting combinations; rule E demonstrates that changes from [low coral, high starfish] states can lead to [high coral, high starfish], while rule u shows [low coral, low starfish] recovering to [high coral, low starfish] Obviously information on the frequency of transitions and the periods required to make them is desirable but the paucity of data makes such estimates unjustifiable.

In Figure 2 we show the derivation sets and it is clear that, in the largest connected set, differences between sequences reflect a change to an **adjacent** sequence. This leads to overlapping of the sequences and thus to a circular structure, indicating some kind of cyclic process. The simple circularity is disrupted by cross connections, but we regard these as a result of gaps in the strings and thus of the paucity of data. It is of course possible that some processes can cause a short cycle to appear, or that alternative paths may exist. Our data are insufficient to distinguish these possibilities with any certainty.

It should also be noted that there are 2 other derivation sets unconnected with this major cyclic process system. They appear to involve an oscillation between low and moderate coral coupled to high and moderate starfish codes. Again the appearance of 2 such derivation sets may simply represent the paucity of our data, but other explanations are possible and we shall discuss the interpretation of these disjoint sets below.

---

[10] This does not mean that the other rules are without any interest. See Appendix for further discussion.

Table 5. Context-free grammar productions. Interesting rules are marked by !!!.
For a definition of interesting, see text.

$S \Rightarrow 2|7|a|b|c|d|i|j|k|l|n|o|p|q|r|s|t|u|v|w|x|$
$\quad y|z|A|B|C|D|E|H|I|J|L|M|O|P|Q$

$Q \Rightarrow b{:}d$	!!!
$P \Rightarrow M{:}4 \mid M{:}6;\ M \Rightarrow D{:}1 \mid D{:}2 \mid D{:}3 \mid D{:}5 \mid 2{:}D \mid 9{:}D$	
$O \Rightarrow N{:}5;\ N \Rightarrow F{:}6;\ F \Rightarrow r{:}j \mid r{:}a$	!!!
$L \Rightarrow K{:}5;\ K \Rightarrow j{:}a$	!!!
$J \Rightarrow y{:}r$	!!!
$I \Rightarrow E{:}9;\ E \Rightarrow l{:}r \mid p{:}l \mid v{:}l$	!!!
$H \Rightarrow G{:}1 \mid G{:}9;\ G \Rightarrow d{:}D \mid v{:}d$	!!!
$D \Rightarrow a{:}5 \mid a{:}8 \mid a{:}9 \mid 1{:}a \mid 2{:}a \mid 3{:}a$	
$C \Rightarrow w{:}1 \mid w{:}7$	
$B \Rightarrow s{:}2 \mid s{:}8;\ s \Rightarrow r{:}1 \mid r{:}2 \mid r{:}5 \mid 5{:}9$	
$A \Rightarrow z{:}3;\ z \Rightarrow k{:}8$	
$y \Rightarrow c{:}7 \mid c{:}9 \mid 3{:}c$	
$x \Rightarrow w{:}j$	!!!
$w \Rightarrow v{:}1 \mid v{:}4 \mid v{:}6 \mid v{:}7 \mid v{:}8 \mid v{:}9$	
$v \Rightarrow 7{:}1 \mid 7{:}2 \mid 7{:}3 \mid 7{:}4 \mid 7{:}5 \mid 7{:}6$	
$u \Rightarrow t{:}d$	!!!
$t \Rightarrow p{:}1 \mid p{:}3 \mid p{:}4 \mid p{:}5 \mid p{:}7 \mid 8{:}p$	
$r \Rightarrow 8{:}1 \mid 8{:}2 \mid 8{:}3 \mid 8{:}7$	
$q \Rightarrow p{:}9 \mid p{:}r$	!!!
$p \Rightarrow 1{:}2 \mid 1{:}3 \mid 1{:}7 \mid 1{:}8 \mid 1{:}9$	
$o \Rightarrow l{:}1 \mid l{:}2$	
$n \Rightarrow m{:}2 \mid 7{:}m;\ m \Rightarrow 5{:}7$	
$l \Rightarrow j{:}1 \mid j{:}2 \mid j{:}3 \mid j{:}6 \mid j{:}7 \mid 1{:}j \mid 7{:}j$	
$k \Rightarrow d{:}1 \mid 1{:}d \mid 7{:}d$	
$j \Rightarrow 9{:}1 \mid 9{:}2 \mid 9{:}3 \mid 9{:}4 \mid 9{:}5 \mid 9{:}7 \mid 9{:}8$	
$i \Rightarrow h{:}4;\ h \Rightarrow g{:}7;\ g \Rightarrow f{:}5;\ f \Rightarrow e{:}7;\ e \Rightarrow 4{:}8$	
$d \Rightarrow 3{:}2 \mid 3{:}7$	
$c \Rightarrow b{:}2 \mid 1{:}b \mid 7{:}b$	
$b \Rightarrow 2{:}1 \mid 2{:}3 \mid 2{:}5$	
$a \Rightarrow 6{:}1 \mid 6{:}2 \mid 6{:}3 \mid 6{:}8 \mid 6{:}9$	

DISCUSSION

Our classificatory analyses coupled with appropriate statistical tests demonstrate that there do exist regularities in the starfish infestations, which we suspect are due to spatio-temporal waves. This provides some empirical validation for models of starfish reproduction and behaviour which result in the production of such waves, e.g. Antonelli *et al.* (1989). This in itself is an important result and the computational cost of obtaining it is relatively low; certainly it is much lower than the cost of inspecting and coding the data originally.

However the grammatical analysis provides further material. The derivation paths indicate that recovery from [high starfish, low coral] states has occurred in the past although we know little about the rapidity with which such recovery can take place. Furthermore this does not necessarily imply that recovery will **always** occur in the future, since such disturbances as fishing of major starfish predators have certainly changed towards higher intensities. This has certainly changed the system and such non-stationarity precludes any direct prediction of the future of the reefs.

The cyclic pattern represents an acute disturbance by invading starfish followed by a recovery when starfish numbers collapse. This is not the only possible kind of system. and the 2 disconnected derivation sets may represent alternative possibilities. Rather than indicating a pattern of infestation and recovery, these 2 sets represent reefs which are **maintaining** moderate to large numbers of starfish. The possibility of such metastable states of chronic infestation has been discussed by Bradbury *et al.* (1985) and Moore (this volume). Such reefs represent a potential source of further infestation since they form centres from which future waves could be generated and as such represent desirable targets for any control measures aimed at reducing the future probability of infestations occurring.

Our conclusions are clearly tentative yet, given the paucity and quality of data, we suggest that the exploratory analyses have provided information of value to understanding the system and possibly towards suggesting possible sources for future infestations in the sense of chronically infested reefs. We do not suggest a continuing series of exploratory analyses, however, since with understanding of the system such as that developed here, we should be able to design more appropriate and more precise quasi-experimental studies to answer more specific questions.

Acknowledgements. To Scott Bainbridge, Johnston Davidson, Craig Mundy and Peter Speare for their efforts in data preparation.

REFERENCES

Abel, D.J. & Williams, W.T. (1981) NEBALL and FINGRP: new programs for multiple nearest neighbour analysis. Aust. Comput. J. 13, 24-26.

Antonelli, P.L., Bradbury, R.H. & Reichelt, R.E. (in press) Multiple time-scale diffusion models of starfish and coral state changes over the whole Great Barrier Reef J. infer. deduc. Biol.

Antonelli, P.L., Kazarinoff, N.D., Reichelt, R.E., Bradbury, R.H. & Moran, P.J. (1989) A diffusion-reaction-transport model for the large-scale waves in the crown-of-thorns starfish outbreaks on the Great Barrier Reef. IMA J. Math. Appl. Med. Biol. 6, 81-89.

Bradbury, R.H. & Antonelli, P.L. (this volume) What controls outbreaks? In: R.H. Bradbury (ed.) The *Acanthaster* phenomenon: a modelling approach. Springer-Verlag, Berlin

Bradbury, R.H., Hammond, L.S., Moran, P.J. & Reichelt, R.E. (1985) Coral reef communities and the crown-of-thorns starfish: evidence for qualitatively stable cycles. J. theor. Biol. 113, 69-81.

Bradbury, R.H. & Mundy, C.N. (1989) Large scale shifts in biomass of the Great Barrier Reef ecosystem. In: K. Sherman & L. Alexander (eds.) Biomass yields and geography of large marine ecosystems. Westview, Boulder. pp 143-167.

Chirathamjaree, C. & Ackroyd, M.H. (1980) A method for the inference of non-recursive context-free grammars. Int. J. Man-Machine Stud. 12,379-387.

Dale, M.B. (1989a) Mutational and nonmutational similarity measures: a preliminary examination. Coenoses 3,121-133.

Dale, M.B. (1989b) Similarity measures for structured data: a general framework and some applications to vegetation data. Vegetatio 81, 41-60.

Dale, M.B. (1989c) Dissimilarity for partially ranked data and its application to cover-abundance data. Vegetatio 82, 1-12.

Dale, M.B. & Barson, M.M. (1989) The use of grammars in vegetation analysis Vegetatio 81, 79-94.

Endean, R. & Cameron, A.M. (1985) Ecocatastrophe on the Great Barrier Reef. Proc. 5th Int. Coral Reef Cong. 5, 309-314.

Friedman, J. & Rafsky, L.C. (1979) Multivariate generalisations of the Wald-Wolfowitz and Smirnov two-sample tests. Ann. Statist. 7, 697-717.

Goodall, D.W. (1966) A new similarity index based on probability. Biometrics 22, 883-907.

Gower, J.C. (1971) A general coefficient of similarity and some of its properties Biometrics 27, 857-871.

Hopley, D. 1982 The geomorphology of the Great Barrier Reef. Wiley-Interscience, New York

Kenchington, R.A. & Morton, B. (1976) Two surveys of the crown of thorns starfish over a section of the Great Barrier Reef: Report of the Steering Committee for the Crown of Thorns Survey, March 1976. Aust. Govern. Printing Office, Canberra.

Lance, G.N. & Williams, W.T. (1967) A general theory of classificatory sorting strategies I. Hierarchical systems. Comput. J. 9, 373-380.

Marczewski, E. & Steinhaus, H. (1958) On a certain distance of sets and the corresponding distance of functions. Coll. Mathem. 6, 319-327.

Mojena, R. (1977) Hierarchical grouping methods and stopping rules: an evaluation. Comput. J. 20, 359-363.

Moore, R.J. (this volume) Persistent and transient populations of the crown-of-thorns starfish, *Acanthaster planci*. In: R.H. Bradbury (ed.) The *Acanthaster* phenomenon: a modelling approach. Springer-Verlag, Berlin.

Moran, P.J. (1986) The *Acanthaster* phenomenon. Oceanogr. Mar. Biol. Ann. Rev. 24, 379-480.

Moran, P.J., Bradbury, R.H. & Reichelt, R.E. (1989) Distribution of recent outbreaks of the crown-of-thorns starfish (*Acanthaster planci*) along the Great Barrier Reef: 1985-86. Coral Reefs 7, 125-137.

Pearson, R.G. (1981) Recovery and recolonization of coral reefs. Mar. Ecol. Prog. Ser. 4, 105-122.

Prim, R.C. (1957) Shortest connection networks and some generalizations. Bell System Tech. J. 36, 1389-1401.

Ratkowsky, D.A. & Lance, G.N. (1978) A criterion for determining the number of groups in a classification. Aust. Comput. J. 10, 115-117.

Raymond, R. (1986) *Starfish wars*. MacMillan, Melbourne.

Reichelt, R.E. (this volume) Dispersal and control models of *Acanthaster planci* populations on the Great Barrier Reef. In: R.H. Bradbury (ed.) The *Acanthaster* phenomenon: a modelling approach. Springer-Verlag, Berlin

Williams, WE.T. (1980) TWONET: a new program for the computation of a two-neighbour network. Aust. Comput. J. 12, 70.

Williams, W.T., Burt, R.L. Pengelly, B.C. & Robinson, P.J. (1980) Network analysis of genetic resources data. I. Geographical relationships. Agro-Ecosyst. 6, 99-109.

Williams, W.T., Lambert, J.M. & Lance, G.N. (1966) Multivariate methods in plant ecology V. Similarity analyses and information-analysis. J.Ecol. 54, 427-445.

Williams, W.T. & Tracey, J.G. (1984) Network analysis of northern Queensland tropical rainforests. Aust. J. Bot. 32, 109-116.

APPENDIX

A grammar is a formal means by which the variations in a (possibly infinite) set of strings can be described. Many grammars are composed from the following parts, although more complex constructions are possible, of which we shall note some later. The parts are:

1. A special initiating symbol representing "any string". We shall call this symbol **S** .

2. A set of symbols which are found in the strings we are trying to describe. These are called terminal symbols because once they appear they are never changed, and we shall use lower case letters for them,

3. A set of symbols which represent structural parts of the strings and which can be replaced, in a manner to be defined below, either by terminal symbols, or by other symbols from this present set. They are called nonterminals and we shall represent them by UPPER CASE letters.

4. A set of rules which show how any non-terminal symbol may be rewritten into 2 (or more) other symbols.

It is perhaps easiest to start with an example of a rule, and we shall use what is known as the Backus-Naur form to write it. This is simply a convenient method of writing the rules explicitly, first developed for computer languages. Take the rule

$$F ::=> A:b \mid A:c \mid d:e$$

This can be read as: The nonterminal symbol F can be replaced by any one of 3 possible strings.The first is the nonterminal symbol A followed by the terminal symbol b. The second is the nonterminal symbol A followed by the terminal symbol c. The third is the terminal symbol d followed by the terminal symbol e.

We could now look for another rule which tells us how to replace the symbol A. There are NO rules for b, c, d or e since they are terminal symbols.

To develop a complete sequence we start from the special symbol **S** and replace it using one of the rules which pertain to it. We next look for any single nonterminal symbol in the result and replace it. We continue replacing individual nonterminal symbols until no further such symbols remain. We shall then have one string generated by the grammar. Obviously by choosing different alternatives we can generate a great variety of strings. If we have rules of the form

$$F ::=> W:F \mid W:f$$

with an example of the nonterminal symbol F on BOTH sides of the rule, we have an example of a **recursive** rule. Such rules permit the generation of infinitely long strings. In our example we specifically prohibit such rules by demanding a nonrecursive grammar.

If in all rules at most 1 nonterminal symbol appears on the right hand side of a rule then the grammar is call **regular**. Such grammars are related to finite state automata and are commonly used in computer studies. If we have rules such as

$$F ::=> W:Y$$

where 2 nonterminals exist on the right hand side of the rule, then we have a context-free grammar. However if we make the application of the rule to the symbol F depend on the symbols surrounding F we have a context-sensitive rule. For example

$$\alpha F\beta ::=> W:Y$$

we would read this as meaning that the nonterminal symbol F **if and only if it is preceded by an** α **and followed by a** β can be rewritten as W:Y. α and β may be any arbitrary sequence of symbols, terminal or nonterminal.

There are other more complex forms of grammars which attempt to capture more complex restrictions based on semantics, as for example the change in pronoun from **him** to **her** if the person referenced is female. Some grammars permit erasing rules where a symbol is just eradicated, others permit several symbols of the same kind to be replaced simultaneously (parallel grammars) rather than individually (serial grammars). Still others specify the order in which rules are applied, as for example in programmed grammars where each rule has associated with it a list of successor rules. Finally we have looked only at grammars for sequences of symbols, but the common foundation of sets of elements and rules for changing them can be applied more widely.

Since our program is capable of inferring a context-free grammar, which thus may contain rules of the form F ::=> W:Y; We shall call these "context-free" rules. The reader will notice that it is just such rules that we define as interesting. If we examine other possible sets of rules such as G::=>A:z ; A::=>B:y ; B::=>w:x, it can be seen that we could as easily write a single rule G::=>w:x:y:z. To do this reduces the complexity of the grammar, but also makes clear that there is no real choice here at all; the sequence w:x:y:z is a single unit. Equally rules where the right hand side consists only of terminal symbols, while important in describing the sequences, do not provide any information of higher level structure. In studying the reefs we submit that we are less interested in such sequences since, unless they appear in context-free rules, they represent the temporal pattern of a single reef only. It is the context-free rules which imply a higher level of structure in our sequences.

STOCHASTIC AND SPATIAL EFFECTS
IN PREDATOR-PREY MODELS
OF *ACANTHASTER*-CORAL INTERACTIONS

J.S. PARSLOW[1]

School of Australian Environmental Studies
Griffith University, Brisbane, Queensland 4111

Abstract. The effects of stochastic variation in starfish recruitment on simple predator-prey models of *Acanthaster*-coral interactions are considered. Single site models with high levels of stochastic variation predict more 'realistic' outbreaks than similar deterministic models in that increases in starfish density occur suddenly from low background densities. In regional models, similar high levels of recruitment variation lead to a breakdown in cycles, and the appearance of severe endemic outbreaks. In models incorporating multiple regions linked downstream, travelling waves of outbreaks are obtained only if parameters in the upstream seed region differ from those in downstream regions. As stochastic variation in recruitment and spatial complexity increases, increasing clearance rates are required for predators of starfish to prevent outbreaks.

INTRODUCTION

Since the early 1960s outbreaks of the starfish *Acanthaster planci* have been widely reported from the Great Barrier Reef. This is primarily a phenomenon at the population level, involving the sudden appearance of large numbers of starfish on individual reefs, and the consequent destruction of major fractions of the coral cover. There has been widespread concern about the possible economic and ecological consequences of these outbreaks, but our understanding in the past has not allowed long-term prediction or the design of management policies. The challenge for modellers is to translate field and experimental results into understanding and prediction at the population and ecosystem level.

During the last three decades, there has been a large set of diverse studies of *Acanthaster* ecology on the Great Barrier Reef and elsewhere. Much of the resulting information is summarized by Moran (1986). It can be divided into information at the population level (that is, information on the distribution of the starfish and interacting species in space and time) and information at the organism level (the basic biology of these same species).

---

[1] Present address: CSIRO Division of Fisheries, Hobart, Tasmania.

Data on the spatial and temporal distribution of *Acanthaster* on the Great Barrier Reef, and accompanying data on coral damage, have been collated by the Great Barrier Reef Marine Park Authority, and analysed by Reichelt (this volume) and by Hundloe & Parslow (1988). Given the scale of the system involved, the considerable difficulties involved in sampling, and the uneven distribution of effort in time and space, it is not surprising that there is much uncertainty associated with the interpretation of the historical record. At a broad qualitative level, the data show two apparent cycles of outbreaks on the Great Barrier Reef, separated by about 15 years. Intensive studies over time of individual reefs (Moran *et al.*, 1985) are still too few, but the data for a number of reefs show an initial outbreak with heavy coral damage, followed by a decline of starfish to very low levels and gradual coral recovery until a second outbreak about 15 years later. There are suggestions that the outbreaks appear suddenly, involving increases in densities of several orders of magnitude, and that the decline in numbers may be less rapid (Hundloe & Parslow, 1988). Despite gaps in the data, analyses of the spatio-temporal pattern of outbreaks suggest a travelling-wave phenomenon, with outbreaks first appearing in a 'seed-region' north of 16°S, and propagating southward over a period of 10 or more years (Reichelt, this volume; Hundloe & Parslow, 1988).

More realistic population models attempt to use biological information at the level of the organism to predict behaviour at the population level. The difficulty with this approach lies in obtaining sufficient information at the lower level, and this is certainly true for *Acanthaster*. The starfish life history can be divided into three stages: planktonic larvae, post-recruits and early juveniles, and late juveniles or adults. Studies in the laboratory and in field enclosures (Lucas, 1975, 1982; Olson, 1987) have established the duration of the planktonic stage to be about 12 to 20 days. Our ignorance concerning the factors affecting the survival of larvae in the field is virtually complete, except that laboratory indications of food limitation appear to have been ruled out by results from field enclosures (Lucas, 1982; Olson, 1987). Studies of both the survival and dispersal of wild larvae are constrained by an inability to sample these populations. Numerical fluid dynamic models have been used to predict larval dispersal, and results are reported in this volume and elsewhere (Black & Gay, 1987, this volume; Gay *et al.*, this volume).

After settlement, early juveniles feed on encrusting algae for about six months, and then switch to hard corals (Moran, 1986). For the first 12 to 18 months, juveniles are extremely cryptic hiding deep in crevices during the day. Sampling of these stages has proved so difficult that their recruitment and survival rates have been virtually unknown, with rare exceptions (e.g.. Yokochi *et al.*, 1988). The late juveniles and adults are less cryptic, particularly during outbreaks, and their feeding rates and preferences, and movement rates, are better studied. Reliable estimates of survival rates are difficult because of tagging problems. The factors affecting survival at low population densities are particularly important for modelling, and are least understood. Adults can reach sexual maturity after two years, and fecundity is very high, with individual females producing up to 70 million eggs (Moran, 1986). Females convert a large proportion of their biomass into eggs. Growth rates depend on food intake (Lucas, 1984) and one might expect fecundity to depend on accumulated food intake. Spawning takes place in aggregations, and there is clearly a possibility of density-dependent effects on fertilization success, although this has not been quantified (Pearson & Endean, 1969; Beach *et al.*, 1975).

This study is concerned with understanding the population dynamics of *Acanthaster*, at the level of the individual reef, and at larger spatial scales of groups or chains of reefs. The approach adopted has been to try to structure models to incorporate relevant biological information, but to use empirical or phenomenological results to patch over the gaps in our knowledge of *Acanthaster* biology. It goes without saying that useful models can be built at a variety of levels, and that results of models constructed at other levels [for example, the movement models of Hogeweg & Hesper (this volume) and Green (this volume)] may in the future be used to fill these gaps in a more satisfying manner.

This study starts from the observation that the cycles of outbreaks on individual reefs look like a predator-prey interaction between *Acanthaster* and corals, and have been modelled as such by a number of authors (Antonelli & Kazarinoff, 1984; Bradbury *et al.*, 1985; Reichelt *et al.*, 1990). The study examines the implications of variability in larval survival for these predator-prey dynamics. High stochastic variability in larval survival and recruitment can be expected both on the grounds of comparison with other marine organisms having high fecundity and planktonic larvae, and on the basis of arguments involving the effects of physical advection subject to wind and tide past complex topography (Parslow & Gabric, 1990).

Given the preceding models of *Acanthaster*, and more general ecological modelling theory, the introduction of stochastic recruitment raises a number of questions:

1. How does stochastic recruitment affect the qualitative nature of the outbreak cycle?

2. What are the implications for control by predators?

3. What are the effects of combining groups of reefs in a mutually-stochastically-recruiting region?

4. How does stochasticity affect the propagation of an outbreak wave in a line of downstream regions?

MODEL FORMULATION

The model is developed in three stages. A basic site model is developed for an individual reef. A regional model is then constructed as a set of mutually recruiting site models. Finally, a linked-regional model is developed by setting up a chain of downstream-linked regional models.

The basic site model uses a simple discrete time structure to allow rapid simulation over long periods, to assess qualitative behaviour. The basic equations are:

$$C_{t+1} = g(C_t) - f(C_t, A_t)$$

$$A_{t+1} = s(R_t).A_t + J_t$$

$$J_{t+1} = r(R_t).s(R_t).A_t$$

$$R_t = f(C_t, A_t) / A_t$$

The site variables are coral cover, C, and *Acanthaster* adults, A, juveniles, J, and ration, R. Coral cover is scaled between 0 and 1.0, so that 1.0 represents maximum coral cover (not necessarily 100%). *Acanthaster* adults includes both 1+ juveniles and older individuals. These are treated as functionally equivalent in the model as they all feed on coral, their survival and reproductive output are assumed to depend on coral ration, and they can contribute to reproduction at the end of the year. The juvenile class J_t represents new recruits at the beginning of year t, with numbers adjusted to allow for density-independent mortality during the first year. All *Acanthaster* densities are scaled to represent units of coral-consuming potential. That is, an adult population density of 1.0, feeding at its maximum consumption rate (not allowing for coral depletion) would reduce coral cover from 1.0 to 0.0 in one year. Given observed consumption rates of about 6 m^2 y^{-1} per adult, and typical levels of coral cover of 50-60% (Moran, 1986), a scaled density of 1.0 corresponds to about 1000 starfish/ha. Because of the scaling, the ration R is a relative ration, with R=1 corresponding to maximum food intake.

In the absence of starfish, the coral growth term g(C) is set equal to $(1+a).C + C_0$. This represents simple exponential growth, with the additional term C_0 representing external recruitment, or having the effect of a refuge from starfish predation, as we discuss later. There is an additional proviso that coral cover C is not allowed to exceed 1.0. The term representing coral consumption by starfish, f(C,A), is set equal to $C.(1 - \exp(-A/(K_c+C)))$. This represents a modification to a simple Type II functional response (Holling, 1965) to allow for the discrete time step. The constant K_c is the coral density at which half-ration is obtained at low starfish densities.

The average annual ration is given by $R = f(C,A)/A$. Survival rates are assumed to depend on ration in a sigmoid fashion, with $s(R) = (1+R_c^2).R^2.s_0/(R_c^2+R^2)$. The parameter s_0 represents the survival rate at maximum ration, while the parameter R_c represents the ration level at which survival drops rapidly. A sigmoid dependence rather than a threshold is used because R represents a population average, and even when R is very small, some individuals may obtain sufficient ration to survive.

In some simulations, the effects of predation on *Acanthaster* adults and late juveniles are represented by multiplying the above survival rate by the factor $\exp(-P.A/(K_A^2 + A^2))$. This can be regarded as representing a fixed population of predators with a Type III functional response (Holling, 1965). Attention on the predators of *Acanthaster* has focused on their potential as control agents, keeping the starfish at low densities (e.g. Endean, 1982; Endean *et al.*, 1987). As McCallum (1987) has pointed out, the most promising agents are generalist predators which are capable of switching on to *Acanthaster* as the starfish densities increase. The parameter P represents the maximum consumption rate by predators, and the parameter K_A determines the switching density, with the maximum mortality rate being equal to $P/(2.K_A)$.

The reproductive output is assumed to increase linearly with ration above the critical level R_c, so that $r = e^z.B.(R-R_c)/(1.-R_c)$. The parameter B represents the mean recruitment rate at maximum ration, allowing for mortality in both the larvae and the first year following settlement. The realized (stochastic) recruitment is obtained by multiplying the mean recruitment for a given ration by a multiplicative, log-normal variate [that is, z is assumed to be i.i.d. $N(0,\sigma^2)$].

The regional model is obtained by running ten site models simultaneously, with the models linked only through larval recruitment. The recruitment linkage is particularly simple, with all reefs contributing larvae to a common larval pool, and recruitment from this pool to individual reefs occurring randomly. The common larval pool is given by

$$L_{t+1} = \Sigma_i \, r^i_t.s^i_t.A^i_t$$

with recruitment to reef i given by

$$J^i_{t+1} = L_{t+1}. \, e^{z^i_{t+1}}$$

The linked regional model contains five regional groups linked in a downstream manner. Recruits from each region are pooled as before, with L^m representing the recruit pool in region m, and J^{mi} representing recruits to reef i in region m. Region 1 acts as a seed region, recruiting both to itself and to region 2, while regions 2 to 4 recruit only to the next region downstream, with recruits from region 5 being lost. Thus:

$$J^{1i}_{t+1} = L^1_{t+1} \cdot e^{z^{1i}_{t+1}}$$

$$J^{mi}_{t+1} = L^{m-1}_{t+1} \cdot e^{z^{mi}_{t+1}} \qquad \text{for } m=2...5.$$

Table 1. List of variables and parameters used in model.

Variables

C_t	.....	relative coral cover in year t.
A_t	.....	relative adult *Acanthaster* density in year t.
J_t	.....	relative juvenile *Acanthaster* density in year t.
R_t	.....	average relative *Acanthaster* ration in year t.
L_t	.....	regional pool of *Acanthaster* larvae in year t.
z	.....	normally-distributed stochastic exponent for *Acanthaster* recruitment.

Parameters

a	.....	exponential growth rate (y-1) for coral cover.
C_o	.....	constant external coral recruitment level.
K_c	.....	coral density producing half-ration at low *Acanthaster* densities.
R_c	.....	critical ration for starfish survival and fecundity.
S_o	.....	*Acanthaster* survival rate at maximum ration.
P	.....	maximum consumption rate by predator of *Acanthaster*.
K_A	.....	inflection *Acanthaster* density for type III functional response for predators of *Acanthaster*.

PARAMETER VALUES

The choice of parameter values is complicated both by our ignorance of crucial aspects of *Acanthaster* biology, and by mismatches between the level of description in the model and the available data. In some cases, parameters can be interpreted and estimated at the population level. For example, if the coral growth parameter a is given a value of 0.30, coral cover recovers from severe depletion (0.03) to full cover (1.0) in about 12 years which is close to minimum estimates of recovery time in the field (Pearson, 1981). Clearly this represents faster growing corals such as *Acropora*; others such as *Porites* grow much more slowly (Done, 1985). The coral input parameter C_o serves to prevent overdepletion of coral and can be regarded either as recruitment from outside the reef, or as having a similar effect to a refuge from predation by starfish. In the model, coral cover cannot drop below C_o which is assigned a value of 0.03, corresponding roughly to minimum observed levels.

The qualitative behaviour of the coral-starfish interaction depends heavily on the parameters K_c and R_c which determine the functional and numerical responses of the starfish. The parameter K_c is best thought about in the context of a small population of starfish on a reef. At what average coral cover will the starfish ration drop to half its maximum value? I am not aware of a published answer to this question, but there are qualitative reasons to expect K_c to be small. The starfish are motile predators feeding on stationary prey. It has been suggested that they can detect corals at a distance using chemosensory receptors (Sloan & Campbell, 1982). Potential complications concern the effects of starfish aggregation and coral spatial distribution, particularly when the latter results from previous patterns of starfish feeding. A level of coral cover of 0.05, for example, could represent small surviving fragments scattered over the entire reef, or a sizeable pocket of live coral which has escaped starfish attention. A default value of 0.2 is assigned to K_c, but some investigation is made of values in the range 0.1 to 0.5.

Similar qualitative arguments suggest that the value of the critical ration, R_c, should also be small. (Recall that R represents a fraction of the maximum ration, and can range from 0 to 1.) *Acanthaster* is a relatively inactive predator with the ability to withstand long periods of starvation (Lucas, 1984), so that one would not expect survival rates to decrease significantly until ration had dropped to relatively low levels. Again, R_c is tentatively set at 0.2, and a range from 0.1 to 0.5 is considered.

Adult starfish in aggregations under conditions of plentiful food are not believed to suffer high mortality rates, and the maximum survival rate s_0 is set at 0.9. There is little evidence that predators exert significant mortality on starfish at outbreak densities, but it is possible that predators play a significant role when starfish densities are low (McCallum, 1987). No attempt is made here to model a particular predator; rather, the analysis will attempt to assess the properties, in the form of P and K_A, required of a predator in order to control starfish densities at low densities under different degrees of recruitment variability.

The remaining parameters are the mean recruits per adult at maximum ration, B, and the stochastic variance in recruitment, σ^2. The parameter B is perhaps the least certain of all parameters. It hides our ignorance of the extremely high loss rates which must take place between female fecundity and recruitment to the 1+ age class on a reef. There is no basis for estimating B from available biological data even to the nearest order of magnitude. All we can do is to look at the effect of different values on the qualitative behaviour of the model.

The principal aim is to study the effects of different levels of stochastic variation in recruitment on model behaviour. Model runs will consider values of σ ranging from 0 to 3. The high fecundity of *Acanthaster* makes relatively large values of σ at least feasible. For σ = 3.0, there is a 10% chance of a recruitment 50 times B, and a 1% chance of a recruitment 360 times B. For moderate values of B, these still represent relatively low survival rates for the original 10 million or more eggs per adult. As σ increases, the probability distribution of recruitment levels becomes flatter, and one might expect that the qualitative behaviour of the model would be less sensitive to the exact choice of B.

MODEL BEHAVIOUR

The behaviour of the model has been analysed by computing numerical solutions, and by comparing the behaviour with previous analytical and numerical results from predator-prey theory. The discrete-time model is easily solved numerically. Because the questions of interest refer to long-term, qualitative behaviour, solutions were normally calculated over periods of 200 years or more, and checked for evidence of convergence to some kind of qualitatively stable behaviour. Except where stated otherwise, the numerical results refer to the second hundred years of solutions.

1. Deterministic Site Model

The deterministic site model, with no predation on *Acanthaster*, was studied first. This model is capable of three qualitative classes of behaviour, depending primarily on the parameters R_c and K_c. If these parameters are high (e.g.. both equal to 0.5), so that functional and numerical responses depend almost linearly on coral cover, a stable equilibrium is approached with coral cover close to 1.0 and moderate starfish densities, just sufficient to crop back annual coral growth. If lower values of K_c and R_c are chosen (e.g.. both 0.2), so that the responses are markedly non-linear, stable limit cycles are obtained. This behaviour is entirely consistent with that of classical Lotka-Volterra predator-prey models involving predator (starfish) satiation and limits to prey (coral) carrying capacity (e.g.. May, 1972). In fact, the time lags introduced in the form of the discrete time step and the age structure in this model have little effect on the qualitative behaviour; simulations run without the age structure showed essentially similar behaviour.

One feature of the classical predator-prey limit cycles is that cycle amplitudes quickly become ridiculously large, and periods extremely long, as the functional and numerical responses become more nonlinear.

This occurs because high starfish densities are maintained down to low coral levels, and in turn reduce coral cover to many orders of magnitude below maintenance levels before dying off. The recovery of coral cover is limited by corals' maximum intrinsic growth rate, and there is consequently a very prolonged period of starfish starvation, which reduces the starfish population to a level many orders of magnitude below its equilibrium level. After coral recovery, there is a further lengthy period of starfish recovery, and an eventual large overshoot in predator population, to start the cycle again. The amplitude of these cycles can easily become such that one must interpret the minimum population levels as extinction of coral and/or starfish.

This behaviour can be avoided if one can prevent unrealistic depletion of coral cover. In this model, this is prevented by the coral input term C_0. At the population level, this can be supported by observing that coral populations on reefs are not depleted below levels of a few percent coral cover. At the biological level, one can invoke both recruitment of coral from outside the reef, and refuges from predation, both by chance due to the complex spatial distribution of predator and prey, and in areas avoided by starfish, such as zones of high wave action.

The introduction of a lower limit to coral cover introduces a third kind of qualitative behaviour in the model. At even lower levels of the parameters K_c and R_c, stable cycles with low amplitude and permanently depressed coral cover are obtained. These correspond to the stable predator-control equilibria produced in classical predator-prey models with a prey refuge: the low amplitude cycle in this model results from the age structure. This corresponds to the situation feared by observers of the first *Acanthaster* outbreaks: that a resident population of starfish would remain to prevent coral recovery. This does not appear to have occurred on the Great Barrier Reef as a whole, although there have been observations of prolonged residual populations on individual reefs (Moran *et al.*, 1985).

A general conclusion from these results is that the qualitative behaviour of large amplitude cycles or outbreaks may be as much or more dependent on behaviour at low densities as at high densities. It is primarily the response of *Acanthaster* to low coral densities which determines the behaviour of this model, and this response is not well studied.

For values of K_c and R_c which produce periodic outbreaks, varying the mean recruitment parameter B has a relatively small effect on the period and amplitude of the cycle. As B is increased from 1.0 to 3.0 to 5.0, the cycle period decreases from 29y to 19y to 17y and the peak

Acanthaster density increases from 0.7 to 1.3 to 1.7. The effect of changing B is much greater if the coral recruitment term C_0 is not included.

For these parameter values which produce periodic outbreaks, one can try to determine what values of the parameters P and K_A are required if predation on *Acanthaster* is to prevent outbreaks. In fact, this can be determined algebraically. If the starfish densities are successfully controlled, coral cover will remain at 1.0 and starfish ration will be maximum. One can therefore determine an intrinsic rate of population increase for starfish which depends on s_0 and B. For the Type III functional response used here, the maximum mortality rate exerted on starfish is $P/(2.K_A)$, occurring at a density K_A. If this mortality rate does not exceed the intrinsic rate of increase, control is impossible. If it does, there will be two values of A, either side of K_A, for which the mortality rate equals the intrinsic rate of increase. The lower value, A_{eq}, represents a stable equilibrium and the upper value, A_T, a threshold for escape from predation control.

For B=3.0, the intrinsic rate of increase is about 0.6 y^{-1}. This means that a maximum clearance rate of at least 0.6 y^{-1} is required from the predator population for control. In order to have a significant margin between A_{eq} and A_T, a value twice this is desirable. As McCallum (1987) has pointed out, most of the study and discussion of potential control agents has focused on the maximum consumption rate P, rather than this maximum clearance rate.

2. Stochastic Site Model

This section will focus on the effects of stochastic recruitment on a site model using parameter values which produce periodic outbreaks for $\sigma = 0$. The effects of increasing σ from 0 to 3 can be seen in Figure 1. The qualitative form of the solution is not altered, in that recognisable periodic or quasi-periodic outbreaks are still produced. There is little effect for $\sigma = 1$, but for $\sigma = 2$ and 3 there is increasing variability in cycle period and amplitude (Table 2), and in the extent of coral recovery between outbreaks. There is a decrease in mean cycle period as σ increases, due to the fact that the effective mean recruitment rate increases, because of the assumed log-normal distribution for variation in recruitment.

Table 2. Effect of σ on cycle period and peak starfish densities for the stochastic site model with B=3.0, K_c=0.2 and R_c=0.2. Values given are means, with standard deviations in parentheses.

σ	1.0	2.0	3.0
Cycle Period (y)	17 (2.0)	17 (2.6)	15 (2.2)
$\ln(A_{peak})$	0.8 (0.4)	2.0 (0.9)	2.4 (2.2)

Another effect which is clearly visible in Figure 1 is the increased steepness of the build-up in starfish numbers for σ = 2 or 3. In the deterministic solution, the starfish population increases smoothly, although exponentially, approximately doubling each year. Despite the holes in the observational record, it is clear for at least some reefs that outbreaks appear much more suddenly (Moran, 1986). For σ = 3, model outbreaks commonly consist of a single large chance recruitment, with starfish populations increasing by one or two orders of magnitude in one year.

It is fairly obvious that increasing variability in recruitment will make regulation of *Acanthaster* by predators more difficult. One way to think about this is to imagine an *Acanthaster* population at the lower equilibrium A_{eq}. In the deterministic model, it will stay at that level for ever. In the stochastic model, chance high recruitments will increase the population above this level. If the increased population is still below the upper threshold, A_T, it will suffer increased predation mortality, and tend to return to A_{eq}. If it exceeds A_T, it will experience reduced predation mortality and continue to increase. One might therefore expect the probability of an escape from predation control to depend on the probability of a chance recruitment pushing the population from A_{eq} to above A_T, which will depend in turn on the ratio A_T/A_{eq} and the value of σ. For the sigmoid Type III functional response used here, the ratio A_T/A_{eq} is proportional to $(P/K_A)^2$; that is, to the square of the maximum clearance rate. One might expect that the maximum consumption rate P would be more important for control in the stochastic environment. However, for a single self-recruiting site, both the existence of the lower equilibrium and its resistance to perturbation depend on the maximum clearance rate.

Simulation shows that the clearance rate of 1.2 y^{-1}, used in the deterministic model, also apparently prevents outbreaks in the stochastic model for σ = 1, but does not prevent outbreaks for σ = 2 or 3. In order to reduce the frequency of outbreaks to less than 1 per 200 y for

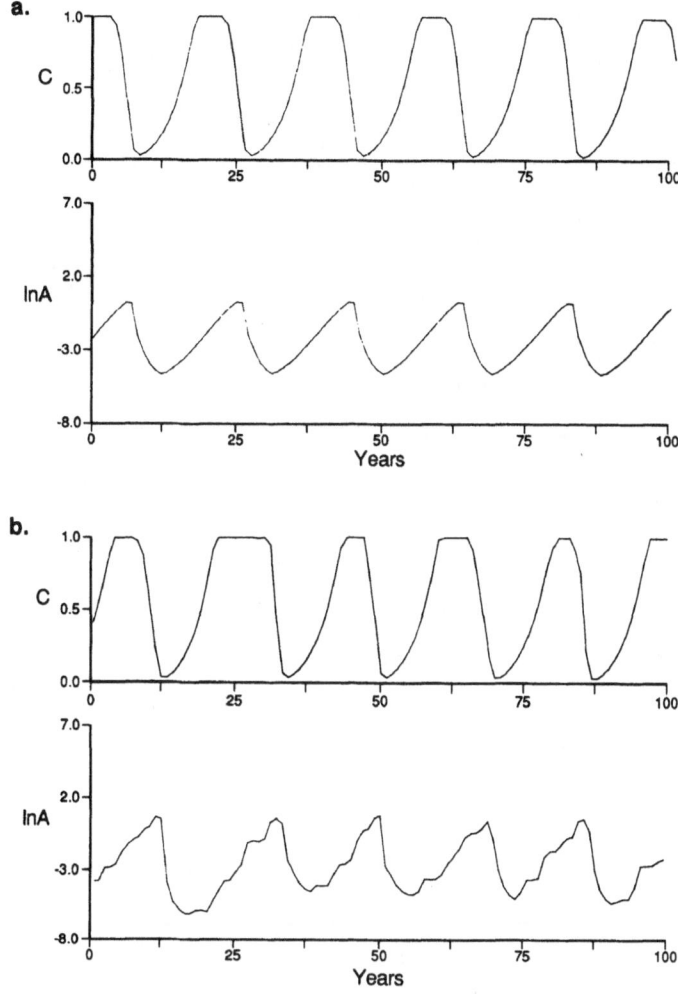

Figure 1. Relative coral cover and log-scaled starfish densities predicted by the site model, with B=3.0, K_C=0.2, R_C=0.2, and σ=0 (a), σ=1 (b).

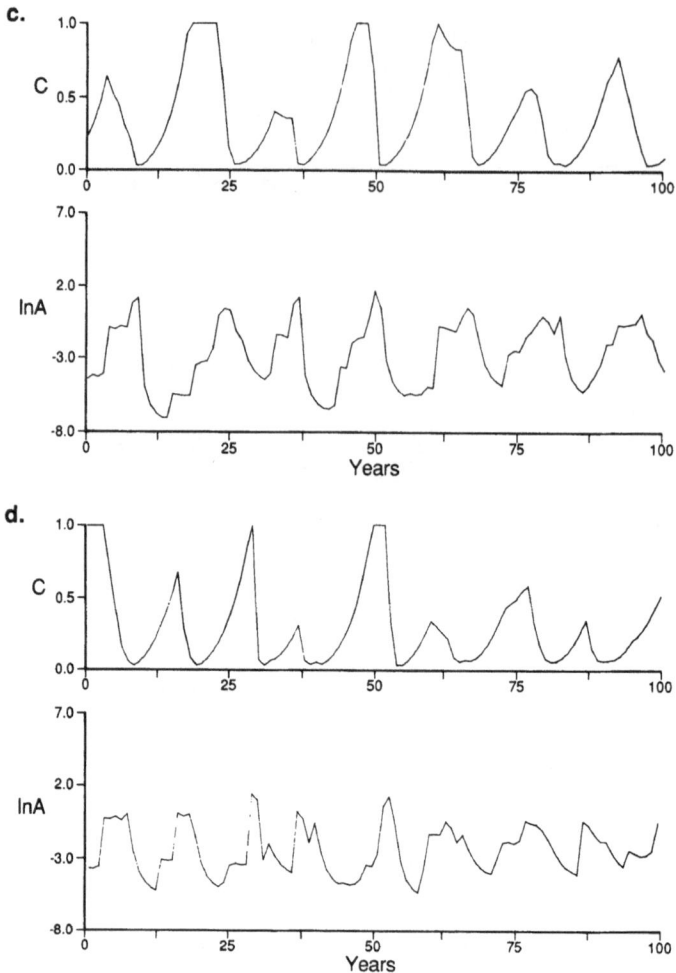

Figure 1 (cont). Relative coral cover and log-scaled starfish densities predicted by the site model, with B=3.0, K_c=0.2, R_c=0.2, and σ=2 (c), σ=3 (d).

$\sigma = 2$, it is necessary to increase the maximum clearance rate by a factor of 4, which increases the ratio A_T/A_{eq} by a factor of 16. Even this is not sufficient to prevent frequent outbreaks for $\sigma = 3$. If recruitment really does have this level of stochastic variability, long-term control by predators will require very high search efficiencies.

3. Regional model

The regional model consists of a set of 10 site models linked via recruitment. All sites contribute to a common larval pool, and recruitment to individual reefs is determined stochastically. For $\sigma = 0$, all reefs see the same recruitment and the regional model behaves identically to a single site model, so the model is only of interest for $\sigma > 0$. Because of the way 'larvae' are pooled and distributed, the mean recruitment B must be divided by 10 in order to accomplish the same mean recruitment level for each reef.

Simulations were used to study the effects of both spatial variation and increasing values of σ on outbreak cycles. For $\sigma = 1$, the regional model maintains regular cycles of synchronous outbreaks on all reefs, with generally good coral recovery between outbreaks (Figure 2a). For $\sigma = 2$ or 3, cycles are much less coherent, and many reefs spend long periods with low coral cover (Figure 2b). Observers of such a system would be more likely to describe the outbreaks as endemic than cyclic.

This model behaviour appears to be due to the role of starfish on some reefs as seed populations. That is, once the variability in recruitment rates is high enough to break down synchronicity, there are always some reefs with moderate or high coral cover and moderate starfish populations. These are capable of supplying high recruitments to any reefs, including those which are in the early stages of recovery.

One might expect predator control of *Acanthaster* to be more difficult in the regional model, and simulations bear this out. For $\sigma = 1$, a maximum clearance rate of 1.2 y^{-1} is no longer sufficient to prevent outbreaks, and levels of 5 y^{-1} are required. Even with these high mortality rates, occasional outbreaks still occur for $\sigma = 2$, although there are extended periods of predator control. Lower clearance rates appear to promote cycle coherence for $\sigma = 2$, and the system then alternates between extended periods of coherent cycles, and periods of endemic outbreaks (Figure 3).

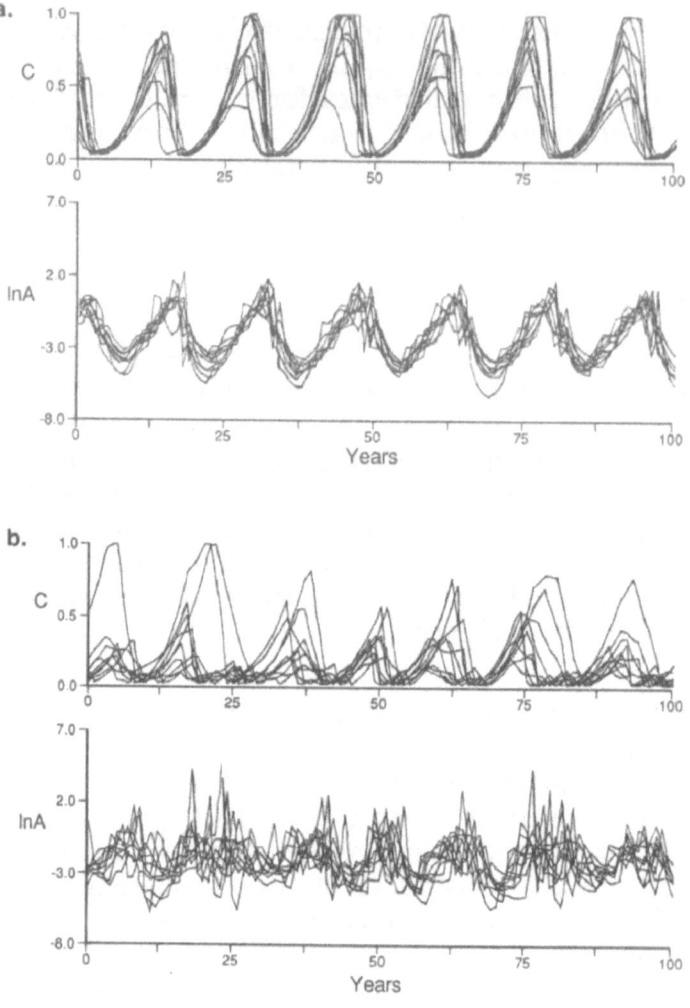

Figure 2. Relative coral cover and log-scaled starfish densities predicted by the regional model, with B=0.5, and σ=1 (a), σ=2 (b).

4. Linked Regions

The linked region model is intended as a simple representation of the Great Barrier Reef according to the picture emerging from spatio-temporal analysis of outbreaks and studies of larval transport. The first or 'northern' region is allowed to recruit to itself, and therefore act as a seed region, as well as recruiting to the second region. The second and subsequent regions recruit only to the next region downstream. This is a very crude model, and it has been used only to look for strong qualitative effects of linking regions in a downstream manner. In fact, a number of strong qualitative phenomena were found in simulations.

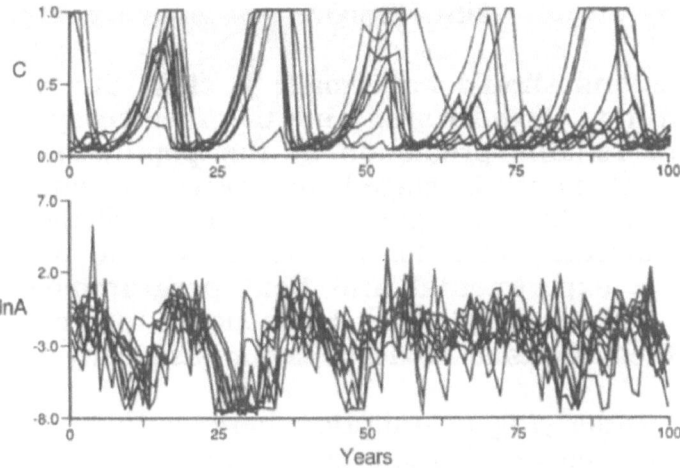

Figure 3. As Figure 2B, but with predation on starfish included, with $P = 0.05$ and $K_A = 0.01$.

There is a strong tendency in this model for stochastic effects to produce a degradation in cycle coherency downstream. For $\sigma = 1$, coherent cycles are maintained in all regions, but for $\sigma = 2$, weak cycles in the seed region collapse into endemic outbreaks downstream. There is also a tendency for outbreak intensities to alternate in successive downstream regions. This is due to the functional and numerical responses in the model, which result in high recruitment from moderate starfish densities on moderate coral cover, and low recruitment from large, starving populations on very low coral cover.

If the parameters in all regions are assumed to be identical, and σ is low enough to allow coherent cycles in all regions, the model predicts zero downstream lag between cycles. At first sight this is counter-intuitive: one would expect the lag implicit in age-structure to be reflected in a phase lag between regions. However, the zero lag does follow logically from the assumption that region 1 and region 2 see the same pool of recruits. This means that region 1 and region 2 have zero lag, and consequently that recruits to region 2 and 3 have zero lag. The same argument extends to all regions. A downstream lag in cycles amounting to two years per region can be produced by assuming a higher mean recruitment rate in the seed region. It then tends to cycle faster than the downstream regions, and to force lagged cycles in the downstream regions.

The results concerning predator control of *Acanthaster* in the linked region model are similar to those for the regional model. High clearance rates are required to prevent outbreaks: lower clearance rates do tend to produce more coherent cycles.

DISCUSSION

The preceding sections should have made it clear that the state of knowledge regarding *Acanthaster* and the reef ecosystem is not sufficient to allow reliable prediction at the population level. The justification for modelling at this stage is to explore the consequences of possible assumptions which span gaps in present knowledge, and to try to provide a logical framework for discussion of the problem, and planning of future experimental and field programmes. In these circumstances, it is incumbent upon the modeller to present the key assumptions and their effects on model behaviour as clearly as possible.

Perhaps the key underlying assumption here is that the outbreak phenomenon can be interpreted primarily as a predator-prey interaction between *Acanthaster* and coral. There is no doubt that *Acanthaster* eats coral, and is capable of reducing coral cover to very low levels during outbreaks. Moreover, there is clearly some numerical response of starfish to coral, in that very large populations cannot be maintained for very long at very low levels of coral cover. However, the models analysed here go further in assuming that the principal deterministic factor controlling *Acanthaster* survival and recruitment is coral cover. In particular, the initiation of the periodic outbreaks predicted by these models is determined by the time scales for coral recovery, and for starfish population build-up, once coral cover is adequate. This assumption is troubling because so little is known of the factors affecting starfish population growth at low population densities. Moreover, in the deterministic single-site model, it leads to a steady exponential build-up in starfish numbers over a prolonged period. There is a general perception that real outbreaks appear suddenly against very low background populations, although the number of intensive studies to support this perception is still quite small.

There are several potential solutions to this problem. From a deterministic point of view, the problem is to keep starfish densities from increasing until coral recovery is almost complete, and then to allow a very high growth rate. An example of this kind of behaviour can be found in the spruce budworm models developed by Holling and co-workers (Holling, 1978). Spruce budworm have a high potential growth rate, and feed on a resource with a long recovery time. In order to predict sudden outbreaks at long periods, keyed to resource recovery, Holling's model needed to include generalist predators on budworm with a Type III functional response. At low to moderate levels of forest cover, these predators regulated budworm at very low densities. This regulation failed at high levels of forest cover, producing a sudden and dramatic increase in budworm density. As well as predation, other ecological processes could play this kind of role for *Acanthaster*. For example, an Allee-type reduction in reproductive success has been suggested (Lucas, 1975). This could in principle produce a coral-dependent population threshold for rapid population growth.

A different solution, explored in this study, is to include stochastic variation in recruitment success. As discussed earlier, some level of interannual variation in recruitment success seems inevitable, if only by analogy with other marine organisms with similar life-history parameters. In a single-site model, very high levels of stochastic variability do give rise to cycles with sudden outbreaks, often representing a single year class. This study considers interannual fluctuations with zero autocorrelation. The amount of autocorrelation in real fluctuations will depend on the mechanism responsible. If the variation arises from small-scale patchiness associated with turbulent transport, one might expect it to have little interannual autocorrelation. If, on the other hand, variation is due to changes in large-scale circulation, it may well show significant autocorrelation or quasi-periodicity. Taken to its extreme, this kind of variation could become equivalent to the familiar hypothesis that outbreaks are triggered externally by unusual large-scale environmental events (Birkeland, 1982). As Reichelt *et al.* (1990) have pointed out, it is hard to reconcile this hypothesis, at least applied on a reef by reef basis, with the observation that, during a large-scale cycle on the Great Barrier Reef, outbreaks are apparently triggered over many consecutive years.

Models of outbreaks must also take into account the degree of exchange of recruits among reefs. There are still some discrepancies among physical models concerning the probability of long-term trapping of larvae on source reefs (Black & Gay, this volume; Gay *et al.*, this volume). If this trapping is present, it seems likely on the grounds of reduced dilution that self-recruitment will outweigh recruitment from other reefs. The regional models studied here take the opposing view and assume that all larvae are broadcast from the parent reef and are equally likely to recruit to that reef or any nearby reef.

There is, of course, a long tradition of studies showing that site models coupled in spatial arrays can show different qualitative behaviour to that from isolated models (e.g., Huffaker, 1958; Hilborn, 1975). Perhaps the most striking result described in this study is that large levels of stochastic interannual and interreef recruitment variation, which produce sudden outbreaks in a site model, lead to a breakdown in cycle coherency and endemic outbreaks in the regional model.

So far, endemic outbreaks maintaining coral at consistently low levels have not been widely reported from the Great Barrier Reef. One could deduce that larvae must be efficiently dispersed over large regions, so that interreef recruitment variation within regions is relatively small. Some other mechanism would then need to be invoked to explain the sudden onset of outbreaks. One can maintain sudden outbreaks by introducing a high level of coherent regional variation in recruitment so that the regional model behaves like a single site model. However, the cycles predicted by a model of this kind involve severe depletion of coral on all reefs simultaneously. In the real world, coral depletion seems to show much more variation than this, with some reefs escaping relatively lightly. This might be due to deterministic aspects of transport between reefs which the model does not account for: perhaps some reefs are only weakly connected to other reefs in the region. However, one is left with the impression that cycles in the real world are more robust under disturbance to outbreak synchronicity than they are in the regional models considered here. It is not clear why outbreaks in the real world which are delayed or prolonged do not act as seed populations, and lead to endemic outbreaks. Again it seems likely that there is some defect in our understanding of juvenile or adult survival on reefs in the early stages of coral recovery.

The problems of cycle coherence and outbreak onset are shifted another level up in the consideration of multiple regions linked downstream. In this model, only the seed region can recruit to itself; all other regions recruit only downstream. The pattern of outbreaks downstream is then determined by the pattern produced in the seed region and the effects of propagation. An interesting conclusion from the model is that the dynamics in the seed region must differ from those in the downstream regions if a travelling wave is to result. In reality, it seems quite likely that the dynamics will differ, if only in the processes affecting larval transport and survival.

If outbreaks are to appear suddenly rather than gradually in this model, it is necessary that they have this character in the seed region, and that it is not lost during cycle propagation. Model results again show that coherence is lost, and cycles degenerate into endemic outbreaks downstream, unless the stochastic variation in recruitment between reefs within regions is small. One might produce a travelling wave of the right shape by exploiting the difference between seed and downstream regions, and invoking a high level of stochastic variability in regional larval survival in the seed region, and a low level downstream. In fact, the argument against invoking episodic, large-scale, environmental triggers disappears if these are assumed to act primarily in the seed region, as the periods of high recruitment in this region may be very short.

The problem of explaining observations of an absence of severe endemic outbreaks despite variability in outbreak intensity and duration among reefs is also shifted up to the seed region in this model. Unfortunately, the suspected seed region north of 16°S has historically been less studied than the central region, especially in the critical periods preceding the discoveries of major outbreaks at Green Island. However, levels of coral damage in this region have been relatively light, compared with the central section. The data are not consistent with a deterministic regional limit cycle involving simultaneous depletion of coral to very low levels on all reefs. Rather, they suggest the maintenance of a low background population, with infrequent regional episodes of high recruitment leading to moderate coral damage, and very high recruitment downstream.

Predation on *Acanthaster* has been discussed here primarily in terms of its potential as an agent for 'permanently' regulating the starfish population at low densities. It has traditionally been associated with arguments concerning the historical precedents for *Acanthaster* outbreaks. If outbreaks are a new phenomenon, and starfish populations have historically been maintained at very low levels for long periods in the presence of abundant coral cover, then it is natural to look to predation. It remains one of the few ecological processes which is in theory capable of maintaining a population at a tiny fraction of its carrying capacity, and then only by invoking a Type III functional response. Other processes such as the Allee effect will establish negative growth rates at low population densities, but lead to extinction, at least in spatially homogeneous, deterministic models.

The model results here suggest that generalist predators with a Type III functional response can prevent outbreaks, even allowing for considerable stochastic variation in recruitment, provided they can exert high enough maximum clearance rates. This of course begs the question as to whether predators of *Acanthaster* with the necessary properties presently exist, or have existed in the past. *Acanthaster* populations in the Red Sea appear to provide an example of a non-outbreaking population (Ormond *et al.*, this volume). Outbreaks have not been observed in the Capricorn-Bunker group of reefs, at the southern end of the Great Barrier Reef, although small starfish populations are found there (Moran, 1986).

To summarize, the qualitative behaviour of population models of *Acanthaster* depends critically upon assumptions about patterns of recruitment among reefs, and interannual variation in larval and juvenile survival. As noted earlier, field data on these stages are almost entirely lacking, due to sampling difficulties. The behaviour of these models also depends on assumptions about survival and reproductive output at low levels of starfish density and coral cover. These affect the modelling of both outbreaking and non-outbreaking populations.

These assumptions have practical implications for management as well as scientific understanding. The management implications of the models described here are addressed by Hundloe & Parslow (1988). The models provide a uniformly gloomy prognosis, at least for direct control of outbreaks by harvesting or poisoning. The problem is that, according to the models, when coral cover is high, starfish are constantly in a state of incipient outbreak. This means that, in order to keep starfish at low densities, high mortality rates must be exerted each year. Given the cryptic nature of the starfish at low densities, the expense of underwater operations, and the spatial scale, completely unrealistic expenditures would be required for control on a regional basis. More limited control of an individual reef is, according to the regional model, likely to produce severe regional endemic outbreaks via the 'seed reef' phenomenon. Again, these conclusions may be overly pessimistic if processes exist which reduce population growth rates at low population densities. However, the conclusions do seem consistent with the Japanese experience with large scale control measures (Yamaguchi, 1986).

Acknowledgements. This research was supported by a grant from the Great Barrier Reef Marine Park Authority. The author is grateful to Peter Moran and Peter Doherty for discussion and advice.

REFERENCES

Antonelli, P.L. & N.D. Kazarinoff. 1984. Starfish predation of a growing coral reef community. J. Theor. Biol. 107: 667 - 684.

Beach, D.H., N.J. Hammond & R.F.G. Ormond. 1975. Spawning phenomena in Crown of Thorns starfish. Nature 254: 135 - 136.

Birkeland, C. 1982. Terrestrial runoff as a cause of outbreaks of *Acanthaster planci* (Echinodermata: Asteroidea). Mar. Biol. 69: 175 - 185.

Black, K.P. & Gay, S.L. (this volume) Reef-scale numerical hydrodynamic modelling developed to investigate crown-of-thorns starfish outbreaks. In: R.H. Bradbury (ed.) The *Acanthaster* phenomenon: a modelling approach. Springer-Verlag, Berlin.

Bradbury, R.H., L.S. Hammond, P.J. Moran & R.E. Reichelt. 1985. Coral reef communities and the Crown of Thorns starfish: evidence for qualitatively stable cycles. J. Theor. Biol. 113: 69 - 80.

Done, T.J. 1985. Effects of two *Acanthaster* outbreaks on coral community structure. The meaning of devastation. Proc. 5th Int. Coral Reef Congress. Tahiti. 5: 315 - 320.

Endean, R. 1982. Crown of Thorns starfish on the Great Barrier Reef. Endeavour, New Ser. 6: 10 - 14.

Endean, R., A. Cameron & H.I. McCallum. 1987. Study of the Crown of Thorns starfish predators on or in the vicinity of reefs of the Great Barrier Reef. Prog. Rept. GBRMPA.

Gay, S.L., Andrews, J.C. & Black, K.P. (this volume) Dispersal of neutrally-buoyant material near John Brewer Reef. In: R.H. Bradbury (ed.) The *Acanthaster* phenomenon: a modelling approach. Springer- Verlag, Berlin.

Green, D.G. (this volume) Cellular automata models of crown-of- thorns outbreaks. In: R.H. Bradbury (ed.) The *Acanthaster* phenomenon: a modelling approach. Springer-Verlag, Berlin.

Hilborn, R. 1975. The effect of spatial heterogeneity on the persistence of predator-prey interactions. Theor. Pop. Biol. 8: 346 - 355.

Hogeweg, P. & Hesper, B. (this volume) Crowns crowding: an individual oriented model of the *Acanthaster* phenomenon. In: R.H. Bradbury (ed.) The *Acanthaster* phenomenon: a modelling approach. Springer-Verlag, Berlin.

Holling, C.S. 1965. The functional response of predators to prey density and its role in mimicry and population regulation. Mem. Entomol. Soc. Can. 45: 60pp.

Holling, C.S. 1978. Adaptive Environmental Assessment and Management. Wiley and Sons, N.Y.

Huffaker, C.B. 1958. Experimental studies on predation: dispersion factors and predator-prey oscillations. Hilgardia 27: 343 - 383.

Hundloe, T. & J. Parslow. 1988. Crown of Thorns risk analysis. Institute of Applied Environmental Research, Griffith University. 271 pp.

Lucas, J.S. 1975. Environmental influences on the early development of *Acanthaster planci* (L). In Proc. Crown of Thorns Starfish Sem., 6 Sept. 1974. AGPS Canberra, pp. 109 - 121.

Lucas, J.S. 1982. Quantitative studies of feeding and nutrition during larval development of the coral reef asteroid *Acanthaster planci* (L). J. Exp. Mar. Biol. Ecol. 65: 173 - 194.

Lucas, J.S. 1984. Growth, maturation and effects of diet in *Acanthaster planci* (L) (Asteroidea) and hybrids reared in the laboratory. J. Exp. Mar. Biol. Ecol. 79: 129 - 147.

May, R.M. 1972. Limit cycles in predator-prey communities. Science 177: 900 - 902.

McCallum, H.I. 1987. Predator regulation of *Acanthaster planci.* J. Theor. Biol. 127: 207 - 220.

Moran, P.J., R.H. Bradbury & R.E. Reichelt. 1985. Meso-scale studies of the Crown of Thorns/coral interaction: a case history from the Great Barrier Reef. Proc. 5th Int. Coral Reef Congr. 5: 321 - 326.

Olson, R.R. 1987. In situ culturing as a test of the larval starvation hypothesis for the crown-of-thorns starfish, *Acanthaster planci.* Limnol. Oceanogr. 32: 895 - 904.

Ormond, R.F.G. , Bradbury, R.H., Bainbridge, S., Fabricius, K., Keesing. J., DeVantier, L., Medlay, P. & Steven, A. (this volume) Test of a model of regulation of crown-of-thorns starfish by fish predators. In: R.H. Bradbury (ed.) The *Acanthaster* phenomenon: a modelling approach. Springer-Verlag, Berlin.

Parslow, J. & A. Gabric. 1990. Advection, dispersal and plankton patchiness on the Great Barrier Reef. Aust. J. Mar. Freshwater. Res. 40: 403 - 419.

Pearson, R.G. 1981. Recovery and recolonization of coral reefs. Mar. Ecol. Prog. Ser. 4: 105 - 122.

Pearson, R.G. & R. Endean. 1969. A preliminary study of the coral predator *Acanthaster planci*. Qld Dept. Harbours and Marine Fisheries Notes 3: 27 - 55.

Reichelt, R.E. (this volume) Dispersal and control models of *Acanthaster planci* populations on the Great Barrier Reef. In: R.H. Bradbury (ed.) The *Acanthaster* phenomenon: a modelling approach. Springer-Verlag, Berlin.

Reichelt, R.E., W. Greve, R.H. Bradbury & P.J. Moran. 1990. *Acanthaster planci* outbreak initiation: a starfish-coral site model. Ecol. Modell. 49, 153-177.

Sloan, N.A. & A.C. Campbell. 1982. Perception of food. In Echinoderm Nutrition. M. Jangoux and J.M. Laurence (eds.), Balkema, Rotterdam, pp 3 - 23.

Yamaguchi, M. (1986) *Acanthaster planci* infestations of reefs and coral assemblages in Japan: a retrospective analysis of control efforts. Coral Reefs 5, 23-30.

Yokochi, H., S. Uemo, M. Ogura, A. Nagai & T. Habe. 1988. Recruitment, diet and growth of juvenile *Acanthaster planci* on reefs in a recovery phase. In: Abstracts, Sixth International Coral Reef Symposium, August 8 - 12 1988. Choat, J.H. and Bellwood, O. eds. Townsville: Organizing Committee of the Sixth Int. Coral Reef Symposium, p. 110.

DISPERSAL OF NEUTRALLY-BUOYANT MATERIAL NEAR JOHN BREWER REEF

STEPHEN L. GAY, JOHN C. ANDREWS[1]

Australian Institute of Marine Science, Townsville, Queensland 4810

AND

KERRY P. BLACK

Victorian Institute of Marine Sciences, Melbourne, Victoria 3002

Abstract. A numerical model of currents, based on the results of an oceanographic experiment at John Brewer Reef, is used to simulate the spawning of passive neutrally-buoyant particles for a four and a half month period in 1987. The simulation shows that the proportion of particles removed from the model grid per hour is proportional to the speed of the free-stream low-frequency current. Also, despite low-frequency current speeds of 0.20 m/s, some particles remained within 10 km of John Brewer Reef after 5 days. By allowing the simulated particles to settle after surviving 10 days in the water column, it was found that, although both the spatial and temporal variability of settlement occurred, some consistencies were observed.

During much of the simulation, particles were not retained near the reef longer than 10 days. However, when the speed of the net free-stream current was small, the number of particles retained was extremely high; in one instance as much as 1/3 of the simulated particles remained in the simulation area for more than 10 days. This paper therefore illustrates that high retention of neutrally-buoyant particles can occur during actual current conditions. These results can be applied to *Acanthaster planci* larvae (considered to be neutrally-buoyant) to show that increased retention of larvae along the Great Barrier Reef will occur when spawning coincides with small net currents.

1. INTRODUCTION

In 1962, an outbreak of *Acanthaster planci* (*A. planci*) occurred at Green Island. The number of reports of infestations has increased considerably over the last twenty years with large proportions of coral being destroyed along many of the reefs between Townsville and Cairns. It has been

---

[1] Present address: Department of Primary Industries, Brisbane, Qld 4001.

suspected that the infestations may be due to the impact of man (Moran, 1986). To resolve this question, the Crown-of-thorns Starfish Advisory Committee was formed in 1986 (Baker & Moran, 1986). Since, then the committee has funded research by over 50 scientists.

An important question which had to be confronted was how the outbreaks spread from one reef to the next. Dight & James (1988) modelled larval dispersal over the Cairns section of the Great Barrier Reef with a large-scale hydrodynamic model and a dispersal model. Their model incorporated the East Australian Current which bifurcates in the northern region near Cooktown. The northern region was said to be seeding both itself and the adjacent southern region. From these results, they inferred that the Cairns section acts as a source region for *A. planci* outbreaks.

Consideration of the currents has raised the question of whether the outbreaks could be due to self-seeding of individual reefs. That is, if small numbers of *A. planci* exist on a reef and then spawn, could large numbers of the larvae be retained near the reef causing an outbreak? This question was first considered by Black (1987). Subsequently, Black *et al.* (in press) simulated dispersal of neutrally-buoyant particles around 18 reefs using currents which consisted of a fixed tidal amplitude and a constant net current as modelled by Black & Gay (1987b). Black showed that *A. planci* larvae could remain near the reefs for periods of time much greater than 10 days. The retention of larvae was found to be highly dependent on the size of the reef and the speed of the low-frequency current.

Black's results had important implications for the occurrence of infestations and it became necessary to extend his work by modelling currents which matched more closely actual conditions. In the work presented here, natural variability of both low frequency and tidal currents is modelled around both John Brewer and Lodestone Reefs. These reefs have both had histories of outbreaks with John Brewer Reef recorded as having two infestations in the last twenty years (Moran *et al.*, 1985). The number of *A. planci* on John Brewer Reef is constantly being monitored by AIMS staff. The reefs are only 7 km apart (Figure 1) so by studying both reefs at the same time, it was possible to consider inter-reefal recruitment as well as self- seeding. 17 current meters and tide gauges were deployed at John Brewer Reef during March to September 1987 to measure actual currents. These measurements were employed both to provide boundary conditions and to test the accuracy of the model.

The results presented in this paper will focus on developing an understanding between retention of neutrally-buoyant particles and currents. Where possible, these results will be applied to dispersal of *A. planci* larvae which are also considered to be neutrally-buoyant

(Olson, *pers. comm.*). Although the time of the field work does not correspond to a spawning season, which is typically during summer, the results presented will provide a basis for estimating the likelihood of larval retention during the spawning season should current meter data become available. This paper will concentrate only on the movement of neutrally-buoyant matter due to currents. The mortality of *A. planci* larvae was considered to be an extremely important factor but was beyond the scope of the present study.

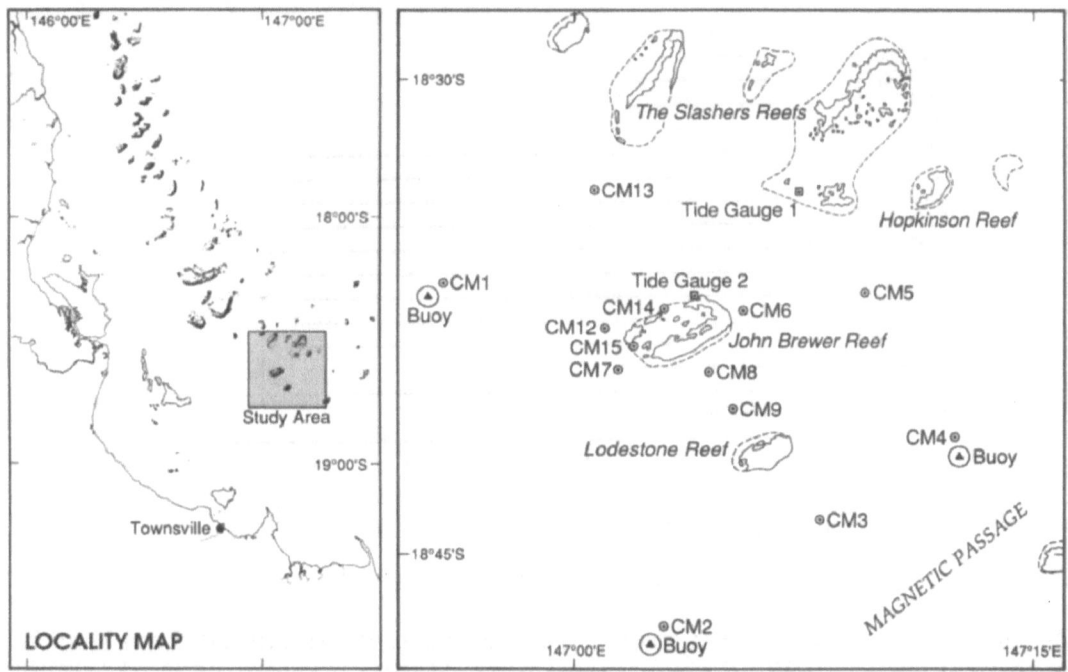

Figure 1. Location of current meter and tide gauge moorings.

2. THE PHYSICAL ENVIRONMENT

John Brewer Reef is about 5 km long and 3 km wide (Figure 1). Half of the reef's area is encompassed by a lagoon which has wide entrances on the western and northern sides. Lodestone Reef is crescentic with the lagoon on the leeward side of the prevailing south easterly trade winds. The size of the reef is about 2 km long and 1 km wide.

Both Lodestone and John Brewer Reefs are within the Central Great Barrier Reef. The tide in this section is predominantly a standing wave flooding and ebbing to and from the coast (Church, Andrews & Boland, 1985; Andrews & Bode, 1987). Tidal current speeds are typically of the order of 0.15 m/s.

Low-frequency currents in the Central Great Barrier Reef are influenced by the strength of the winds and the southward flowing East Australian Current (Andrews, 1983). During summer, with the onset of the monsoons, the winds are predominantly northerly with typical speeds of 5 m/s (Pickard et al., 1977). These winds enhance the southward flow of the East Australian Current near the coast. In winter, south-easterly winds with speeds of 5 m/s cause the coastal current to reverse with speeds of up to 0.15 m/s (Andrews, 1983).

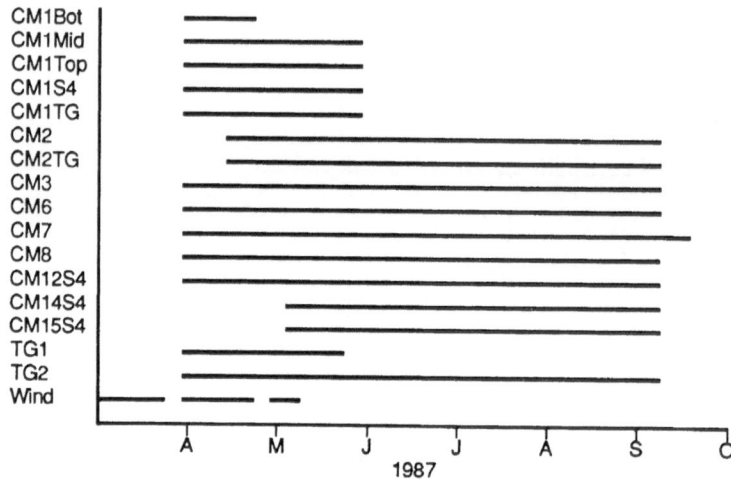

Figure 2. Reliable time periods of the data

3. FIELD STUDY

The field work for the John Brewer Reef experiment occurred from March to September, 1987. The instruments deployed included thirteen Aanderaa current meters, four S4 current meters and four Aanderaa tidal data recorders. Fourteen sites were selected to deploy the instruments (Figure1). To study the changes in current velocity with depth, four current meters were deployed on mooring CM1. Most current meters were deployed two fifths of the way down the water column. At this depth, current measurements are best estimates of the depth-averaged current.

The program utilized data from the Australian Institute of Marine Science wind gauge placed in the lagoon of John Brewer Reef and wind data from Cairns, Cardwell, Townsville and Mackay supplied by the Bureau of Meteorology.

Unfortunately, many of the current meter moorings had been vandalized or had been accidentally trawled while in the water. The locations where current meter data were not usable are CM4, CM5, CM9 and CM13. Despite these losses, the amount of reliable data retrieved was sufficient to allow the currents near John Brewer Reef to be modelled. The times at which reliable data exists for the current meters, tide gauges and wind gauge are shown on Figure 2.

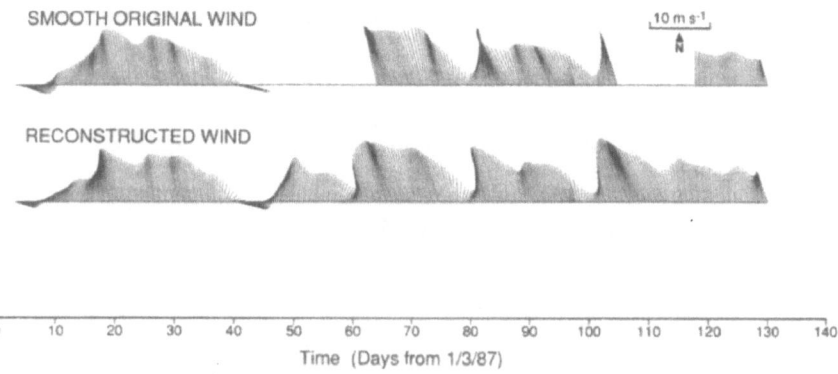

Figure 3. Original and reconstructed low-frequency components of the wind.

4. DATA ANALYSIS

The data were filtered to obtain low-frequency and tidal components. The low-frequency components were obtained by applying a Godin 24,24,25 moving average filter (Godin, 1972). The low-frequency components were then subtracted and the tidal components were found using Foreman's tidal analysis package (Foreman, 1977, & 1979). All current meters were analyzed in this way (Gay & Andrews, 1989). The remaining unfiltered data consisted largely of noise and were not included in the model.

The John Brewer Reef wind gauge retrieved only sparse data for the duration of the experiment due to technical problems (Figure 2). These data were insufficient for the modelling. Accordingly, multilinear regression was applied to the low-frequency components for the 1986 John Brewer Reef wind data and the coastal station data supplied by the Bureau of Meteorology. We were then able to reconstruct the low-frequency components of the 1987 John Brewer Reef wind data using linear interpolation applied to the coastal station records. The reconstructed wind vectors were compared to the actual smoothed data retrieved during 1987 (Figure 3). The similarity of the data can be clearly seen.

5. MODELLING METHODS

Currents were simulated using a depth-averaged numerical model. To determine the suitability of applying a depth-averaged model, 3 current meters located at CM1 at different depths were analyzed. Visual inspection reveals that for both the tidal and low-frequency components, the currents were very similar throughout the water column (Figure 4).

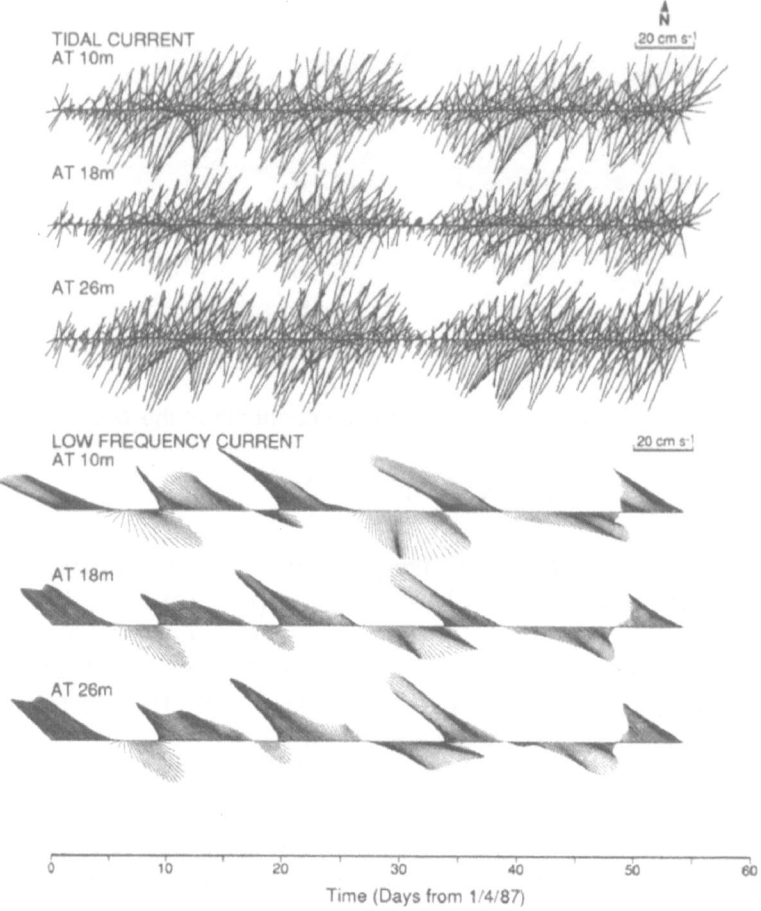

Figure 4. Vertical distribution of current at mooring CM2.

The equations used to determine the depth-averaged 2-Dimensional Flow (assuming incompressible flow and constant density) are:

Momentum

$$\frac{\partial u}{\partial t} + u\frac{\partial u}{\partial x} + v\frac{\partial u}{\partial y} - fv = -g\frac{\partial \xi}{\partial x} - g\frac{u\,(u^2 + v^2)^{1/2}}{H\,C^2} + \frac{\rho_a\,\gamma\,W_x\,|W|}{H\,\rho}$$
$$+ A_H\left(\frac{\partial^2 u}{\partial x^2} + \frac{\partial^2 u}{\partial y^2}\right)$$

$$\frac{\partial v}{\partial t} + u\frac{\partial v}{\partial x} + v\frac{\partial v}{\partial y} + fu = -g\frac{\partial \xi}{\partial y} - g\frac{v\,(u^2 + v^2)^{1/2}}{H\,C^2} + \frac{\rho_a\,\gamma\,W_y\,|W|}{H\,\rho}$$
$$+ A_H\left(\frac{\partial^2 v}{\partial x^2} + \frac{\partial^2 v}{\partial y^2}\right)$$

Continuity

$$\frac{\partial \xi}{\partial t} + \frac{\partial (Hu)}{\partial x} + \frac{\partial (Hv)}{\partial y} = 0$$

where u, v are the mean vertically averaged velocities in the x and y directions; f is the coriolis parameter; ξ is the sea level above datum, H is the sea-level above the bottom; ρ is the water density; ρ_a is the density of air, W is the wind speed at 10 m above sea-level, γ is the wind resistance coefficient, C is Chezy's C and A_H is the horizontal eddy viscosity coefficient. The value of A_H was chosen to be 10 m^2/s which is compatible with the values of A_H utilized by Black & Gay (1987a).

Chezy's C is a function of both depth and bottom roughness and is described by the formula:

$$C = 18 \log_{10} (\frac{0.37 {}^* H}{z_0})$$

where the roughness length, z_0, represents the bottom roughness.

Following the recommendations of Black & Gay (1987b), z_0 was represented in the model as a function of bed type so that:

z_0 =.001 for sand.
z_0 =.01 for rubble or reworked sand.
z_0 =.08 for coral.

To determine z_0 at each grid-cell, the nature of the sea bed for each cell had to be estimated. By perusal of the Australian Survey Office bathymetric charts and consultation with scientists familiar with John Brewer Reef, the bed was taken as being sand away from the reef, reworked undulating sand and bommies in the lagoon region, and coral along the reef flat and slope.

The momentum and continuity equations are solved using an Alternating Direction Implicit (ADI) scheme which has been described by Falconer (1976). Very briefly, the ADI method alternates between solving all the u components and then all of the v components of velocity. Sea-level is also updated when either u or v are solved. The u components are solved implicitly one row at a time using Gaussian elimination. In the same manner, the v components are solved along each column. The computer program to solve this scheme is called the Corspex model (Wolanski *et al.*, 1989). We have slightly adapted this program by using quadratic bottom friction instead of linear friction and by applying boundary conditions specifically developed for the study. These boundary conditions have been fully described in the Appendix.

Grid representation

The grid was oriented at 33°T so that Lodestone Reef was directly below John Brewer Reef. At this orientation, low-frequency currents directed up the coast moved slightly left of the top boundary and flood tides travelled toward the bottom of the left boundary.

To ensure that the wake effects of currents were significantly reduced at the boundaries, the grid was extended to be about 8 km from the closest reef. Hence, the size of the model region became 20 km (or 50 grid cells) by 22 km (or 54 grid cells) with the size of each grid cell being 400m by 400m.

Verifying the results

The boundary conditions for the model were developed using the measured free-stream currents. All measurements from the current meters located away from the reefs, apart from CM2, were not reliable so the boundary conditions were developed by using the low-frequency and tidal currents and tidal heights measured at CM2 only. The model simulated currents in hourly intervals for all the available CM2 data. That is from 8 Apr 87 until 27 Aug 87.

Two tests were made to verify the model. The first test compared the model results with the boundary data. The former were analysed by averaging the velocity from the boundary cells located midway along each boundary. The second test compared the model results with all the actual low-frequency and tidal currents. The statistical measure of error was:

$$\varepsilon^2 = \frac{\sum\limits_{i=1}^{N} (U_i - u_i)^2 + \sum\limits_{i=1}^{N} (V_i - v_i)^2}{\sum\limits_{i=1}^{N} (U_i^2 + V_i^2)}$$

where u_i, v_i are the model results and U_i, V_i are the addition of the hindcast tidal currents and low-frequency currents from the current meter.

Hence, ε^2 can vary from 0 for perfect results to infinity for extremely bad results. If $\varepsilon^2 = 1$, then the u_i and v_i are no better estimates than using (0,0).

Particle dispersal

Particles are tracked individually using a second-order tracking equation developed by Black & Gay (this volume) which makes allowance for the curvature of the trajectories. If a particle was removed from the model region, the value of the free-stream velocity specified in the boundary data was assigned to the particle. The particle was monitored until it had moved a model-length's distance (20km) from the boundary. As the currents reversed, some particles re-entered the model region. This approximation ignores the presence of downstream reefs so these particles must be presumed to represent larval translation around an isolated reef pair.

Integrated retention

Particles were released continuously from the reefs for the full four and a half month model period at a rate of 110 per hour. One of the purposes of this paper was to discuss likely settlement positions for *A. planci* larvae should the larvae behave passively before settling. These settlement positions were found by monitoring the positions where particles were situated from 10 to 25 days after release. After 25 days, the particles were no longer monitored. Hence, predicted settlement distributions correspond to the integral over time, for each grid cell, of the number of particles that have remained in the model region after 10 days and are referred to in this paper as "integrated retention".

6. RESULTS

Results of the field experiment

The speed of the low-frequency current exceeds 0.25 m/s (Figure 5) only 10% of the time. The maximum measured current was 0.40 m/s and the median was 0.11 m/s. Dominant directions were toward either the north-west or south-east (parallel to the coastline). However, the current measured in winter of 1987 was predominantly north-westerly. Lodestone Reef was therefore mainly upstream of John Brewer Reef.

The winds (Figure 5) were principally from the south south-east with an average speed of 5 m/s. Comparison of the wind with the low-frequency current showed that during strong south-easterly winds, the current was more likely to be toward the north-west. When the winds either subsided or became northerly, the current was generally toward the south-east.

The tidal currents (Figure 5) were elliptical with the principal direction being toward the coast. Average major axis amplitudes of the tidal currents were about 0.15 m/s.

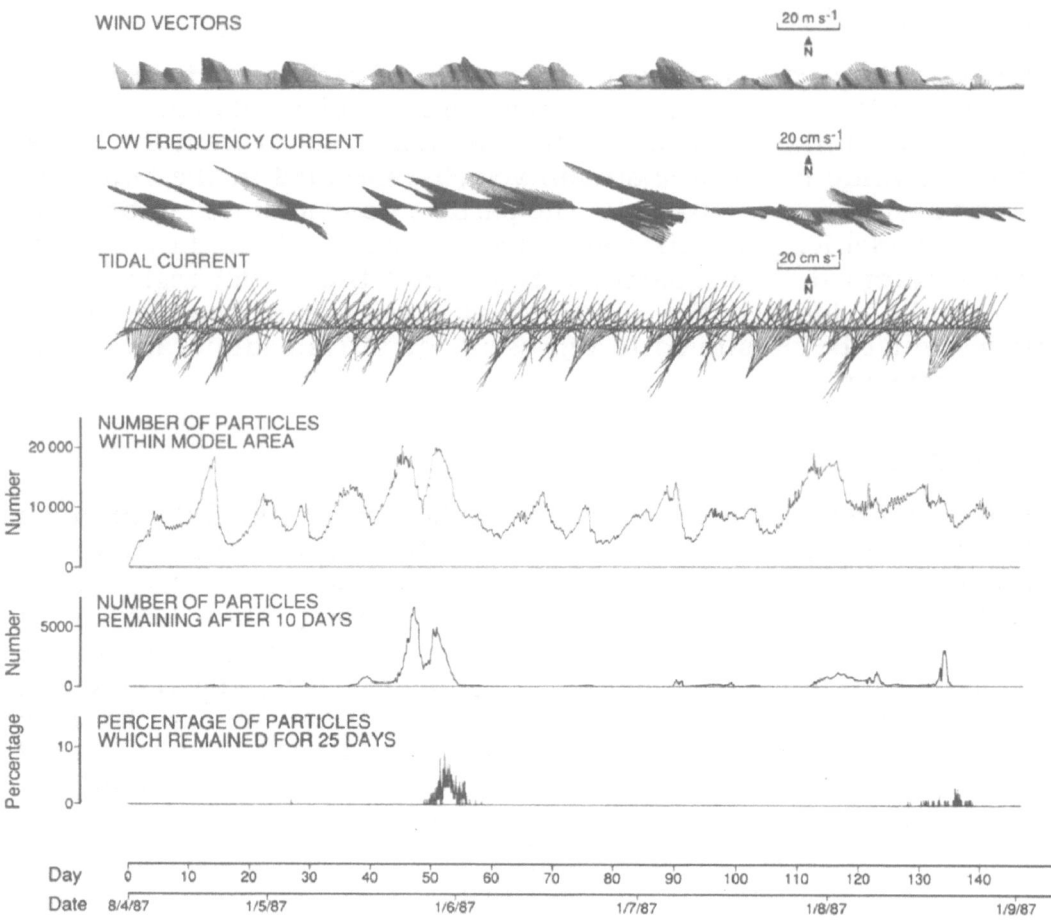

Figure 5. Results of the dispersal simulation.

Verifying the hydrodynamic model

Two tests were used to verify the model. The first test compared the model results with the specified boundary conditions. The second test compared the model results with the measured currents at each available mooring.

The boundary currents were found to be identical to CM2 indicating that no anomalies were occurring at the boundary. The model results at each mooring compared reasonably well with the measured currents as well (Table 1). The current measurements which best matched the model was CM2 with $\varepsilon^2=0.11$ and the average value of ε^2 was 0.2. The worst match between the model results and the measured currents was for CM8 with $\varepsilon^2=0.38$. CM8 is located in between Lodestone and John Brewer Reefs and is very sensitive to any minor change in the free-stream current. It is therefore not surprising that this location has the worst comparison between the measured currents and the model currents. The predicted model results for CM8 were compared directly with the measured current (Figure 6a and Figure 6b). The northward component of the modelled current (Figure 6a) is far better correlated with the actual data than the eastward current (Figure 6b). The results for CM8 are still reasonable with the close agreement between the model results and the actual measurements clearly visible. However, better results would have been achieved if other free-stream current meter data had been incorporated in the boundary conditions.

Location	ε^2
CM2	0.11
CM3	0.14
CM6	0.18
CM7	0.18
CM8	0.38
CM12	0.20
Avge.	0.20

Table 1. Error of the predicted currents at each current meter location.

Larval dispersal

Figures 7a to 7f show a simulation of particles being dispersed during a 66 hour period with an initial release at 24:00 hours on 25 May 87 (day 55 on Figure 5). This interval corresponds to a time when a strong current directed grid-easterly was abating (Figure 5, Table 2) and for the remainder of the period, the current was very small. The average current during the period was 5 cm/s directed grid-south (Table 2).

Figure	Time (Hours)	Current (cm/s) Up	Across
7a	0	0.9	-13.6
7b	18	-3.6	-4.9
7c	48	-5.7	4.7
7d	54	-4.7	4.1
7e	60	-3.4	2.9
7f	66	-2.0	1.2
Average	0-66	-4.5	-0.8

Table 2. Low-frequency currents during the dispersal simulation (time 0 is 0 hrs 26/5/87)

The particles were initially released around the reef perimeter (Figure 7a). During the first 18 hrs (Figure 7b), the strong current caused particles to disperse towards the left boundary. However, many particles were retained near the reefs with more particles remaining near John Brewer Reef than at Lodestone Reef. A large concentration of particles can be seen left of Lodestone Reef. This concentration moved towards the bottom of the grid in the next 12 hours due to the southerly current and is removed from the model area (Figure 7c). A second large concentration still remains near Lodestone Reef while a third cloud of particles from

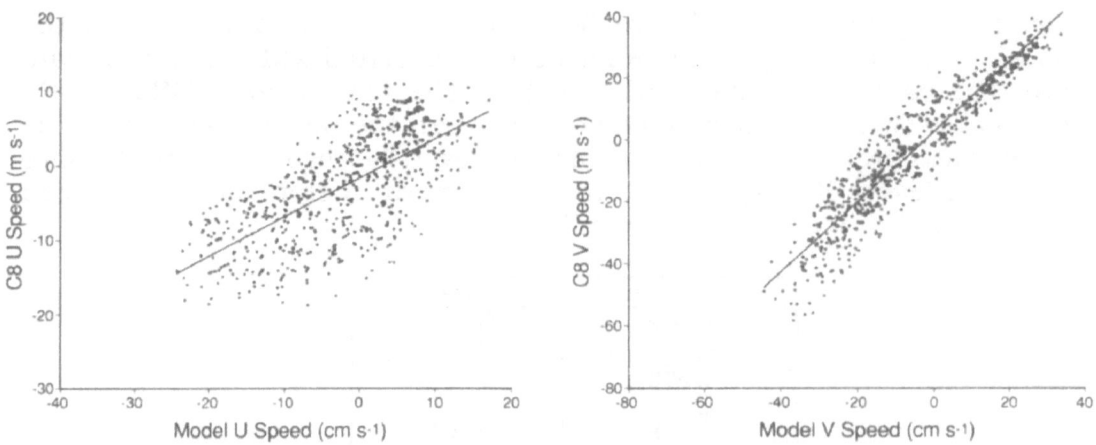

Figure 6. Comparison of the model-predicted speeds at CM8 with the measured speeds: **a.** CM8 U speed versus model-predicted U speed, **b.** CM8 V speed versus model-predicted V speed.

John Brewer Reef approaches Lodestone Reef (Figure 7d). At 60 hrs (Figure 7e), the approaching cloud split as it hit Lodestone Reef. The major concentration resumed its drift to the bottom boundary. By 66 hrs (Figure 7f), this cloud became completely separated from the particles retained near Lodestone Reef to eventually move out of the grid region. All of the particles that were removed from the grid were removed through the bottom boundary on the left side of Lodestone Reef.

Two patches of particles located to the right of John Brewer Reef moved to the right and slightly toward the top boundary against the free-stream current (Figures 7c-7f). Similarly, particles retained at Lodestone Reef show no clear sign of moving toward any boundary. At the end of the simulation (Figure 7f) about 1/2 of the initial particles still remain in the model region while about 1/10 of those remaining are within 1 km of a reef. More particles are located near John Brewer Reef than Lodestone Reef.

Particle retention

Temporal variability of retention is studied by considering the number of particles retained in the model after 0, 10 and 25 days (Figure 5).

The number of particles retained for 10 days (Figure 5e) is highly variable. Most of the time, there is little, if any retention. However, in some cases, the proportion of particles remaining is quite high. The largest retention is on day 46 with some 6000 particles retained. Over a 10 day cycle, 26400 particles are released, so this is just over 20% retention. The total number of particles within the grid is about 16000. Hence, more than 1/3 of the particles in the model are more than 10 days old.

There are only two time intervals during which particles are retained for more than 25 days (Figure 5f). These are from days 50 to 58 where the average retention rate is 5% and days 128 to 148 where the average

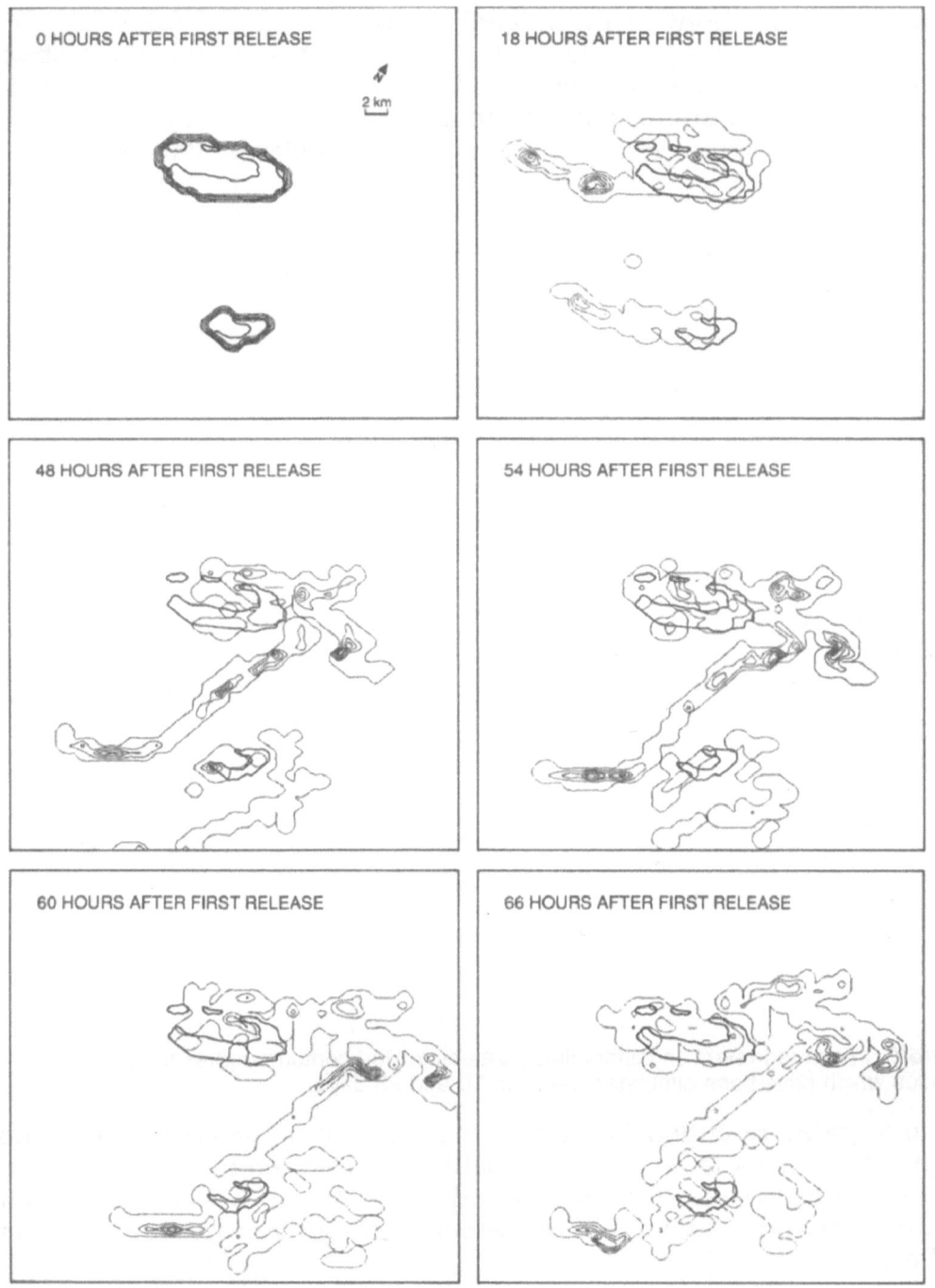

Figure 7. Numerical simulation of larvae initially released from John Brewer and Lodestone Reefs (time 0 corresponds to 0 hrs 26/5/87).
Each contour corresponds approximately to 20% of the maximum value in a grid-cell. The first contour corresponds to greater than 0 particles. The sixth contour corresponds to the grid-cell with the largest number of particles.

retention rate is about half a percent. Particles retained from day 30 to day 55 were released just after a 0.25 m/s north-westerly current was abating. After this current decreased, the currents remained generally small with the maximum current being 0.15 m/s. On day 55, a strong north-westerly current removed most of the particles by day 58.

Figure 8. Integrated retention distributions. Values for the contours represent the number of particles which have been simulated between 10 and 25 days.

The integrated retention for passive particles over the entire simulation (Figure 8a) shows large spatial variability. Large retention occurs 4 km west of John Brewer Reef, in a band extending to 5 km east of John Brewer Reef and 3 km west of Lodestone Reef. Settlement occurs at every grid-cell.

Integrated retention was then considered within 3 time intervals to determine whether the areas of high retention were the same for each time period. The time intervals are:

(i)   0-48 days;
(ii) 48-107 days;
(iii) 107-148 days.

In time interval (i), currents to the north-west dominate. However, during the period of maximum retention, the currents varied between north-west and south-east (Figure 5). Maximum retention occurs about 4 km SE of John Brewer Reef (Figure 8b).

Retention was maintained at high levels from the end of (i) to the start of (ii) as the currents shifted from south-easterly to north-westerly (Figure 5). Maximum retention then occurred 4 km west of John Brewer Reef (Figure 8c). As the north-westerly currents strengthened, retention subsided (Figure 5). The current remained generally strong for the remainder of (ii) until strong southeasterly currents developed toward the end of the period.

In (iii), retention occurred in the middle and towards the end of the period. Both of these intervals coincided with south-easterly currents (Figure 5). The maximum retention occurs 4 km SE of John Brewer Reef. Large retention also occurs near the bottom left corner.

For all of these time intervals (i) to (iii), at least a small amount of retention occurs at every grid-cell.

Relationship of low frequency current to larval numbers

The number of particles remaining in the model region (Figure 5d) were compared with the speed of the low-frequency current (Figure 5c), the speed of the wind (Figure 5a) and the tidal currents (Figure 5b). No simple relationship was found. Instead of using the number of particles remaining, the percentage of particles removed per hour was calculated. However, this percentage fluctuated markedly from one hour to the next. By applying the Godin 24,24,25 moving-average filter, these fluctuations were removed. Figure 9 shows the percentage of particles removed from the grid region per hour (smoothed) and the speed of the low-frequency current. The correlation between these two curves was found to be as high as 0.8. This relationship can be expressed analytically as

$$\delta N/\delta t = -\alpha \, S \, N \tag{1}$$

where N is the number of particles in the model region,
S is the speed of the low-frequency current,
α is the loss rate per 1 m/s and
t is time.

Using linear regression, α is found to be 6.9% per hour per 1 m/s.

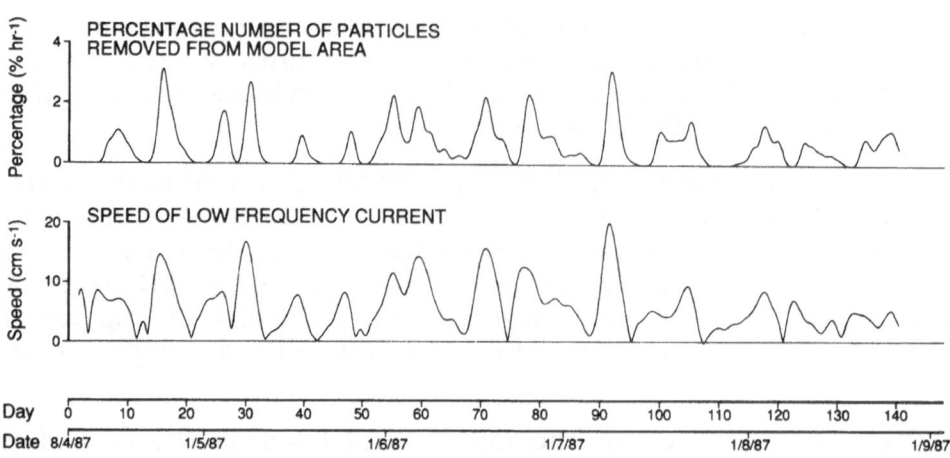

Figure 9. Comparison of the number of particles removed per hour (smoothed) with the speed of the low-frequency current at CM2.

7. DISCUSSION

Retention was examined over two time-scales, 10 days and 25 days. The largest retention of particles over a 10 day period occurred on day 46 of the simulation. In this case, 20% of particles were retained. This peak corresponded to both small low frequency currents and reversals in current direction which were able to return particles into the model region. The largest retention of particles over a 25 day period occurred from day 30 to day 55 of the simulation with about 5% of the particles retained.

Hence, high retention will occur when the speed of the low frequency currents are small. An analytical solution was developed which showed that the flushing of particles away from the reef is directly proportional to the speed of the low-frequency current. It can be shown that for a single release with a constant low-frequency current, eqn (1) becomes

$$N = N_0 \, e^{-\alpha S t}$$

where N_0 is the number of particles released.

This result is in agreement with Black *et al.* (in press) who showed that retention of neutrally-buoyant matter decreases exponentially with time. Black *et al.* went further to show that the rate of exponential decrease is a function of the shape of the reef, the speed of the low-frequency current and the amplitude of the tides; both larger reefs and larger tidal currents enhancing retention. They also found that the decay rate was proportional to the speed of the low frequency current which is in agreement with our results.

Prior to the peak in settlement on day 46 of the simulation, both the increase in the total number of particles and the increase in the number of particles older than 10 days were, at times, greater than the release rate (110 per hour). Hence, the increased retentions were due, partly, to particles returning to the model region. However, as the number of particles which are lost from the model was found to be highly correlated to the current speed, the reversals are not considered to be dominant. The exact contribution that reversals contribute to retention can only be determined after further investigation.

The winds were important in increasing retention by changing the speed of the low-frequency current. When the winds were negligible, the current was toward the south-east. As south-easterly winds developed, the winds caused the current to reverse and flow to the north-west. Maximum retention will therefore be achieved when the wind is sufficient to cause a low-frequency current of negligible speed.

The time of this experiment was when the south-east trade winds dominated. *A. planci* spawn during summer (Moran, 1986). During this season, there are more northerly winds causing the low-frequency current to be mainly south-easterly. However, strong south-easterly winds which will cause north- westerly currents still exist in this season (Andrews *et al.* 1988). Hence, it is very likely that during summer, there will be times when the low- frequency currents will allow high retention. The results presented in this paper therefore show that outbreaks of *A. planci* may be a direct result of high retention caused by small low-frequency currents after spawning. It is therefore essential that summer currents be measured and that this data be analyzed in direct comparison with present *A. planci* surveys.

The movement of particles around the reef proved to be highly complex. Comparison of John Brewer Reef with Lodestone Reef reveals that more particles were retained near John Brewer Reef as John Brewer Reef is larger. This result is in agreement with the results of Black *et al.* (in press). When the current was large at the start of the release, most particles were advected away from the reef, yet a large proportion of the particles were still able to remain. Hence, the complex circulation around the reef causes the particle dispersion over a wide area. This feature was also observed when a cloud of particles approaching Lodestone Reef split into 2 clouds with 1 cloud moving toward the left of the reef and through the bottom boundary and the second cloud remaining near Lodestone Reef. This movement indicates that the current flow near Lodestone Reef was much different to the movement within the free-stream.

Integrated retention patterns were found to be highly variable both temporally and spatially. Patterns for any one of the time intervals (Figures 8b-8d) were not representative of the settlement patterns for the entire period (Figure 8a). However, some clear temporal and spatial consistency was evident. Black (1988) showed that the interaction of tidal and steady currents cause the zones of maximum retention to occur on the downstream side of the reef relative to the steady current, approximately 45⁰ anti- clockwise of the coastal current. In reversing coastal flows, this puts the integrated retention maxima at the north-east and south-west of a notional reef subjected to a north/south oriented tidal current and east/west coastal current.

The results obtained in this study are consistent with this general behaviour except that the expected maximum on the east of Lodestone Reef is only evident in Figure 8d. Maximum retention otherwise coincides with the general predictions (Figure 8a). Thus, while there is variability in strength and direction of the coastal current, the predominance of north-west and south- east flows ensures some consistency. The zone of maximum integrated retention does not occur in the wake of the coastal current in the lee of the reef.

8. CONCLUSIONS

This paper has shown that retention of particles is found to be inversely proportional to the speed of the low-frequency currents and also that passive neutrally-buoyant particles can be retained at reefs of the Great Barrier Reef for up to 25 days. Applying these results to the dispersal of A. planci larvae indicates that there is a strong possibility that A. planci outbreaks may be due to high retention of larvae at reefs when low frequency currents are small.

Acknowledgements. The authors would very much like to thank Marc Jeffrey, Craig Steinberg and the crews of the Lady Basten and Pacific Voyager for their assistance in deploying and retrieving current meters and tide gauges. Special thanks are given to Ray McAlister for his dedication in supervising all field work and to Eric Wolanski for allowing use of the CORSPEX Model. This work was funded by the Great Barrier Reef Marine Park Authority.

REFERENCES

Andrews J.C. (1983) Lagoon-ocean interactions. Proceedings, Inaugural Great Barrier Conference, Townsville, JCU Press. p. 402-408.

Andrews J.C. & L. Bode (1987) The tides of the Central Great Barrier Reef. Continental Shelf Research 8: 1057-1085.

Andrews, J.C., S.L. Gay & P.W. Sammarco, 1988. Influence of circulation pattern on self-seeding at Helix Reef - Great Barrier Reef. Proceedings 6th International Coral Reef Symposium, Townsville, Qld., Australia, 2: 469-474.

Baker J.T. & P.J. Moran (1986). A summary of research on ecological aspects of the crown-of-thorns starfish funded by COTSAC and coordinated by the Australian Institute of Marine Science. Crown-of-thorns Study Report No. 3, Australian Institute of Marine Science, Townsville.

Black K.P. (1987) Dispersal of matter on and around a coral reef - important physical mechanisms, retention times and rates, and the relative concentrations of free-floating larvae. ANZAAS Congress, crown-of-thorns starfish session, Townsville, Australia.

Black K.P. (1988) The relationship of reef hydrodynamics to variations in the number of planktonic larvae on and around coral reefs. Proceedings 6th International Coral Reef Symposium. Townsville, 2: 125-130.

Black K.P. & S.L. Gay (1987a) Eddy formation in unsteady flows. Journal of Geophysical Research, 92 (C9): 9514-9522.

Black K.P. & S.L. Gay (1987b) Hydrodynamic control of crown-of-thorns starfish larvae. 1. Small-scale hydrodynamics on and around coral reefs. Victorian Institute Marine Sciences Technical Report No. 8, 62 pp.

Black K.P. & S.L. Gay (this volume) A numerical scheme for determining trajectories in particle models. In: R.H. Bradbury (ed) The *Acanthaster* phenomenon: a modelling approach. Springer-Verlag, Heidelberg.

Black K.P., S.L. Gay & J.C. Andrews (in press) Residence times of neutrally-buoyant matter such as larvae, sewage or nutrients on coral reefs. Coral Reefs.

Church J.A., Andrews, J.A. & Boland, F.M. (1985) Tidal currents in the central Great Barrier Reef. Continental Shelf Research. 4: 515-531.

Dight I.J. & M.K. James (1988). Simulation of large-scale population dynamics of crown of thorns starfish in the Great Barrier Reef system. James Cook University of North Queensland Research Bulletin No. CS36. 38 pp.

Falconer R.A. (1976) Mathematical modelling of jet-forced circulation in reservoirs and harbours. PhD. thesis, University of London, England.

Foreman M.G.G. (1977) Manual for tidal heights analysis and prediction. Institute of Ocean Sciences, Patricia Bay, Victoria B.C.

Foreman M.G.G. (1979). Manual for tidal currents analysis and prediction. Institute of Ocean Sciences, Patricia Bay, Victoria B.C.

Gay S.L. & J.C. Andrews (1989). Field survey: physical oceanography, John Brewer Reef, March to September, 1987, submitted as Technical Report to the Australian Institute of Marine Science.

Godin G. (1972) The analysis of tides. Liverpool University Press.

Moran P.J. (1986). The *Acanthaster* phenomenon. Oceanography and Marine Biology Annual Review. 24: 379-480.

Moran P.J., R.H. Bradbury & R.E. Reichelt (1985). Mesoscale studies of the crown-of-thorns/coral interaction: A case history from the Great Barrier Reef, Proceeding of the Fifth International Coral Reef Congress, Tahiti, 5: 321-326.

Pickard G.L., J.R. Donguy, C. Henin & F. Rougerie (1977). A review of the physical oceanography of the Great Barrier Reef and Western Coral Sea. AIMS Monograph Series, Vol. 2. 135 pp.

Thomas G.B. & R.L. Finney (1979). Calculus and Analytical Geometry. Addison-Wesley Publishing Company, 870 pp.

Wolanski E., D. Burrage & B. King (1989). Trapping and dispersion of coral eggs around Bowden Reef, Great Barrier Reef, following mass coral spawning. Continental Shelf Research, 9, No. 5: 479-496.

APPENDIX: BOUNDARY CONDITIONS
FOR THE JOHN BREWER REEF SIMULATION

The method used to determine boundary conditions for the simulation of currents was developed specifically for this study. In principle, the values of the normal velocities along the boundaries are adjusted so that the combined average of the normal velocities from opposing boundaries equals a specified input velocity. Furthermore, the normal velocities are constrained to ensure that the average sea-level throughout the grid region equals the sea-level specified in the boundary data.

The whole model region was considered as a rectangle as shown below.

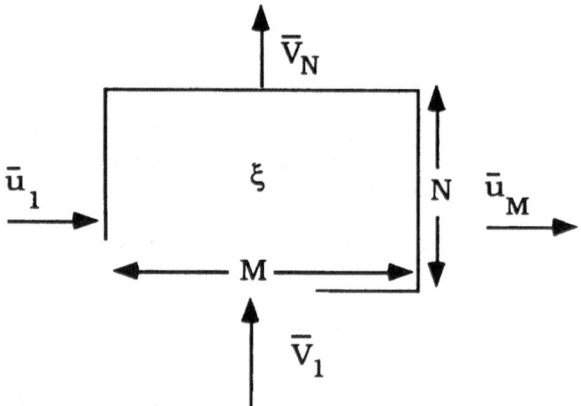

$\bar{u}_1$, $\bar{u}_M$, $\bar{v}_1$, and $\bar{v}_N$ are the averages of the normal velocities along each boundary. L and W are the distances in the x and y directions from the middle of the first grid-cell to the middle of the last. Hence, if M is the number of cells in the x-direction and δx is the length of a grid cell, then

$$L=(M-1)*\delta x$$

Similarly, if N is the number of cells in the y-direction and δy is the width of a grid cell, then

$$W=(N-1)*\delta y$$

Now if u_b and v_b are the values of the velocity specified in the boundary data, then:

$$1/2 \ (\bar{u}_1 + \bar{u}_M) = u_b \tag{1}$$

$$1/2 \ (\bar{v}_1 + \bar{v}_N) \ = v_b \tag{2}$$

In the case of John Brewer Reef, the depth at all the boundary points was taken as being 50m. Hence, the continuity equation becomes:

$$\xi_t = H \ (\frac{\bar{u}_1 - \bar{u}_M}{L} + \frac{\bar{v}_1 - \bar{v}_N}{W}) \tag{3}$$

where H is water depth and ξ_t is the wanted rate of change of the average value of sea-level within the model region.

This can be calculated by:

$$\xi_t = \frac{(\xi_b - \xi)}{\delta t}$$

where δt is the time step between iterations,

ξ_b is the value of the sea-level specified in the boundary data and

ξ is the current value of sea-level averaged from throughout the model region.

So far, there are 4 unknowns $(\bar{u}_1, \bar{u}_M, \bar{v}_1,$ and $\bar{v}_N)$, but only 3 equations (1, 2 and 3) from which these unknowns can be determined. Hence, there are an infinite number of possible solutions. Of course, many of these solutions would not be realistic. The boundary solution we choose is the one which satisfies eqns (1), (2) and (3) and is most similar to the current averages of normal velocity at the boundaries. These averages are represented by
$\bar{u}'_1, \bar{u}'_M, \bar{v}'_1,$ and $\bar{v}'_N$

More formally, it is desired to minimize S^2 where:

$$S^2 = 1/2 \ [\ N \ (\bar{u}_1 - \bar{u}'_1)^2 \ + N \ (\bar{u}_M - \bar{u}'_M)^2 \ + M(\bar{v}_1 - \bar{v}'_1)^2 \ + M \ (\bar{v}_N - \bar{v}'_N)^2 \ ] \tag{4}$$

constrained by equations (1) to (3)

Using Lagrange multipliers (Thomas & Finney, 1979), equation (4) is minimized when

$$\bar{u}_1 = \bar{u}'_1 + \frac{\lambda_1}{2N} + H \frac{\lambda_3}{L \ N} \tag{5}$$

$$\bar{u}_M = \bar{u}'_M + \frac{\lambda_1}{2N} - H \frac{\lambda_3}{L \ N} \tag{6}$$

$$\bar{v}_1 = \bar{v}'_1 + \frac{\lambda_2}{2M} + H \frac{\lambda_3}{W \ M} \tag{7}$$

$$\bar{v}_N = \bar{v}'_N + \frac{\lambda_2}{2M} - H \frac{\lambda_3}{W \ M} \tag{8}$$

Without loss of generality, we can write

$$\lambda_1 = \frac{\lambda_1}{2N} \qquad \lambda_2 = \frac{\lambda_2}{2M} \qquad \lambda_3 = H \lambda_3$$

Eqns (5), (6), (7) and (8) are now:

$$\bar{u}_1 = \bar{u}'_1 + \lambda_1 + \frac{\lambda_3}{LN} \tag{9}$$

$$\bar{u}_M = \bar{u}'_M + \lambda_1 - \frac{\lambda_3}{LN} \tag{10}$$

$$\bar{v}_1 = \bar{v}'_1 + \lambda_2 + \frac{\lambda_3}{MW} \tag{11}$$

$$\bar{v}_N = \bar{v}'_N + \lambda_2 - \frac{\lambda_3}{MW} \tag{12}$$

Manipulation of eqns (9) and (10) gives:

$$\frac{\lambda_3}{LN} = \frac{1}{2} [(\bar{u}_1 - \bar{u}_M) - (\bar{u}'_1 - \bar{u}'_M)] \tag{13}$$

and

$$\lambda_1 = u_b - \bar{u}' \tag{14}$$

where

$$\bar{u}' = (\bar{u}'_1 + \bar{u}'_M)/2$$

Similarly, manipulation of eqns (11) and (12) gives:

$$\frac{\lambda_3}{MW} = \frac{1}{2} [(\bar{v}_1 - \bar{v}_N) - (\bar{v}'_1 - \bar{v}'_N)] \tag{15}$$

and

$$\lambda_2 = v_b - \bar{v}' \tag{16}$$

where

$$\bar{v}' = (\bar{v}'_1 + \bar{v}'_N)/2$$

Algebraic manipulation of (13) and (15) gives

$$2 H \lambda_3 \left(\frac{1}{L^2 N} + \frac{1}{W^2 M}\right) = H \left(\frac{\bar{u}_1 - \bar{u}_M}{L} + \frac{(\bar{v}_1 - \bar{v}_M)}{W}\right) - H \left(\frac{(\bar{u}'_1 - \bar{u}'_M)}{L} + \frac{(\bar{v}'_1 - \bar{v}_M)}{W}\right)$$

Using eqn (3), this reduces to

$$2 H \lambda_3 \left(\frac{1}{L^2 N} + \frac{1}{W^2 M}\right) = \xi_t - \xi'_t$$

where ξ'_t is the previous value of ξ_t.

Thus,

$$\lambda_3 = \frac{W^2 M L^2 N}{L^2 N + W^2 M} \quad \frac{(\xi_t - \xi'_t)}{2H}$$

For large N and M,

$$W M \propto L N$$

Substituting eqns (14), (16) and (17) into eqns (9)-(12) yields:

$$\bar{u}_1 = \bar{u}'_1 + (u_b - \bar{u}') + \frac{L\,W}{L+W} \quad \frac{(\xi_t - \xi'_t)}{2\,H} \tag{18}$$

$$\bar{u}_M = \bar{u}'_M + (u_b - \bar{u}') - \frac{L\,W}{L+W} \quad \frac{(\xi_t - \xi'_t)}{2\,H} \tag{19}$$

$$\bar{v}_1 = \bar{v}'_1 + (v_b - \bar{v}') + \frac{L\,W}{L+W} \quad \frac{(\xi_t - \xi'_t)}{2\,H} \tag{20}$$

$$\bar{v}_N = \bar{v}'_N + (v_b - \bar{v}') - \frac{L\,W}{L+W} \quad \frac{(\xi_t - \xi'_t)}{2\,H} \tag{21}$$

From these equations, it can be seen that if the desired rate of change of sea-level is the same as the current rate of change of sea-level, the velocity gradient from one opposing boundary to the other must not change. This result is realistic. To determine whether these boundary conditions will accurately match physical processes, consider what would happen if there were a 1m tidal oscillation over a 12-hour cycle (semi-diurnal) caused by currents flooding and ebbing to the left and right boundaries.

The sea-level is given by:
$$\xi = \xi_0 \sin (2\pi\, t/T) \tag{22}$$

where $\xi_0 = 1m$ and
$\quad$ T = 12 *3600 s.

During ebb, the increase in u_b will cause the values of u at the left and right boundaries to increase. This pushes water into the left boundary and out through the right boundary causing a sea-level gradient to develop across the model. This slope will correspond to the expected sea-level slope due to the tide. Specifying u_b will not cause the average sea-level throughout the model to change.

Sea-level rises are caused independently by water entering the model through each boundary. Taking L=W, then the amount of water entering from each boundary will be the same. Using eqn (20), the velocity directed internally through all of the boundaries is given by:

$$v = L/4H * \xi_t$$

Hence, v will be 0 at low-tide ($\xi = -\xi_0$) and will be largest when $\xi=0$.

Substituting eqn(22)

$$v_{max} = L/4H * 2 \pi/T \xi_0$$

For John Brewer Reef, L is about 20 000m and H is 50m.

Thus,
$$v_{max} = 1.5 \text{ cm/s}$$

Hence, a 1.5 cm/s current will flow through the top and bottom boundaries. For the actual process, the flow through the top and bottom boundaries will be 0. So the model causes a discrepancy of 1.5 cm/s. This value is small compared to the maximum tidal currents (15 cm/s). Hence, the agreement between the modelled process and the actual process is reasonable. A major advantage of this boundary condition method is that it is not necessary to split the current and sea-level into its various tidal and low-frequency constituents within the boundary conditions.

So far, no allowance has been made for variations in the normal velocities along a boundary. For example, in the lee of a reef, the current is expected to be reduced. To compensate for this effect, we want the normal velocities along each boundary cell to be as close as possible to adjacent values found internally. At the same time, the average velocities should be defined as in eqns (18) to (21).

The method used to compensate for local variations is explained by considering the calculation of the new value of $u(1,n)$. This value corresponds to the normal velocity at the nth boundary cell along the left boundary. Defining the previous value of u one grid-cell internally as $u'(2,n)$, it is desired to have the difference between $u(1,n)$ and $u'(2,n)$ to be as small as possible. However, the average velocity along the left boundary must be as calculated in eqn (18).

Using Lagrangian multipliers again, the equation to minimise is now

$$S^2 = 1/2 \sum_{n=1}^{N} [u(1,n) - u'(2,n)]^2$$

constrained by

$$\bar{u}_1 = \frac{1}{N} \sum_{n=1}^{N} u(1,n)$$

This has a solution:

$$u(1,n) = u'(2,n) + \bar{u}_1 - \bar{u}'_1$$

where

$$\bar{u}'_2 = \frac{1}{N} \sum_{n=1}^{N} u'(2,n)$$

Applying the same method to all the boundary cells, it is found that:

$$u(1,n) = u'(2,n) + \bar{u}_1 - \bar{u}'_2 \qquad n=1..N$$

$$u(M,n) = u(M-1,n) \quad + \bar{u}_M \; - \; \bar{u}'_{M-1} \qquad\qquad n=1..N$$

$$v(m,1) = v(m,2) \quad + \; \bar{v}_1 \; - \; \bar{v}'_2 \qquad\qquad m=1..M$$

$$v(m,N) = u(m,N-1) + \; \bar{v}_N \; - \; \bar{v}'_{N-1} \qquad\qquad m=1..M$$

where $\bar{u}'_1$, $\bar{u}'_M$, $\bar{v}_1$ and $\bar{v}'_N$ are found from eqns (18)-(21).

The boundary conditions developed for this paper are only applicable for a model region where the depths at all the boundaries are the same. Boundary conditions which can be used more generally are presently being developed.

REEF-SCALE NUMERICAL HYDRODYNAMIC MODELLING DEVELOPED TO INVESTIGATE CROWN-OF-THORNS STARFISH OUTBREAKS

KERRY P. BLACK

Victorian Institute of Marine Sciences, Melbourne, Victoria 3002

AND

STEPHEN L. GAY

Australian Institute of Marine Science, Townsville, Queensland 4810

Abstract. The direct application of numerical hydrodynamic models to develop an understanding of biological phenomena is described. The paper reviews the gains in knowledge which have arisen from the reef-scale numerical modelling study of Great Barrier Reef hydrodynamics, developed to investigate crown-of-thorns starfish outbreaks. The models indicated an unusual and complex hydrodynamic environment, but one in which a number of patterns recur.

INTRODUCTION

Numerical studies of coral reef hydrodynamics, which specifically aim to examine a biological phenomenon, have occurred only recently on Australia's Great Barrier Reef (GBR). They arise out of an improved calibre of numerical simulation, with better numerical techniques and computer power. In addition, we now have a broader knowledge of the phenomena to be modelled. This is reflected by the increased number of physical and biological measurements that are now available (e.g. Moran & Johnson, 1990). A fundamental knowledge of the environment to be simulated is a modelling pre-requisite. However, the model then may provide insights into the physics and inter-related forcing phenomena, which are either difficult to measure or which have not been examined fully in the prototype. The model also provides a tool for application to a wider range of regions and conditions, often beyond the capacity attainable with available funding.

This summary paper examines the findings of the numerical modelling study initiated to seek relationships between crown-of-thorns starfish (COTS) outbreaks and small-scale reef hydrodynamics. The paper highlights the results which arose out of the application of the numerical procedures, and aims to demonstrate the impact of the modelling on the development of a clearer understanding of coral reef dynamics, for subsequent application to planktonic larval dispersal.

Prior to the commencement of this investigation in 1986 there had been no simulations of small-scale inter-tidal circulation on and around GBR coral reefs. Falconer *et al.* (1986) had modelled tidal circulation around an island (Rattray Island in the southern central GBR), and Frith & Mason (1986) had modelled One Tree Island under wind-forcing for a selection of mean water levels. The tides were otherwise not included. Thus, the provision of a comprehensive simulation of the hydrodynamics at reef scales was an initial priority of this investigation. This became more important when it was found that the hydrodynamics on and in the immediate vicinity of coral reefs had not been thoroughly measured or explained (Pickard, 1986). The complexity, scale and spatial variation of the circulation meant that large field programs were needed to specify the dynamics. However, attempts were being made, with concurrent computer modelling programs, to measure the circulation (Falconer *et al.*, 1986; Sammarco & Andrews, 1988; Andrews & Black, in Baker & Moran, 1987; Coral Reef Spawning Experiment, GBRMPA). Data of this kind greatly improve the capacity for accurate numerical simulation, and its verification.

The second priority of this investigation was to provide a reef-scale view of the dispersal processes. There had been no experiments to measure long-term dispersal characteristics of individual reefs, primarily owing to the lack of suitable experimental techniques. This meant that only some limited short-term dye experiments had been made (e.g. Ludington, 1981). Numerical modelling after calibration provided the capacity for comprehensive description of the reef hydrodynamic environment with the result that the modelling was able to provide the first view of the long-term dispersal around reefs on the GBR (Black, 1988; Black *et al.*, in press). Simultaneously, informative attempts to infer the medium-term dispersal around Bowden Reef were made using coral spawn as a biological tracer (Willis & Oliver, 1988) and correspondence with the modelling was reported by Black (1988).

In this paper, detailed numerical procedures are presented in Appendices, i.e. the tidal boundary condition (Appendix 1), the 3-dimensional model (Appendix 2), the model calibration (Appendix 3) and a new radiative boundary condition (Appendix 4).

MODELS EMPLOYED

For the hydrodynamic simulations, the 2-dimensional model 2DD of Black (1983) was chosen. This was subsequently upgraded during the study to a 3-dimensional model (Appendix 2) to examine the 3-dimensional circulation around Davies Reef (Black, *et al.*, submitted). Both models included the features necessary for simulation of inter-tidal reefs. The vertically-averaged momentum and continuity equations for horizontal flows in a homogeneous medium are solved using an explicit time-stepping procedure (Appendix 2).

PHYSICAL ENVIRONMENT

Flow measurements on the GBR indicated three dominant current components. These components are caused by tides, the East Australian current (EAC) and wind. In the central GBR, the tides pass across the shelf perpendicular to the isobaths (Church *et al.*, 1985). The low frequency EAC currents are coast parallel and directed poleward in the Central GBR. These are modulated, and often reversed, by local winds at around 10 day periods (Andrews, 1983; Andrews & Furnas, 1986). The three current components have similar magnitudes (Andrews, 1983, Church *et al.*, 1985) indicating that separate simulation of each component would be unsatisfactory as the non-linear interactions would be expected to modify the intensity of each. For example, for the case of steady coastal currents and perpendicular tidal currents, the frictional resistance is increased by the factor 1.4 if both components are equal, as sometimes occurs on the GBR.

Table 1 shows that, of the terms in the momentum equation (Appendix 2), the non-linear advective momentum and eddy viscosity terms are large near the reef, but both are small away from the reef. The sea gradient and Coriolis force are the dominant forces in the free stream. The bed friction is very large on the reef flat when currents are at their maximum.

MORPHOLOGIES AND CONDITIONS MODELLED

Eighteen actual reefs were simulated (Figure 1; Table 2) for a range of tide, wind and steady current conditions (Black & Gay, 1987a). The reefs were selected in collaboration with marine biologist, Dr Peter Moran, who had conducted surveys of starfish numbers and outbreaks on the GBR. Reefs from a variety of morphological categories were chosen when information about their starfish history was available. The reefs came from the morphological categories of planar, lagoonal, fringing, crescentic and ribbon (Hopley, 1983). Reefs as small as Helix Reef (0.6 x 0.6 km) up to reefs as large as Big Broadhurst (15 x 7 km) were simulated. Both individual reefs and reef groups within the GBR reef matrix were considered.

Table 1. Value of terms in the momentum equation ($\times 10^6$) across a section (22,31,22,22) through the middle of the square reef from the reef centre.

U terms at 60.5 hrs. $(-U\partial U/\partial x - V\partial U/\partial y)$, fV, $-g\partial z/\partial x$, $gU(U^2+V^2)^{1/2}/C^2d$, $A_H(\partial^2 U/\partial x^2 + \partial^2 U/\partial y^2)$

Cell	Depth	U Speed	V Speed	Non-lin	Coriolis	Sea Grad	Friction	Eddy Dif
22.0	2.1	0.01	0.01	-0.1	-0.5	11.8	-3.1	-0.2
23.0	2.1	0.01	0.01	-0.4	-0.5	19.6	-0.4	-0.3
24.0	19.0	0.03	-0.12	19.8	5.4	23.6	-0.4	10.0
25.0	41.5	0.08	-0.27	25.8	12.1	-31.4	-1.0	-0.3
26.0	47.1	0.08	-0.28	14.4	12.7	-17.7	-0.9	-0.7
27.0	47.1	0.07	-0.27	8.1	12.0	-7.9	-0.7	-0.1
28.0	47.1	0.07	-0.26	5.4	11.7	-7.8	-0.6	0.0
29.0	47.1	0.06	-0.25	3.9	11.5	-4.0	-0.6	0.0
30.0	47.1	0.06	-0.25	2.8	11.3	-3.9	-0.6	0.0
31.0	47.1	0.06	-0.25	2.1	11.1	-4.0	-0.6	0.0

V terms at 60.5 hrs. $(-U\partial V/\partial x - V\partial V/\partial y)$, $-fU$, $-g\partial z/\partial y$, $-gV(U^2+V^2)^{1/2}/C^2d$, $A_H(\partial^2 V/\partial x^2 + \partial^2 V/\partial y^2)$

Cell	Depth	U Speed	V Speed	Non-lin	Coriolis	Sea Grad	Friction	Eddy Dif
22.0	2.1	0.02	-0.01	-0.9	0.9	19.6	2.5	-0.2
23.0	2.1	0.01	0.00	-0.2	0.6	29.4	0.1	-0.3
24.0	13.4	0.07	-0.01	15.9	3.1	5.9	0.2	-17.1
25.0	41.5	0.13	-0.29	26.6	5.6	-4.0	3.8	16.7
26.0	47.1	0.11	-0.29	14.5	4.9	7.9	3.2	3.4
27.0	47.1	0.09	-0.27	5.6	3.9	15.7	2.7	-0.2
28.0	47.1	0.07	-0.26	1.6	3.3	17.7	2.5	-0.1
29.0	47.1	0.07	-0.26	0.0	3.1	19.6	2.4	0.0
30.0	47.1	0.07	-0.25	-0.6	3.1	17.7	2.4	0.0
31.0	47.1	0.07	-0.25	-0.9	3.0	17.7	2.3	0.0

V terms at 63 hours.

Cell	Depth	U Speed	V Speed	Non-lin	Coriolis	Sea Grad	Friction	Eddy Dif
22.0	2.7	-0.01	0.09	0.5	-0.2	74.6	-75.8	-0.2
23.0	2.7	0.01	0.09	0.9	0.2	70.6	-10.0	0.1
24.0	13.9	-0.01	0.09	3.6	-0.3	90.2	-1.4	6.6
25.0	42.0	0.01	0.26	41.1	0.4	23.5	-2.8	-11.1
26.0	47.7	0.06	0.20	25.8	2.9	37.3	-1.5	-0.9
27.0	47.7	0.09	0.14	12.0	4.2	37.3	-0.8	1.0

Table 2. Actual reefs simulated.
The key to reef types is C, crescentic; L, lagoonal; R, ribbon; P, planar; F, fringing.

Reef	Code	Type	Area (ha)	Latitude
Hicks	14086	R	2200	14°27'
Day	14089	R	2170	14°30'
Carter	14137	R	1370	14°33'
Yonge	14138	R	1110	14°36'
No-name	14139	R	700	14°39'
Lizard Is.	14116	F/L	1809	14°40'
Green Is.	16049	F/P	710	16°46'
Cayley	17023	C	990	17°29'
Glow	18071	P	310	18°32'
Helix	18076	P	60	18°38'
J. Brewer	18075	L	1750	18°38'
Lynch's	18091	P	3690	18°47'
Wheeler	18095	P	190	18°48'
Davies	18096	C/L	1380	18°50'
Big Broadhurst	18100	C/L	8240	18°55'
Little Broadhurst	18106	C/L	1250	18°58'
Bowden	19019	C/L	940	19°02'
Rattray	19110	F	60	19°59'

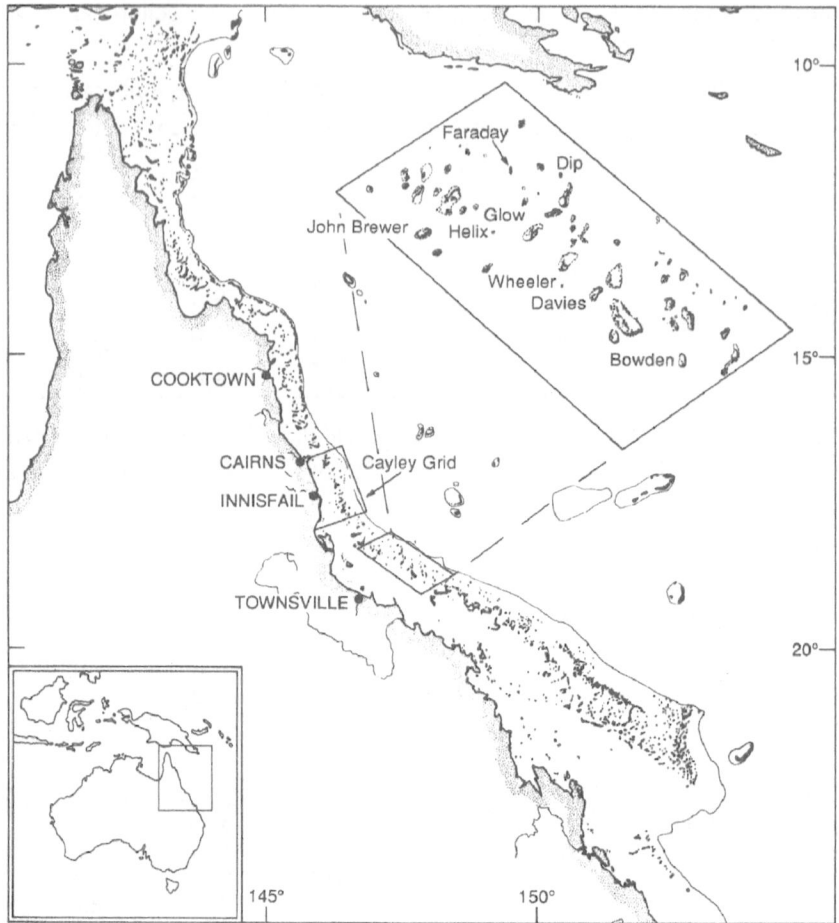

Figure 1. Reef locations on Australia's Great Barrier Reef (modified from Great Barrier Reef Authority). The Cayley grid in the southern section of the Cairns regions is shown.

To further clarify the hydrodynamic processes, a series of 18 schematized reefs were also modelled (Black *et al.*, in press). In total, 77 separate simulations of GBR and schematized reefs were made. This encompassed a wide variety of tidal and steady coastal current and wind strengths, selected to include measured values. The tidal boundary conditions are presented in Appendix 1.

The experimental design proved to be an informative structure for the investigation. The choice of a wide range of morphologies and flow conditions ensured that some broad generalisations as well as specific conclusions could be developed. General results were sought to establish the necessary framework required to specify and understand the characteristics and components of GBR reef-scale hydrodynamics. More specific modelling was then conducted at John Brewer Reef using measured currents and tides to examine matters such as the unsteadiness of the coastal flows (Gay *et al.*, this volume). It was found that, because of the low frequencies of the coastal current oscillations,

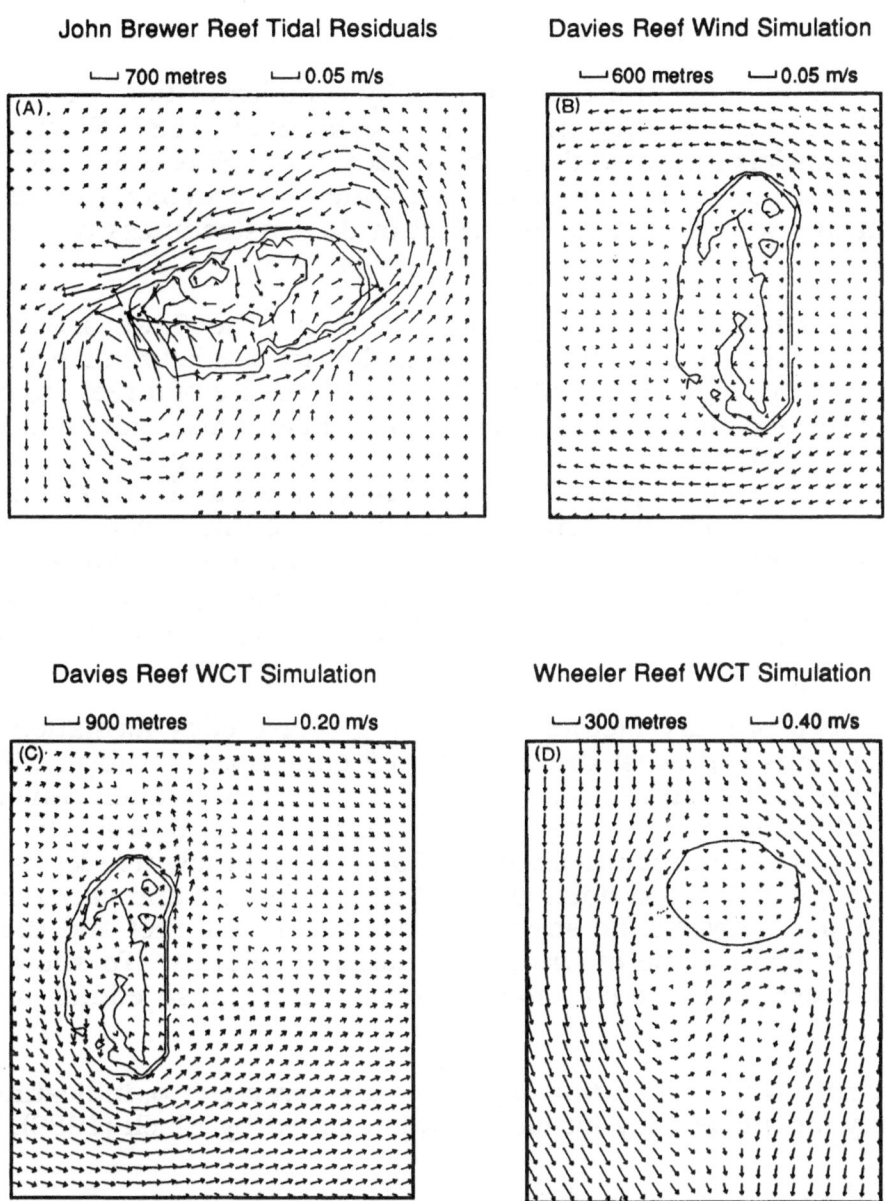

Figure 2. Circulation around a selection of reefs. Residual tidal current at John Brewer Reef (top left), wind-driven circulation at Davies Reef (top right), residual current for simulation including a 5 ms^{-1} trade wind, 0.15 ms^{-1} south-easterly current, and tides at Davies Reef (bottom left) and an instant in time after peak flow at Wheeler Reef (bottom right).

simulations of a variety of steady coastal currents and time-varying tidal flows were satisfactory and informative reproductions of the GBR hydrodynamics (Black *et al.*, in press; Gay *et al.*, this volume). These provided the data needed to break down the complex flow into its components.

RESULTS

Spatially the hydrodynamics are unusually complex at tidal frequencies (Figure 2), far more so than in most estuaries (Black 1987; Black *et al.*, 1989). The models indicated that "phase eddies" (Black & Gay, 1987b) are the most distinctive feature of the dynamics, and demonstrated that these features were primarily the result of the unsteadiness of the tidal currents.

In the free stream, commonly 30-50 m deep, the flows are mostly undamped by bed friction and continue to flow for 2-3 hours after the sea gradients change sign each half tidal cycle. In the lee of a reef, however, the wake formed during the accelerating stage of the half tidal cycle accelerates quickly after the change in sea gradients, and the observer sees recirculation develop which has the appearance and dimensions of a steady-state eddy. The important indicator of the phase eddy, however, is the strengthening of the recirculation (relative to the free stream) after peak flow in the free stream. This is contrary to steady-state conditions, where the eddy strength increases with current speeds at the point of flow separation.

There are two primary factors determining phase eddy formation. The first is flow unsteadiness; the well-defined eddy does not form without temporal variability, although a wake and limited recirculation can occur. The second is due to the sudden depth transition of coral reefs which rise sharply from the continental shelf (30-50 m deep) to low tide levels (0 m deep). This produces a sudden gradient in bed friction and variations in velocity phase between the sheltered flow behind the reef and the free stream. The plan shape of the reef is mostly irrelevant. To show this, Figure 2 presents dissimilar morphologies, and phase eddies always appear.

In accordance with this finding, a phase eddy parameter φ, indicating the propensity for phase eddy formation, was defined by Black (1989). That is,

$$\varphi = 1 - (2/\pi) \tan^{-1}(r_0 T/2\pi H) \tag{1}$$

where r_0 is the bed friction coefficient, T the tidal period and H the shelf depth. This varies as $0<\varphi<1$ with phase eddies becoming more apparent as φ approaches 1. φ was found to be 0.32 at Rattray Island. A typical value on the GBR around coral reefs is about 0.70.

Clearly, phase eddies are expected to be common on the Great Barrier Reef and eddies have been sighted on a number of occasions (e.g. Wolanski *et al.*, 1984). The eddy forms and dissipates each half tidal cycle in deep water off the reef and this has important consequences for larval dispersal (Black, 1988).

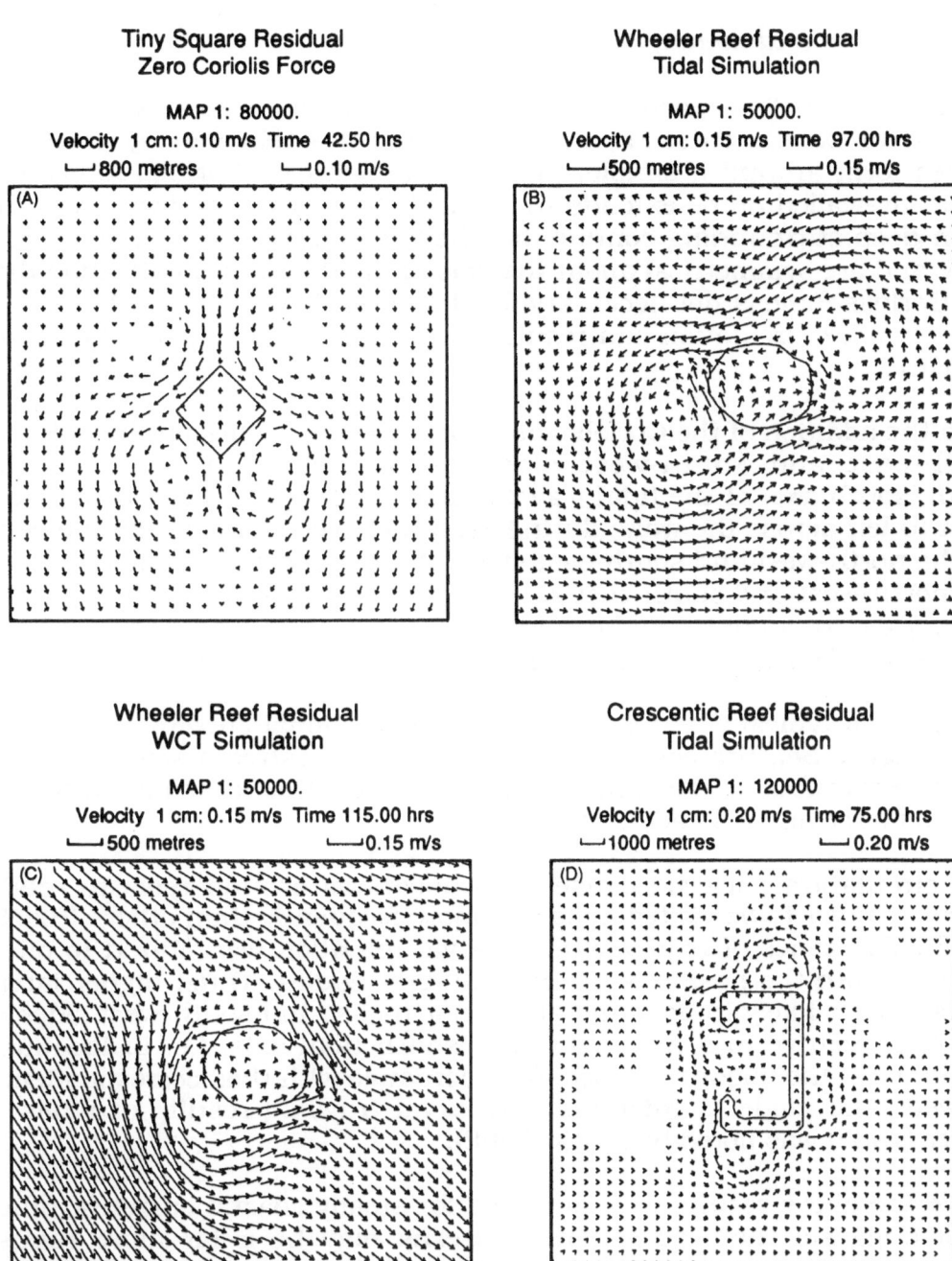

Figure 3. Residual currents (a) without Coriolis force around a square reef (Note, the simulation is yet to converge to a symmetrical image about the E/W axis) (b) with the addition of Coriolis force at Wheeler Reef (c) after the addition of a steady current and Coriolis force at Wheeler Reef and (d) around the schematised crescentic reef, for tidal currents only.

Coriolis force and steady current

Using the model's capacity to isolate particular influences by controlled numerical experiments, Figure 3a demonstrates that four symmetrical eddies form around a simple, planar dome with no Coriolis force. The vertically-walled dome extends to 1.4 m below mean sea level from the 45 m deep surrounding continental shelf. The eddies develop each half cycle and are symmetrical about the axis of the tidal wave, indicating that the Coriolis force is not necessary for the formation of phase eddies, and that complex circulation can result around the simplest plan shapes on the GBR.

With the addition of the Coriolis force, only the two eddies on the notional north-east and south-west sides of the island (relative to the north/south axis of the tidal wave) remain (Figure 3b). Flow turns from right to left across the wake during each half cycle, eliminating the eddy on the north-west and south-east sides. If the reef is wider, the eddies on the north-west and south-east sides may remain. However, these eddies are not as pronounced as those to the north-east and south-west (Black & Gay, 1987a, Horizontal bar tidal residual current, C series figures).

Finally, if a typical Great Barrier Reef unidirectional current is included next (Figure 3c), the eddy on the unsheltered upstream side no longer forms. From four eddies initially, only one remains after including the Coriolis force and a steady current.

The circulation patterns cause some locations around reefs to retain more larvae or to be visited by larvae more frequently than others (Black, 1988). The zones of relatively high larval retention were found to often coincide with the positions of the dominant eddies. Thus, higher larval numbers are likely to be found at the north-east and south-west sides of the reef, relative to a notional north/south tidal axis, i.e. in locations clockwise of the major tidal axis. In accordance with this, correspondence between the model predictions and larval trawling results of Willis & Oliver (1988) was noted by Black (1988).

While the above results were obtained for steady coastal currents, other work suggests that the general patterns may be a far more robust occurrence. Gay et al. (this volume) identified highest concentrations of larvae clockwise of the tidal axis at John Brewer Reef when they modelled measured flows which included the unsteadiness of the coastal currents over a 4.5 month interval during the winter of 1987. Although the coastal currents varied in strength and direction, the consistency of the position of highest larval numbers was said to occur because the coastal currents flow predominantly up or down, rather than across, the continental shelf, while the tidal current major axis was directed across the shelf.

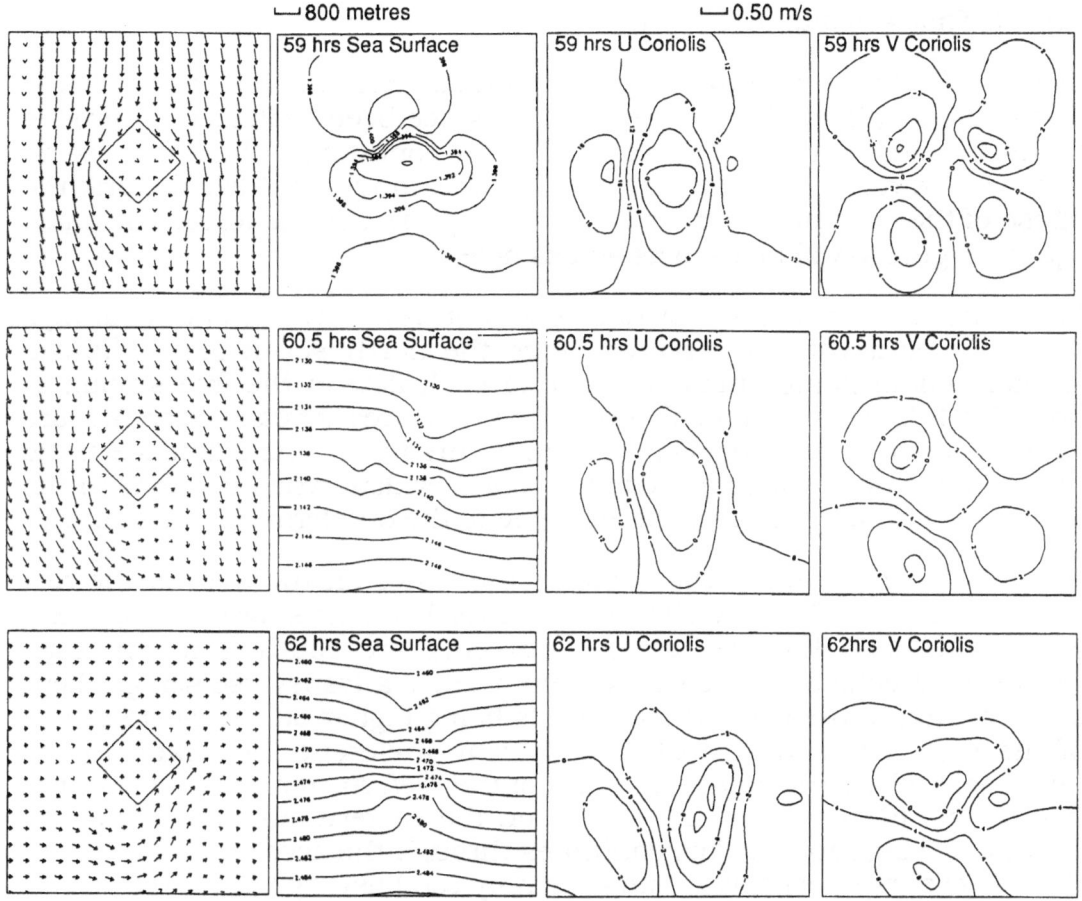

Figure 4. Flow patterns, sea surface contours, U Coriolis force and V Coriolis force around the square reef at hours 59, 60.5 and 62, encompassing the 3 hours from peak flood flow. A positive force acts in the positive current direction. Positive U is to the left and positive V is up the page.

Application to a more complicated bathymetry

Numerical experiments were conducted on the common cres-centic/lagoonal reef to specify the major interactions between the flow and bathymetry around a more complex shape. The residual circulation averaged over a tidal cycle (Figure 3d) can be decomposed as follows. Phase eddies form at the north-east and south-west sides of the reef. A small, matching residual eddy occurs on the south-east face inducing a net southerly current against the reef face in this region. The equivalent eddy at the north-west forms at the opening at the rear of the reef and enters the lagoon to drive a net northerly circulation in the northern half. The lagoon circulation implies that the southern half of GBR lagoons will have longer residence times than the northern half. Circumscribing these patterns is a net anti-clockwise circulation in deep water around the reef.

Anti-clockwise residual current

The anti-clockwise residual current which encircles the reef, as predicted by the model (Black & Gay, 1987a), has since been indicated by drogue paths measured at Bowden Reef in calm conditions (Wolanski, *pers. comm.*). The result is most easily understood in terms of two factors; the phase of the current and the rotation due to the Coriolis force. The small square (Figure 4) will be used as an example.

In this simulation, the sea surface gradient in the free stream is always positively or negatively directed along the north/south axis. The sea gradient changes sign near the time of peak flow in the free stream at hour 59 (Figure 4). At peak flow then, the free stream currents are directed south and minor recirculation occurs in the immediate lee of the reef. At hour 60.5, the phase eddy becomes evident when currents in the wake flow with the sea gradient, opposite to the decelerating free stream.

At peak flow (59 hours), the currents are faster on the west of the reef than on the east. This results in a stronger U Coriolis force to the west. Simultaneously, the free stream currents are rotating in an anti-clockwise direction. As the free stream rotates, a transverse sea gradient develops across the reef with sea levels being higher on the western side of the reef. The eddy circulation thereby preferentially heads to the north-east, along the south-east face of the reef. At minimum flow in the free stream (62 hours), the currents on the eastern side of the reef are already moving rapidly. This continues through the accelerating phase of the tide over the next three hours, thereby establishing the conditions for faster flow on the eastern side of the reef before the process repeats on the other half cycle of the tide.

Thus, the combination of spatial variations in phase and the rotation to the left due to the Coriolis force drives flow in the wake from right to left (looking in the direction of the flow). This partially or wholly obliterates the eddy on the south-east face and establishes a net anti-clockwise residual current.

Pingree & Maddock (1979) describe a similar result derived from numerical simulations of flow around an island. They found that a clockwise residual (northern hemisphere) occurred with the addition of Coriolis force in their model and suggested that frictional torque in the shallow waters near the island generated four symmetrical residual eddies, and that the Earth's rotation produces small asymmetries by deflection of flow near the island in a clockwise direction. While similar to the cases here, the bed frictional resistance is relatively unimportant in deep water where the eddies form (Table 1). The important aspect on the GBR is the presence of the reef creating shelter in its wake. Various combinations of these features are seen in simulations of actual reefs (Figure 2).

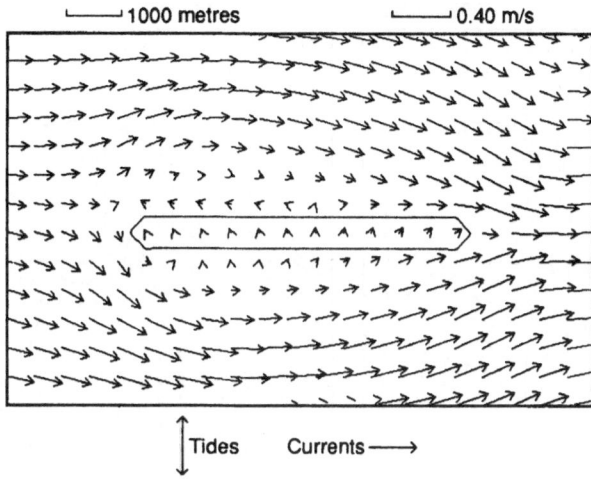

Figure 5. Residual velocities around a horizontal bar shaped reef from a simulation including north/south directed tidal currents and a coastal current to the east.

Interactive eddies

A second type of eddy, identified with the modelling, was denoted as "interactive". It was depicted when the major axis of the tidal currents was aligned with the steady current. The eddy formed when the steady current exceeded the tidal flow in the free stream. Currents then reversed only in the lee of the island, which was sheltered from the steady current. This resulted in oppositely-directed flows or an "eddy" during much of the half tidal cycle when tides and currents were opposed.

Current/tide interaction

The interactions of current components strongly modify the steady flows, causing net currents to be diverted around the reef rather than over it. An example of this is seen in the horizontal bar case (Figure 5) where residual flows near the reef are directed "upstream" to the current. Without the tides, the current would be unattenuated and directed opposite to this, being essentially unaffected by the long, narrow reef. The interaction of tides and currents is vitally important as a means of retaining larvae near a reef (Black *et al.*, in press).

Inertia and ribbon reefs

Inertia is important on the shelf, as seen in the formation of phase eddies. Another interesting case occurred at Carter Reef, a ribbon reef offshore from Lizard Island (Figure 1). A jet forms in the gaps between adjacent reefs, and extends some 10 km across the shelf (Figure 6a).

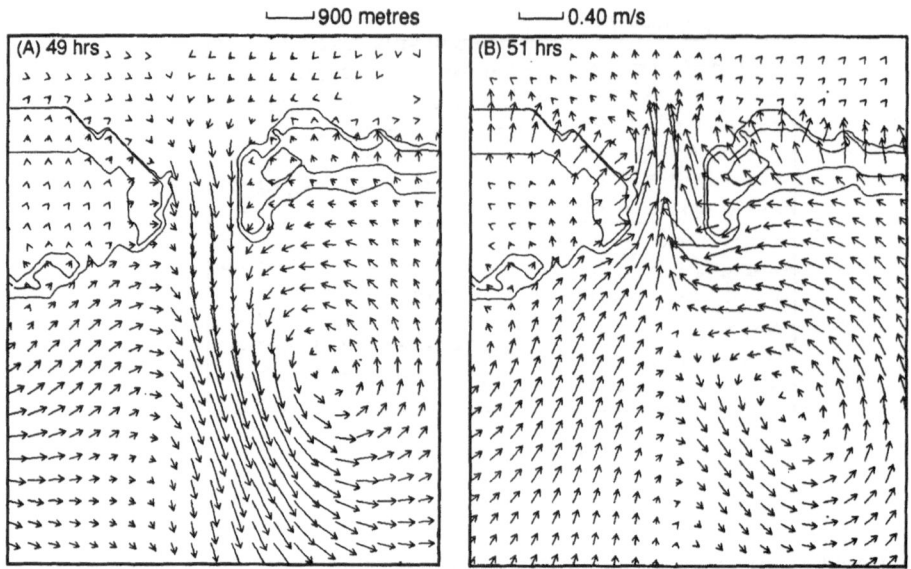

Figure 6. The jet between Carter and Yonge Reefs, ribbon reefs offshore of Lizard Island.

Coriolis force causes the eddy to the left of the flow to predominate (see also Figure 7a of Black, 1989). The shelf friction is low and the jet takes time to decelerate and dissipate, resulting in two interesting phase differences. First, flow on the reef flat is to the north, while the jet continues to flow in the opposite direction (Figure 6a). Second, Figure 6b shows the jet propagating freely across the shelf after marginal ebb flows, initially associated with eddy circulation in the lee of the reef, separate the jet from the entrance. Horizontal eddy viscosity would help to dissipate the jet on the shelf but no measurements have been made to study this phenomenon, which the model has indicated.

The inertia of the tidal jet must strongly influence the continental shelf flushing rates. Because the water in the jet travels a long way across the shelf, it tends not to be flushed on the ebb cycle of the tide. Instead, water from the lee of the reefs is lost. The volume exchanged each tidal cycle is therefore large, relative to cases where a jet does not form.

Reef groups

The dynamics around reef groups was also examined by modelling the southern half of the Cairns section of the GBR (Figure 1). Denoted as the "Cayley" grid, 100 km of coastline from the north of Dunk Island was modelled. The grid size of 2.25 km resolves most of the reefs in the region but it does not adequately resolve the detailed flow on and around the reefs, as discussed below.

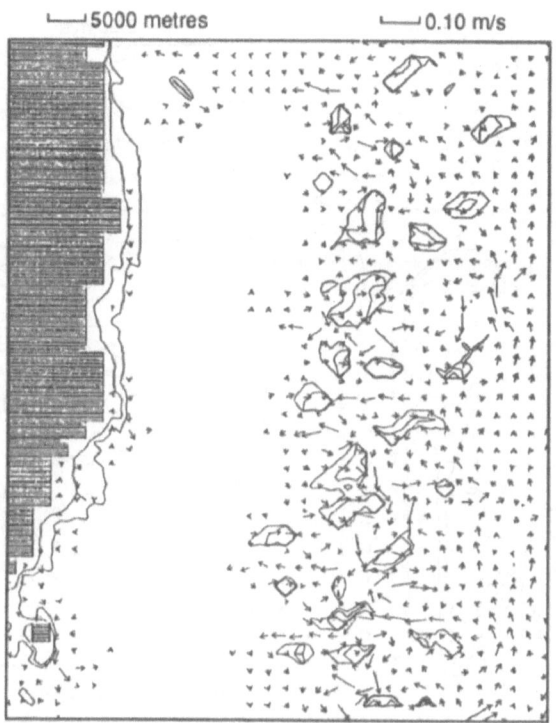

Figure 7. Residual tidal velocities in the Cayley grid.

The circulation is greatly modified in the reef matrix. The residual tidal flow (Figure 7) is essentially net zero in the lagoon (closed orbits in the absence of strong bathymetric effects) but extremely varied among the reefs. Some of the disjointed variability in the vectors is likely to be a result of poor grid resolution. To examine this, the circulation around Davies Reef with its neighbours included (Big and Little Broadhurst and Lynch's Reefs) was compared with the simulations of Davies Reef alone, both at a 400 m grid scale. The residual circulation (Figure 8) provides the most informative comparison, as it includes the flows over the full cycle.

We find (Figure 2 *cf* Figure 8) that the flow on the shelf between Davies Reef and its neighbours was more complex with the neighbouring reefs included. However, the general patterns of tidal circulation closer to Davies Reef were similar in both cases, e.g. the southerly flow at the rear of the reef, northerly flow in the lagoon and on the reef face, and the easterly flow at the north. One difference was the weaker eddy at the south in the isolated case. However in general, by retaining the fine-scale grid size, the previously identified flow patterns on the isolated reef mostly re-emerged, even in this case where the neighbours were relatively large and close. Evidently the individual reef tidal dynamics can remain

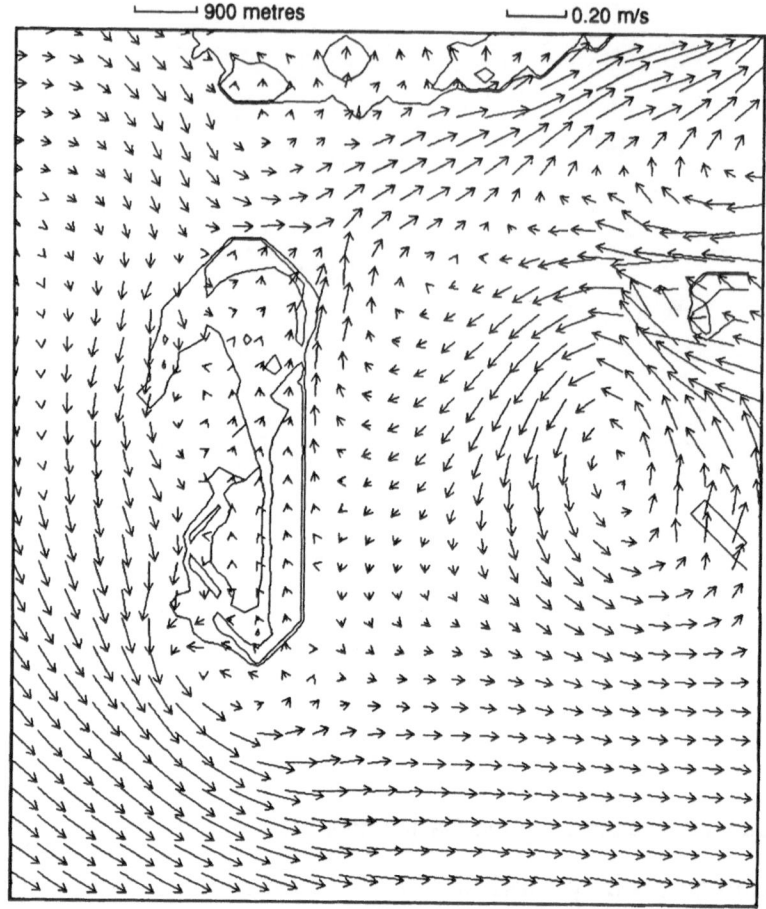

Figure 8. Residual velocities around Davies Reef and its neighbours. The simulation included a 5 ms^{-1} trade wind, 0.15 ms^{-1} south-easterly current and tides.

essentially intact, although there will be interaction in the "free stream" between reefs and it is noted below that wind-driven coastal currents are strongly modified by the reef matrix. On-going research is presently examining reef/reef interactions in more detail with particular emphasis on quantifying the effect on reef residence times.

Wind-driven circulation

The boundary conditions described in Appendix 4 were applied to model the wind-driven circulation on the "Cayley" grid. Most apparent in the simulation (Figure 9) is the effect of reef sheltering which reduces the mean flow in the reef matrix, compared with flows in the inner lagoon and

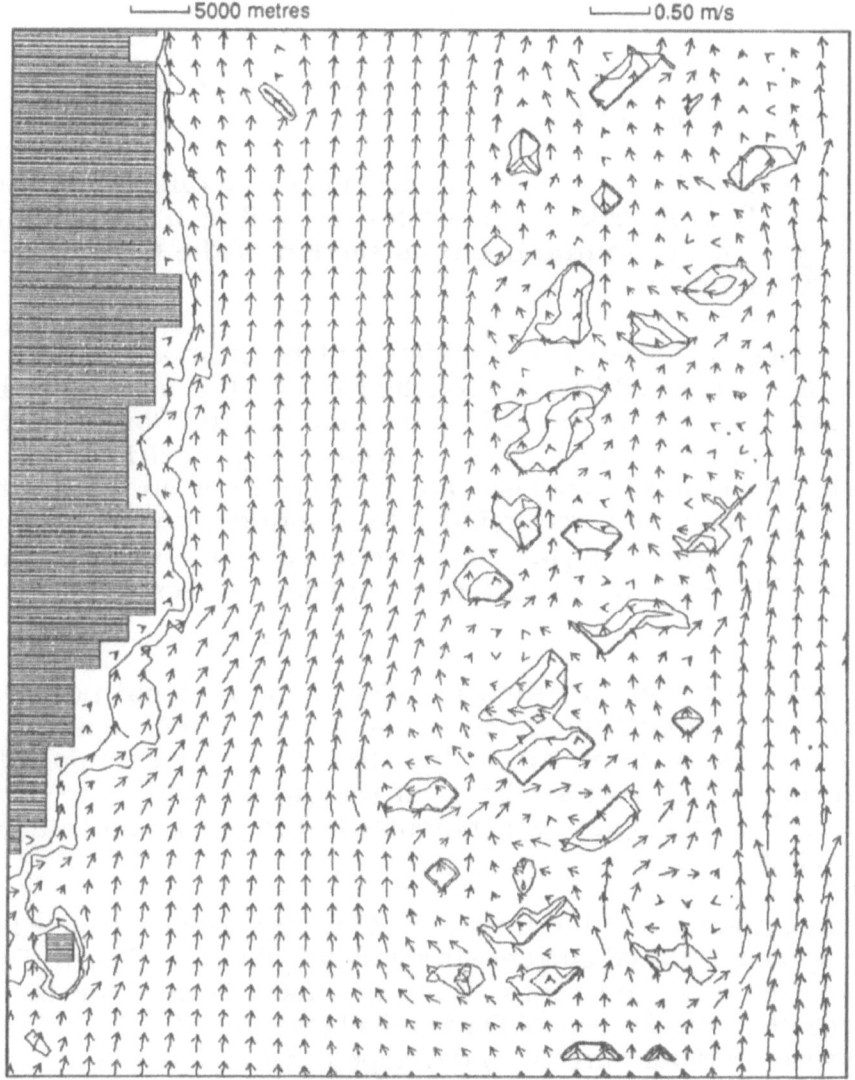

Figure 9. Wind-driven circulation on the Cayley grid. 10 ms^{-1} trade wind.

offshore. Consequently, larval drift speeds will be significantly reduced in the matrix. When this is combined with the capacity of reefs to trap some larvae as they pass, either temporarily or until settlement (Black *et al.*, in press), the net drift of the centre of mass of the larval cloud will be reduced. Second, horizontal mixing of larvae at a regional scale will also be dependent on the density of the reef matrix. Tidally-driven cross-shelf transfer in combination with the potential for larvae to be temporarily trapped on reefs will cause particles to spread over a much wider region which expands as the reef density increases (Parslow & Gabric, 1989), particularly when the reef spacing is less than the tidal excursion.

CONCLUSIONS

(1)   These studies provided the first general view of the fine-scale circulation around coral reefs.

(2)   More specifically, the modelling provided a detailed record of the circulation at 18 GBR reefs at fine-scales, as well as a simulation on a meso-scale grid of the southern Cairns section of the GBR from Dunk Island to 100 km north.

(3)   The modelling revealed that eddies which commonly form around GBR reefs are phase eddies, established by spatial variability in the velocity phase. These were found to be modified by the Coriolis force and the presence of low frequency currents. Around a symmetrical reef, it was shown that the dominant eddies occur in deep water on the north-east and south-west sides of a reef relative to a notional north-south major tidal axis, a result of the asymmetry introduced by the Coriolis force. In the Central GBR, this corresponds with dominant eddies to the east and west of the reefs.

(4)   For crescentic reefs oriented parallel to the notional north/south direction, net tidal circulation in the lagoon was near neutral at the south end and northerly at the north. In the GBR, this corresponds with the north-east and south-west segments of the lagoon, when the reefs are aligned with their long axes perpendicular to the prevailing trade winds (e.g. Davies Reef).

(5)   Net circulation around individual reefs under tidal flow alone is anti-clockwise. Typical residual current strengths are about 0.04 m/s. Thus, in the absence of other currents, it would take 5 days for larvae to circumnavigate a reef the size of Davies Reef (assumes 6.4 x 2.8 km). The anti-clockwise residual current is developed in the wake of a reef each tidal cycle and is associated with negative Coriolis force in the southern hemisphere.

(6)   A number of weather conditions were modelled. These demonstrated that eddy circulation depends on the current intensity. In fast currents, the eddy on the upstream side of the reef is dissipated. The eddy in the lee, however, can be strengthened by the tide/current interaction. Non-linear interaction causes more of the flow associated with low-frequency currents to pass around rather than over the reef, when the tides were present. Any strengthening of the lee eddy reinforces this process. If free stream currents pass around rather than over the reef, local residence times of crown-of-thorns larvae or pollutants can be expected to increase.

(7)   Interactive eddies were identified as forming in conditions when tidal currents were parallel to but less than the free stream current. In these cases, the currents only changed sign in the lee of the reef sheltered from the free stream current during the opposed half tidal cycle, and an eddy forms.

(8)   Ribbon reefs were characterised by large eddies forming adjacent to the tidal jet. At Carter Reef, a jet was seen to split off from the entrance and to propagate across the shelf, after being separated by convergent marginal currents flowing along the lee of the reef towards the entrance. The flow patterns were similar to those observed in the entrance gorge and marginal channels of tidal inlets. The deflection by the Coriolis force caused the eddy on the left hand side (looking in the tidal flow direction) to be more pronounced.

(9)   Tidal circulation around Davies Reef was seen to be partially unaffected by its neighbours, the closest of which was 2 km away. Wind-driven currents were severely attenuated by the matrix of reefs around Cayley Reef. This would cause a reduction of the centre of mass speed of a larval cloud and a greater propensity for regional dispersion under the combined action of cross-shelf tidal flows and net coast-parallel currents.

(10) The results and numerical procedures developed are applicable to biological, ecological and engineering studies. A tool for management applications has been established. With this knowledge of the hydrodynamic environment, the results are being applied to assessment of crown-of-thorns larval dispersal, reef flushing times and specification of likely starfish outbreak sites.

Acknowledgements. The authors would like to acknowledge the support of Dr Peter Moran, Australian Institute of Marine Science, and of the staff at the Victorian Institute of Marine Sciences. This study was partly funded by COTSAC and the Victorian Institute of Marine Sciences. David Hatton worked as a research assistant during much of this program.

REFERENCES

Andrews, J.C.: Lagoon-ocean interactions. In: Baker, J.T., R.M. Carter, P.W. Sammarco, & K.P. Stark (eds), Proc. of the inaugural Great Barrier Reef Conference. Townsville, JCU Press, Townsville. pp. 403- 408. (1983)

Andrews, J.C. & M.J. Furnas: Subsurface intrusions of Coral Sea water into the Central Great Barrier Reef. I. Structures and shelf-scale dynamics. Continental Shelf Res. 6, 491-514. (1986)

Arnold, R.J.: An improved open boundary condition for a tidal model of Bass Strait. In: Numerical modelling: applications to Marine Systems. ed. J. Noye. North-Holland Mathematics Studies. Elsevier Science Publishers. p.145-158. (1987)

Baker, J.T. & P.J. Moran: A summary of research on ecological aspects of the crown-of-thorns starfish funded by COTSAC and coordinated by the Australian Institute of Marine Science. Crown-of-thorns starfish study report No. 3, Australian Institute of Marine Science, Townsville. (1987)

Black, K.P.: Sediment transport and tidal inlet hydraulics. PhD thesis, Univ. of Waikato, Hamilton, N.Z. Vols. 1 and 2, 331 pp. (1983)

Black, K.P.: A numerical sediment transport model for application to natural estuaries, harbours and rivers. In: J. Noye (ed.) Numerical modelling: Applications to marine systems. Elsevier Science Pub., pp. 77-105. (1987)

Black, K.P.: The relationship of reef hydrodynamics to variations in numbers of planktonic larvae on and around coral reefs. Proc. Int. Coral Reef Symp., Townsville. (1988)

Black, K. P. Numerical simulation of steady and unsteady meso-scale eddies. Proc. 9th Australasian Coast. Eng. Conf. p. 204-208. (1989)

Black, K.P. & S.L. Gay: Hydrodynamic control of the dispersal of crown-of-thorns starfish larvae. 1. Small-scale hydrodynamics on and around schematized and actual reefs. Victorian Institute of Marine Sciences Technical Rep. No. 8, 67 pp. (1987a)

Black, K.P. & S.L. Gay: Eddy formation in unsteady flows. Jour. Geophys. Res. 92(C9): 9514-9522. (1987b)

Black, K. P., Gay, S. L. & Andrews, J. C. (in press). A method to determine residence times of neutrally-buoyant matter such as larvae, sewage or nutrients on coral reefs. Coral Reefs.

Black, K. P. & Healy, T. R. Sediment transport investigations in a New Zealand tidal inlet. Proc. 18th Int. Coastal Eng. Conf. (ASCE). Cape Town. 2436-2457. (1982)

Black, K. P. & Healy, T. R. The sediment threshold over tidally-induced megaripples. Mar. Geol. 69, 219-234. (1986)

Black, K. P. & Healy, T. R. Formation of ripple bands in a wave-convergence zone. Jour. Sed. Petrol. 58, 195-207. (1988)

Black, K. P., Healy, T. R. & Hunter, M. G. Sediment dynamics in the lower section of a mixed sand and shell-lagged tidal estuary, New Zealand. J. Coast. Res. 5(3), 503-521. (1989)

Bode, L. & Sobey, R.J.: Initial transients in long wave computations. J. Hydr. Eng. 110, 1371-1395. (1983)

Chapman, D.C.: Numerical treatment of cross-shelf open boundaries in a barotropic coastal ocean model. J. Phys. Ocean. 15, 1060-1075. (1985)

Chen, M.H.T.: Tsunami propagation and response to coastal areas. Hawaii Inst. Geophys. Rep. NOAA-JTRE-95, HIG-73-15, 60 p. (1973)

Church, J.A., J.C. Andrews, & F.M. Boland: Tidal currents in the central Great Barrier Reef. Cont. Shelf Res., 4: 515-531. (1985)

Falconer, R.A. & Mardapitta-Hadjipandeli, L. Bathymetric and shear stress effects on an island's wake: a computational model. Coast. Eng. 11, 57-8. (1987)

Falconer, R.A., E. Wolanski & L. Mardapitta-Hadjipandeli: Modelling tidal circulation in an island's wake. Jour. Waterway Port, Coast. Ocean Eng. (ASCE) 112(2), 234-254. (1986)

Frith, C.A. & L.B. Mason: Modelling wind-driven circulation in One Tree Reef, southern Great Barrier Reef. Coral Reefs, 4, 201-211 (1986)

Gay, S.L., Andrews, J.C. & Black, K.P. (this volume) Dispersal of neutrally-buoyant material near John Brewer Reef. In: R.H. Bradbury (ed.) The *Acanthaster* phenomenon: a modelling approach. Springer- Verlag, Berlin.

Hopley, D.: Morphological classifications of shelf reefs: a critique with special reference to the Great Barrier Reef. Chap 11 In: Barnes, D.J. (ed) Perspectives on Coral Reefs. Australian Institute of Marine Science, Brian Clouston Pub. 277 pp. (1983)

Lee, T. T. & Black, K. P. The energy spectra of surf waves on a coral reef. Proc. 16th Int. Coastal Eng. Conf. 1, 588-608. (1978)

Leentertse, J.J. & Liu, S-K.: Modeling of three-dimensional flows in estuaries. 2nd Annual Symposium on Modelling Techniques. Waterways, Harbours and Coastal Engineering (ASCE). p. 625-642. (1975)

Lettau, H. Note on aerodynamic roughness-parameter estimation on the basis of roughness-element description. J. Appl. Meteor. 8, 828-832. (1969)

Longuet-Higgins, M. S. & Stewart, R. L. Radiation stresses in water waves: a physical discussion with applications. Deep Sea Res. 11, 529-562. (1964)

Ludington, C.A.: A study of flushing and exchange in reef lagoons using fluorescent dye. Proc. Conf. Environmental Eng. Townsville, Australia, 1981, 102-106 (1981)

Moran, P.J. & Johnson, D.B. Final report on the results of COTSAC ecological research: December 1985 - June 1989. Crown-of-thorns starfish study report 11. Australian Institute of Marine Science. Townsville. 40 pp. (1990)

Orlanski, I.: A simple boundary condition for unbounded hyperbolic flows. J. Computational Phys. 21, 251-269. (1976)

Parslow, J.S. & Gabric, A.J. Advection, dispersal and plankton patchiness on the Great Barrier Reef. Aust. J. Mar. Freshwater Res. 40(4): 403-419. (1989)

Pearson, R.A.: Consistent boundary condition for numerical models of systems that admit dispersive waves. J. Atmos. Sci. 31(6), 1481-1489. (1974)

Pickard, G.L. Effects of wind and tide on upper-layer currents at Davies Reef, Great Barrier Reef, during MECOR (July-August 1984). Aust. J. Mar. Freshwater Res. 37, 545-566. (1986)

Pingree, R.D. & L. Maddock: Tidal Flow around an Island with a regularly sloping bottom topography. J. mar. biol. Ass. U.K. 59, 699-710. (1979)

Sammarco, P.W. & J.C. Andrews: Localised dispersal and recruitment in Great Barrier Reef corals: The Helix experiment. Science 239, 1422- 1424. (1988)

Swart, D. H. Offshore sediment transport and equilibrium beach profiles. Delft Hydr. Lab. Pub. No. 131. (1974)

Willis, B.L. & J.K. Oliver: Distributions of coral eggs and larvae in the central section of the Great Barrier Reef marine park following the annual mass spawning of corals. Report to the Great Barrier Reef Marine Park Authority, Townsville. 49 pp. (1988)

Wolanski, E., J. Imberger & M.L. Heron: Island wakes in shallow coastal waters. Jour. Geophy. Res. 89(C6): 10553-69. (1984)

APPENDIX 1: TIDAL BOUNDARY CONDITIONS

The tide in the central GBR mostly consists of M_2-S_2 standing waves reflecting off the coast (Church *et al.* 1985). The reefs were modelled with tides of M_2 frequency and constant amplitude when it was found that long-term dispersal and the dispersal patterns were unaffected by diurnal inequalities. However, a number of different tidal ranges were modelled to examine the influence of tidal current intensity. Both the input and reflected tidal wave were included to simulate the standing wave conditions on the shelf. The relative phase ϕ was set to create free stream currents of a selected maximum magnitude U_m, taken from field measurements. For a standing wave in a frictionless environment,

$$\tan \phi = U_m / \xi\sqrt{g/d} \qquad (1)$$

where

$$\xi = 2a \cos \phi \qquad (2)$$

d is the water depth on the shelf, and a is the tidal amplitude at the site. Application of this method using known values of velocity, tidal amplitude and depth ensures that the free stream current in the model approximates the known speed. The frictionless assumption proved to be acceptable.

APPENDIX 2: MODEL CALIBRATION

The model 2DD had been applied to a number of locations where good quality data was available for its verification (e.g. Black 1987; Black *et al.* 1989). Thus, the computer code and its capacity to model hydrodynamic circulation had been thoroughly verified. The main concern with the GBR modelling was the specification of physical coefficients representing bed friction and horizontal eddy viscosity. The latter is discussed by Black (1989).

Bed roughness

The roughness in the model was estimated in three independent ways. First, we utilised wave heights measured at a number of locations in the direction of wave travel across the reef flat on Oahu, Hawaii by Lee & Black (1978). The wave models

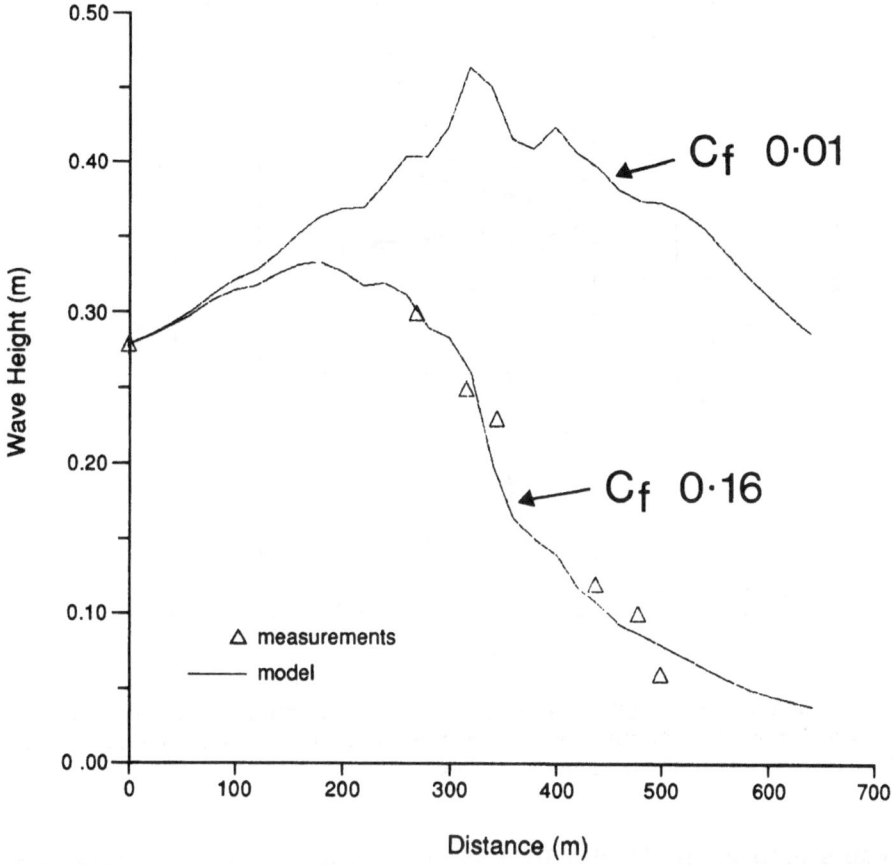

Figure A2.1. Comparison of measured wave heights across a coral reef flat, Oahu, Hawaii, with predictions from a wave transformation numerical model. The upper curve shows the prediction for a typical beach roughness and the lower curve gives the best fit to the data with a much higher friction coefficient of C_f=0.16.

of Black & Healy (1988) were employed to simulate the height attenuation due to bed friction and breaking. The bed roughness coefficient which gave the best correspondence with the measurements was C_f=0.16 (Figure A2.1). This compares with beach values over sandy beds of 0.01 suggested by Longuet-Higgins & Stewart (1964). The calculated roughness coefficient was converted to a roughness length z_0 using the formula of Swart (1974) and was found to be z_0=0.08 m.

Second, Reichelt (unpub. data) measured bed undulations on the reef flat of Davies Reef of the order of 0.1-0.3 m per meter traversed, with RMS value of 0.27 m. Applying these measurements in the form drag formula of Lettau (1969), we obtain

$$z_0 = h^2/2L = 0.04 \text{ m} \tag{1}$$

where h is the height and L the wavelength. The very high skin friction around individual corals would be expected to increase this value significantly.

Third, measurements of the vertical current profile on a patch reef in the lagoon of Little Broadhurst Reef provided a direct specification of the roughness length, defined as the notional height above the bed where the velocity goes to zero. At Little Broadhurst, z_0 was measured as being in the range of 0.08 to 0.15 m (Figure A2.2), which is in accordance with the previous results.

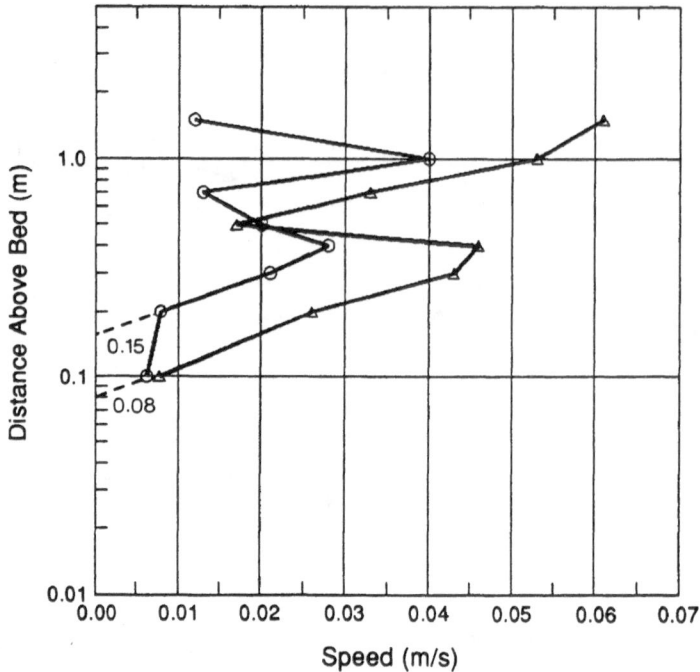

Figure A2.2. Vertical velocity profiles measured at Little Broadhurst Reef on 12/12/88. The linearity at the lower levels indicates a logarithmic boundary profile. The intercepts of these curves at U=0 give z_0=0.08 and 0.15 m respectively.

Calibration

A direct calibration of the Davies Reef simulation was then undertaken onto the measurements of Pickard (1986), selected because current meters had been placed off the reef face, on the reef flat and in the lagoon. No calibration of the free stream conditions was required because these were given boundary conditions (Appendix 1).

Pickard (1986) measured current speeds of 0.08 ms^{-1} on the reef flat at Davies Reef in calm conditions. Predicted velocities from the model were in agreement with this (Figure 3) when a roughness of z_0=0.08 m was employed. Demonstrating the critical importance of the roughness length, the currents from a simulation using the marginally lower roughness of 0.04 m were much faster than the measurements. Notably, initial tests with a typical estuarine roughness of 0.001 m (Black & Healy, 1982, 1986) resulted in model speeds which were highly excessive.

Speeds in the lagoon were similarly well predicted with z_0=0.08 m on the reef flat, but too fast with the low roughness. As the reef flat is a major barrier to the passage of water through the lagoon, an under-estimate of the reef flat friction was expected to result in a poor specification of lagoon speeds. No similar strong dependence on the local bed friction in the lagoon was identified and lagoon roughness was taken as 0.01 m to represent a reworked sand bottom with occasional patch reefs. In deep water off the reef face, Pickard found that tidal velocity was given by V=0.19R where R is the tidal range in metres. We expect velocity to be 0.33 ms^{-1} for the modelled R=1.72 m which agrees with the model predictions of 0.32–0.35 ms^{-1} (Figure A2.3). As the water is approximately 50 m

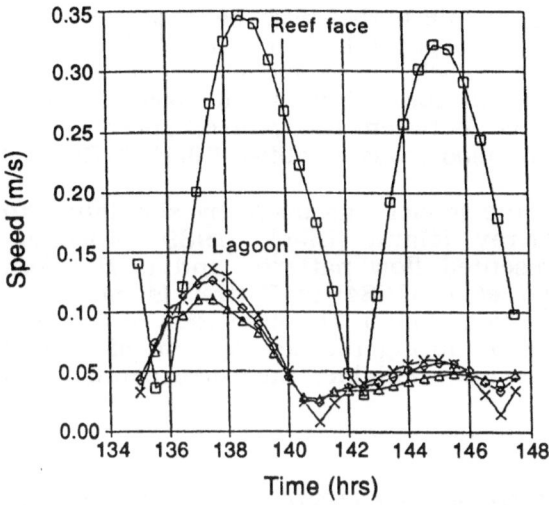

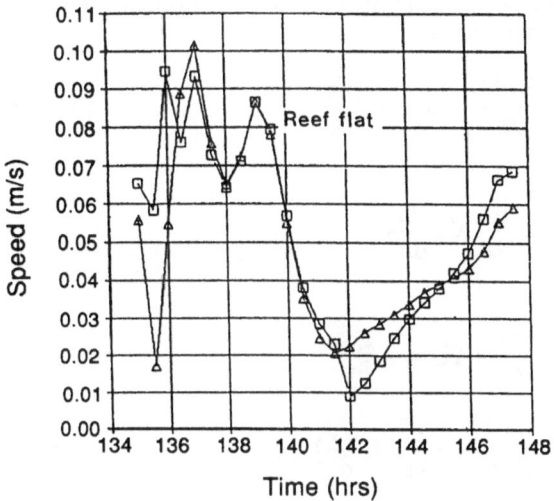

Figure A2.3. Model predictions of water speeds on Davies Reef for a roughness length of 0.08 m at the 6 sites measured by Pickard (1986).

deep, the frictional retardation is a minor term in the hydrodynamic equation off the reef face, and z_0 was taken as 0.001 m to reflect the sand and detritus bed. While some error may arise, it was assumed that average coral roughness would be similar on different reefs and the roughness values for the reef flat, lagoon and reef face were applied universally in all reef simulations.

Verification

Verification of the model was made by comparison of model results with drogue experiments at John Brewer Reef and currents measured at Rattray Island. Drogues were released near slack water at John Brewer Reef in calm conditions (Black & Gay, 1987b) and allowed to drift for about 6 hours. Their speeds were calculated from the straight line path length between observations. Observation times were converted to a tidal phase by comparison with the tidal phases at the closest port of Townsville, to which the model was also related. These results were then interpolated to half hourly intervals for comparison with the model. A consistent correspondence between the phase, speed and direction of the measured currents and the model was identified (Black & Gay, 1987).

Further confirmation of the model was made by simulating the hydrodynamic conditions around Rattray Island (Black, 1989. Falconer & Mardapitta-Hadjipandeli (1987) presented flow patterns and tidal information from the experiments of Wolanski et al. (1984) at Rattray Island in sufficient detail to enable a simulation to be undertaken. The prediction of the model was compared with the field data collected during the development of an eddy and, although accurate bathymetry was not available, the correspondence was good (Black, 1989).

APPENDIX 3: THREE-DIMENSIONAL NUMERICAL MODEL

The model 3DD is a 3-dimensional extension of the vertically-averaged model 2DD (Black, 1983) for vertically homogeneous water bodies. The vertically-averaged equations are solved in each layer.

$$\frac{\partial U}{\partial t} + \frac{U\partial U}{\partial x} + \frac{V\partial U}{\partial y} + \frac{W\partial U}{\partial z} - fV = -\frac{g\partial \xi}{\partial x} \tag{1}$$

$$+ A_H \left(\frac{\partial^2 U}{\partial x^2} + \frac{\partial^2 U}{\partial y^2}\right) + \frac{\partial}{\partial z}(N_z \frac{\partial U}{\partial z})$$

$$\frac{\partial V}{\partial t} + \frac{U\partial V}{\partial x} + \frac{V\partial V}{\partial y} + \frac{W\partial V}{\partial z} + fU = -\frac{g\partial \xi}{\partial y} \tag{2}$$

$$+ A_H \left(\frac{\partial^2 V}{\partial x^2} + \frac{\partial^2 V}{\partial y^2}\right) + \frac{\partial}{\partial z}(N_z \frac{\partial V}{\partial z})$$

$$W = -\frac{\partial}{\partial x} \int_h^z U \, dz - \frac{\partial}{\partial y} \int_h^z V \, dz \tag{3}$$

At the surface $z=\xi$,

$$\frac{D\xi}{Dt} = W \tag{4}$$

t is the time, U, V are mean vertically-averaged velocities in the x,y directions respectively, W the vertical velocity in the z direction (positive upward) at the top of each layer, h the depth, g the gravitational acceleration, ξ the sea level above a horizontal datum, f the Coriolis parameter, A_H the horizontal eddy viscosity coefficient, and N_z the vertical eddy viscosity coefficient.

Surface boundary conditions at $z=\xi$ are

$$\rho(N_z \frac{\partial U}{\partial z}) = \tau_x{}^s \qquad \rho(N_z \frac{\partial V}{\partial z}) = \tau_y{}^s \qquad (5)$$

where $\tau_x{}^s$, $\tau_y{}^s$ denotes the components of wind stress and

$$\tau_x{}^s = \rho_a \gamma |S| S_x/\rho \qquad \tau_y{}^s = \rho_a \gamma |S| S_y/\rho \qquad (6)$$

ρ is the water density, S the wind speed at 10 m above sea level while S_x and S_y are the x and y components, γ is the wind drag coefficient, ρ_a the density of air,

At the sea bed, z=h, we have

$$\rho(N_z \frac{\partial U}{\partial z}) = \tau_x{}^h \qquad \rho(N_z \frac{\partial V}{\partial z}) = \tau_y{}^h \qquad (7)$$

where $\tau_x{}^h$, $\tau_y{}^h$ denotes the components of bottom stress. Applying a quadratic law at the sea bed,

$$\tau_x{}^h = gU_h(U_h{}^2 + V_h{}^2)^{1/2} / C^2 \qquad (8)$$

$$\tau_y{}^h = gV_h(U_h{}^2 + V_h{}^2)^{1/2} / C^2$$

with U_h, V_h being the bottom currents and C is Chezy's C. For a logarithmic profile,

$$C = 18 \log_{10}(0.37 \ h/z_o) \qquad (9)$$

where z_o is the roughness length.

The depth-averaged convective momentum terms can be scaled by a coefficient applied to correct for non-uniform velocity distributions. However, the correction is small in most cases and is normally taken as 1.0. The form of the horizontal eddy viscosity term results when the depth is presumed constant before taking the derivative of the horizontal shear stresses. The term, as presented, behaves as a velocity smoothing algorithm. The horizontal eddy viscosity coefficient is a variable in space in the model.

A staggered finite difference grid is utilised similar to that applied by Leentertse & Liu (1975) which places the V and U components on "north and east" walls respectively. W is located in the centre of the "top" wall. The sea level replaces W in the top layer. The solution is found by time stepping with an explicit solution, as employed in 2DD (Black, 1983; 1989).

APPENDIX 4: GEOMETRICAL BOUNDARY CONDITION

A geometrical boundary condition developed for the GBR study is described in this appendix. Various procedures for treatment of tidal waves at an open boundary have been described in the literature, but they fall essentially into 3 main categories of clamped, sponge and radiative. The clamped condition is disadvantaged by the fact that transients are locked into the grid, which can lead to a complete collapse of the numerical solution (Bode & Sobey, 1983; Arnold, 1987). The sponge condition is disadvantaged by the large number of cells required around the grid where wave amplitudes are progressively damped to zero.

In a review of boundary conditions, Chapman (1985) found that the accuracy and long-term stability of radiation conditions varied significantly. For example, the well-known Orlanski (1976) condition behaved inadequately in his tests unless a sponge condition was applied simultaneously. We examined this problem and obtained similar results. We found that the finite difference approximation applied to calculate wave phase speed in the Orlanski condition was inadequate for shorter waves and for the transient waves associated with the cold start in the model.

Radiation conditions assume either perpendicular (e.g. Pearson, 1974; Orlanski, 1976) or incident (e.g. Chen, 1973) wave approach. Chen (1973) expressed the angle of incidence of the wave as a function of velocity. This meant that sea level and velocity were both required from the previous time step. A more desirable condition would find sea level only, independent of velocity. The method we developed presumes that the wave travels in a direction, n, normal to the wave crest. Then,

$$n = \frac{s.\nabla\xi}{|\nabla\xi|} \tag{1}$$

where s is the sign specifying rectilinear travel direction and ξ is the sea level. Define n_B as normal to the boundary and directed outwards. n is also directed outwards for a wave radiating out, then

$$n.n_B > 0 \tag{2}$$

and

$$s = \text{sign} \, (\nabla\xi.n_B) \tag{3}$$

For a wave with phase speed c, the orthogonal distances moved by the crest in a small time increment δt are, from equation (1),

$$\delta x = c.\delta t.s \frac{\partial \xi}{\partial x} / \, [(\frac{\partial \xi}{\partial x})2 + (\frac{\partial \xi}{\partial y})2]1/2 \tag{4a}$$

$$\delta y = c.\delta t.s \frac{\partial \xi}{\partial y} / \, [(\frac{\partial \xi}{\partial x})2 + (\frac{\partial \xi}{\partial y})2]1/2 \tag{4b}$$

For a tidal model, the phase speed is obtained from the shallow water approximation, c=√gh, although any selected dispersion relation can be applied. The geometry of the sea surface is approximated using a Taylor series expansion to first order into the grid and second order along the boundary. The sea level is unknown outside the grid and the derivative into the grid is reduced to an offset approximation.

For a boundary parallel with the y-axis, the sea level at the point

$\xi(x+\delta x,\ y+\delta y)$ is

$$\xi(x+\delta x,\ y+\delta y) = \xi + \delta x\frac{\partial \xi}{\partial x} + \delta y\frac{\partial \xi}{\partial y} + \frac{\delta y^2}{2}\frac{\partial^2 \xi}{\partial y^2} \qquad (5)$$

where $\delta x, \delta y$ are distances moved by the wave in time δt, in a model with grid sizes $\Delta x, \Delta y$. The calculated sea level provides the new sea level at the location (x,y) at the updated time $t = t + \delta t$. Standard finite difference approximations are used to find the derivatives, i.e.

$$\frac{\partial \xi}{\partial x} = (\xi_{i+1} - \xi_i)/\Delta x \ ; \ \frac{\partial \xi}{\partial y} = (\xi_{j+1} - \xi_{j-1})/2\Delta y$$

$$\frac{\partial^2 \xi}{\partial y^2} = (\xi_{j+1} + \xi_{j-1} - 2\xi_j)/\Delta y^2 \qquad (6a)$$

For shorter wavelength waves with greater surface curvature, a higher order offset derivative may be applied, e.g.

$$\frac{\partial \xi}{\partial x} = (-11\xi_i + 18\xi_{i+1} - 9\xi_{i+2} + 2\xi_{i+3})/6\Delta x \qquad (6b)$$

In cases with an input wave entering the grid, the same equations with an opposite sign for the wave travel direction are used to find the sea level at the internal cells from the known input levels and phases at the boundary. The output levels are the residuals obtained by subtracting the input waves from the total sea level.

In a one-dimensional model, the above equations simplify after all the y-derivatives (along the boundary) are neglected. The equations then reduce to the standard Sommerfeld condition.

To test the boundary condition, a flat-bottomed channel of constant 5 m depth was modelled. Waves of 0.1 m amplitude and 25 s period were introduced at the upstream open end while the radiation sea level boundary conditions were applied downstream. Time series were obtained of velocities and sea levels at Cells 25 and 50, i.e. at the channel mid-point and at the downstream end respectively.

Amplitude and relative phase was well predicted throughout the simulation (Figure A4.1). With this condition imposed, 4% of the amplitude was lost and there was a small phase shift (<1 s) between sea level and velocity. The amplitude loss is due to dispersion in the model, rather than the boundary condition. Subsequent tests, with a 4th order approximation for the gradient of the sea level in the model, eliminated this problem.

The behaviour in 2-dimensions was also found to be satisfactory with the following test. Four sinusoidal waves with different orientations were simultaneously passed through the grid (Figure A4.2). Each input wave was represented at the two sides on the upstream open boundaries, while outputs were handled by the radiation condition. All four boundaries treated both input and output waves simultaneously.

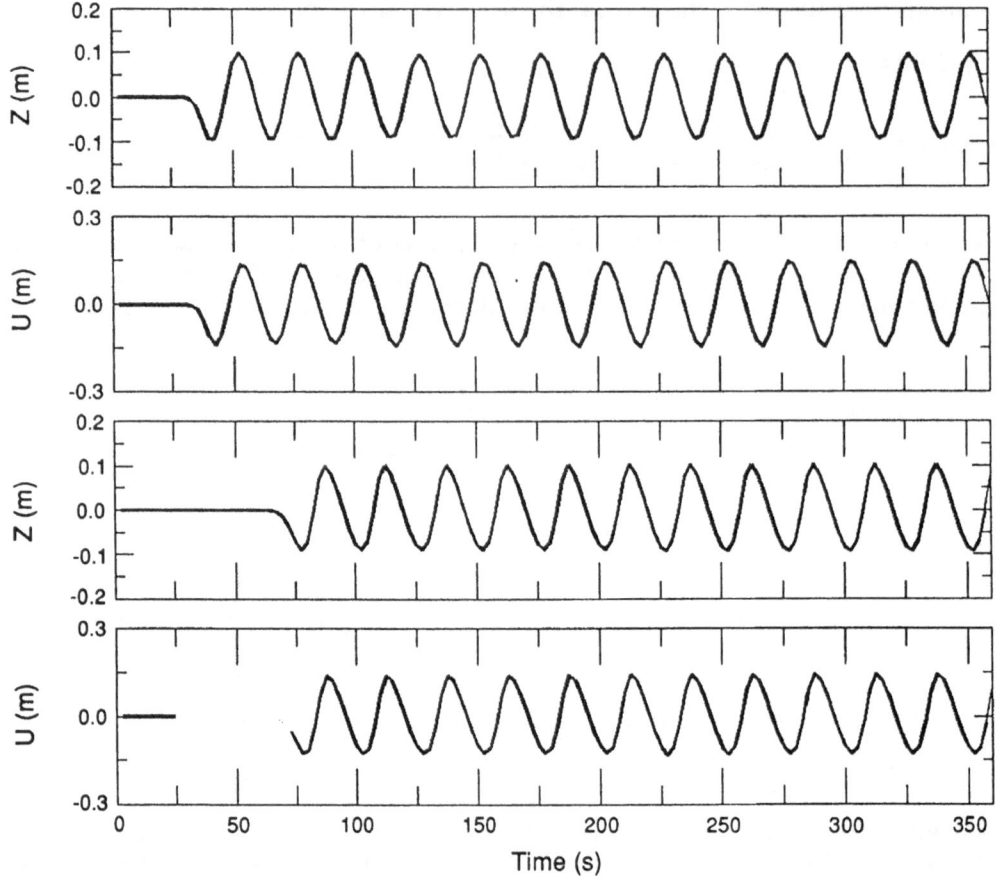

Figure A4.1. Sea levels and velocities at the mid-point and downstream end of a 1-dimensional channel. The geometrical radiation condition was applied at the downstream end of the channel.

The expected sea level at the grid midpoint, calculated assuming a 2-dimensional frictionless case, is compared with the model simulation in Figure A4.3. The amplitude and phase are accurately predicted. There is, however, a slow drift in mean sea level in the model. All boundaries are radiative, which lets the mean level change unhindered, if no boundaries are clamped. A low-frequency damping function is needed to eliminate this problem (e.g. Arnold, 1987).

Finally, the simple test of a steady wind blowing onshore against a straight coastline was applied. The offshore boundary was clamped to zero sea level, while the geometrical condition was applied to the boundaries perpendicular to the coast. The contours of sea level are seen to be straight and parallel (Figure A4.4) and velocities are negligible, as expected.

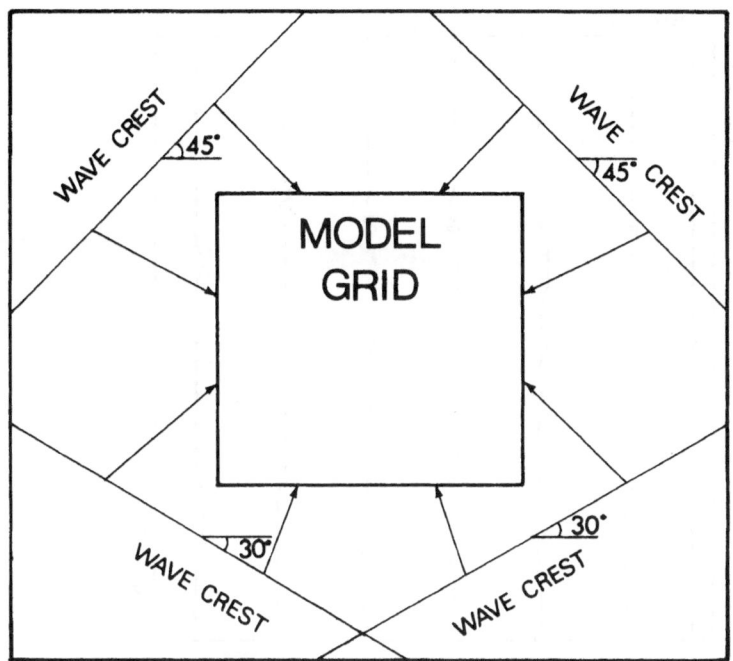

Figure A4.2. Four sinusoids pass into the model grid with the orientations shown. The geometrical radiation boundary condition treats both the input and output waves.

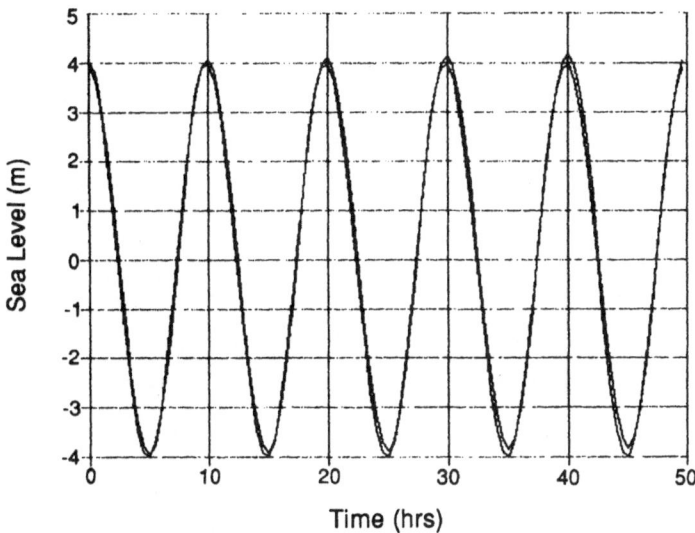

Figure A4.3. Comparison of modelled sea level with the expected value at the centre of the grid in Figure A4.2. The modelled levels drift upwards slowly, but amplitude and phase are well predicted.

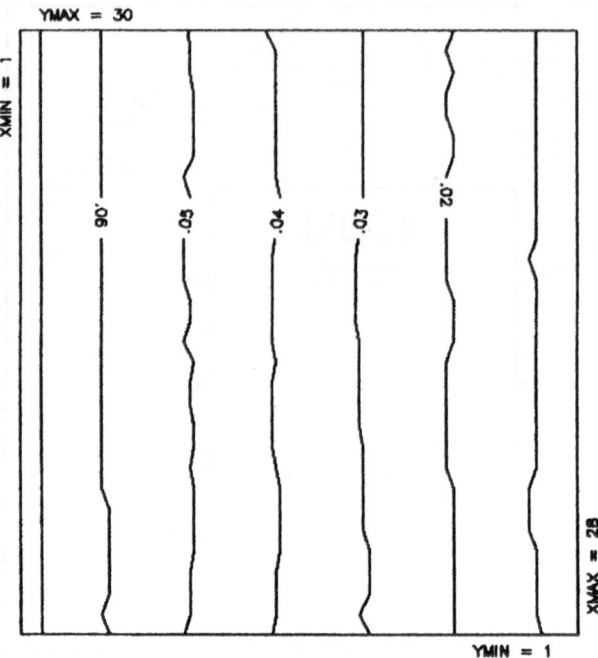

Figure A4.4. Sea level contours for an onshore wind of 10 ms^{-1}. The offshore boundary was clamped to zero sea level, while the geometrical boundary condition was applied at the perpendicular boundaries.

A NUMERICAL SCHEME FOR DETERMINING TRAJECTORIES IN PARTICLE MODELS

KERRY P. BLACK

Victorian Institute of Marine Sciences, Melbourne. Victoria 3002

AND

STEPHEN L. GAY

Australian Institute of Marine Science, Townsville, Queensland 4810

Abstract. A scheme for modelling of trajectories of particles in Lagrangian advection/dispersion models is presented. The second-order accurate method is numerically tractable and eliminates the tendency for particles to spiral outwards on curved streamlines. The method is compared with Euler and Taylor series solutions.

INTRODUCTION

With the increase in modern computing capacity (e.g. faster and/or parallel processors), a resurgence in interest in particle tracking techniques in oceanography is evident. The particle models operate by tracking a large number of individual particles as they traverse a velocity grid associated with a finite difference hydrodynamic model. These Lagrangian particle models have been applied recently to diverse phenomena such as the dispersion of sewage and larvae on Australia's Great Barrier Reef (Black, 1988; Dight *et al.*, 1988), abalone larval dispersal (McShane *et al.*, 1988) and estuarine circulation problems (Hunter, 1987).

Lagrangian particle methods, although requiring a greater computer capacity, eliminate the smearing of concentrations (numerical viscosity) throughout the grid which occurs with an Eulerian scheme, particularly at sharp concentration gradients. This important advantage will often be more important than the cost of extra computer time (Hunter, 1987).

For tracking particles in a grid-cell (where orthogonal velocities U and V are known and derivatives U_x, U_y, V_x, and V_y can be easily calculated) it was found that for circular orbits the use of a second order Taylor's scheme resulted in the trajectories slowly spiralling away from the centre. By slightly varying this simple scheme the problem was overcome. This paper presents a method to track particles which is truly second-order accurate, in the sense that particles will remain on the arc of a circle indefinitely, excluding round-off error.

NUMERICAL SCHEME FOR DETERMINING TRAJECTORIES

Consider the following ordinary differential equation:

$$\frac{dx}{dt} = u(x,t) \tag{1}$$

Starting at x_0, t_0 the first-order estimate of the next value of x after a time step δt is:

$$x = x_0 + u(x_0, t_0)\, \delta t \tag{2}$$

This scheme is known as Euler's method.

Higher order methods take account of the fact that $u(x,t)$ varies during the length of the trajectory. There are numerous higher order schemes used to track trajectories, some of the more well known are Runge-Kutta, Taylor, Midpoint, Trapezoidal and Milne's (Atkinson, 1978). The scheme chosen by a modeller depends on the desired accuracy of the trajectories, the type of equations and computational constraints. Of the schemes which presently exist, the most suitable for our purposes appeared to be Taylor's scheme.

To second-order, Taylor's scheme finds $\delta x/\delta t$ as the average of the initial and final values of u. Putting $h = \delta t$, the equation for tracking a particle may be written as:

$$\frac{\delta x}{\delta t} = 1/2 \left(u(x_0, t_0) + u(x_0 + h\frac{\delta x}{\delta t}, t_0 + h) \right) \tag{3}$$

Using Taylor's Theorem $u(x_0 + h\, \delta x/\delta t, t_0 + h)$ can be approximated as:

$$u(x_0, t_0) + h\frac{\delta u}{\delta t}(x_0 + h\frac{\delta x}{\delta t}, t_0 + h) \tag{3a}$$

$$= u + h\frac{\delta x}{\delta t} u_x + h\, u_t \tag{3b}$$

where u implies $u(x_0, t_0)$

Hence, Eqn (1) becomes:

$$\frac{\delta x}{\delta t} = u + 1/2\, h\, (\frac{\delta x}{\delta t}\, u_x + u_t) \tag{4}$$

In the standard Taylor's scheme, Eqn (1) is used to replace $\delta x/\delta t$ on the right hand side of the above equation with $u(x_0, t_0)$. Hence, Eqn (4) becomes:

$$\frac{\delta x}{\delta t} = u + 1/2\, h\, (u_t + u\, u_x) \tag{5}$$

In our scheme however, $\delta x/\delta t$ of the right hand side of Eqn (3) is assumed to be the same as the left hand side. In this way $\delta x/\delta t$ is taken as being the slope of the trajectory rather than the slope at the start of the trajectory. Therefore, it is slightly more accurate.

Rearrangement of Eqn (4) now yields:

$$\frac{\delta x}{\delta t} = \frac{u + 1/2\, h\, u_t}{1 - 1/2\, h\, u_x}$$

For tracking of particles in 2-dimensions, the equations are slightly more complicated.

$$\frac{\delta x}{\delta t} = u(x, y, t) \tag{6}$$

$$\frac{\delta y}{\delta t} = v(x, y, t) \tag{7}$$

Applying Taylor's Theorem causes the tracking equations to become:

$$\frac{\delta x}{\delta t} = u + 1/2\, h\, (\frac{\delta x}{\delta t}\, u_x + \frac{\delta y}{\delta t}\, u_y + u_t) \tag{8}$$

$$\frac{\delta y}{\delta t} = v + 1/2\, h\, (\frac{\delta x}{\delta t}\, v_x + \frac{\delta y}{\delta t}\, v_y + v_t) \tag{9}$$

substituting $\delta x/\delta t$ with u and $\delta y/\delta t$ with v under the standard Taylor Scheme yields:

$$\delta x = (u + 1/2\, h\, (u\, u_x + v\, u_y + u_t))\, h \tag{10}$$

$$\delta y = (v + 1/2\, h\, (u\, v_x + v\, v_y + v_t))\, h \tag{11}$$

Using our scheme and denoting $\delta x/\delta t$ by $\bar{u}$, $\delta y/\delta t$ by $\bar{v}$, and $u + \dfrac{hu_t}{2}$ by u', causes Eqns (8) and (9) to become:

$$\bar{u} = u' + 1/2\,h\,(\bar{u}\,u_x + \bar{v}\,u_y)$$

$$\bar{v} = v' + 1/2\,h\,(\bar{u}\,v_x + \bar{v}\,v_y)$$

Solving these equations simultaneously for $\bar{u}$ and $\bar{v}$ and replacing $\bar{u}$ and $\bar{v}$ gives the changes in x and y as:

$$\delta x = \frac{(\,u' + (u_y v' - v_y u')\,h/2\,)\,h}{(1 - u_x h/2)\,(1 - v_y h/2) - u_y\,v_x\,h^2/4} \tag{12}$$

$$\delta y = \frac{(\,v' + (v_x u' - u_x v')\,h/2\,)\,h}{(1 - u_x h/2)\,(1 - v_y h/2) - u_y\,v_x\,h^2/4} \tag{13}$$

The reasons for adopting this scheme as compared to Euler's scheme and Taylor's scheme will now be considered.

Consider what would happen if u and v were known to satisfy:

$$u = \quad \Omega y$$
$$v = -\,\Omega x$$

A particular trajectory of a particle has a solution

$$x = \sin\,(\Omega t)$$
$$y = \cos\,(\Omega t)$$

Hence, this trajectory defines a circle of radius 1 and angular frequency Ω

If Euler's scheme is adopted starting at $(0,1)$ then

$$u \neq \Omega$$
$$v = 0$$

after time step h,

$$\delta x = \Omega h$$
$$\delta y = 0$$

Therefore, $x = \Omega h$
$$y = 1$$

The value of x^2+y^2 at $[(x_0+\delta x),(y_0+\delta y)]$ is $1+\Omega^2 h^2$. So the trajectory spirals away from the centre.

Furthermore, it can be shown that by using this method on the above equations, all of the trajectories will spiral away from the centre in a similar manner to this example. Clearly then, Euler's method is not adequate, particularly if the purpose of monitoring trajectories is to determine whether particles can be retained near the same area.

Consider now what happens if Taylor's scheme is adopted.

Again $u=\Omega$, $v=\Omega$ but also $u_x=0$, $u_y=\Omega$, $v_x=-\Omega$, $v_y=0$, $u_t=0$, $v_t=0$.

Hence, Eqns (10) and (11) become:

$$\Omega x = \quad \Omega h$$

$$\Omega y = - 1/2\, \Omega^2 h^2$$

So, the new value of x^2+y^2 is:

$$(\Omega h)^2 + (1 - 1/2\, \Omega^2 h^2)^2$$

$$= 1 + 1/4\, \Omega^4 h^4$$

So, in this case, the particle again spirals away from the centre. However, if $\Omega h < 1$, which is normally the case, the rate of spiralling is much slower than for Euler's scheme.

Using Eqns (12) and (13) however gives:

$$\delta x = \frac{\Omega h}{1 + \Omega^2 h^2/4}$$

$$\delta y = \frac{- \Omega^2 h^2/2}{1 + \Omega^2 h^2/4}$$

x^2+y^2 now equals 1.

In this case, no spiralling occurs.

CONCLUSIONS

An accurate solution to second-order is presented for application to particle models which utilize velocities from the finite difference grid of a hydrodynamic model. The method eliminates the outward spiral error on curved streamlines which occurs in Euler's or Taylor's scheme solutions.

Acknowledgements. The authors would like to thank John Hunter and Lance Bode for their comments on this manuscript. The work was jointly funded by the Crown-of-Thorns Starfish Advisory Committee and the Victorian Institute of Marine Sciences.

REFERENCES

Atkinson, K.E. (1978) An Introduction to Numerical Analysis, Publisher; John Wiley and Sons, New York.

Black, K.P. (1988) The relationship of reef hydrodynamics to variations in numbers of planktonic larvae on and around coral reefs. Proceedings 6th International Coral Reef Symposium 2, 125-130.

Dight, I.J., James, M.K. & Bode, L. (1988) Models of larval dispersal within the central Great Barrier Reef: patterns of connectivity and their implications for species distributions.
Proceedings 6th International Coral Reef Symposium 3, 217-224.

Hunter, J.R. (1977) The application of Lagrangian particle-tracking techniques to modelling of dispersion in the sea. In: J. Noye (ed) Numerical Modelling: Applications to Marine Systems. Elsevier Science Pub. pp 257-270.

McShane, P.D., Black, K.P., & Smith, M.G. (1988) Recruitment processes in *Haliotis rubra* (Mollusca: Gastropoda) and regional hydrodynamics in south-eastern Australia imply localised dispersal of larvae. Journal of Experimental Marine Biology and Ecology 124, 175-203.

CELLULAR AUTOMATA MODELS
OF CROWN-OF-THORNS OUTBREAKS

DAVID GEOFFREY GREEN

Ecosystem Dynamics Group, Research School of Biological Sciences,
Australian National University, Canberra 2601

Abstract. Spatially explicit simulation of within-reef movement and activity of the crown-of-thorns starfish provides a tool with which to test hypotheses about the nature of outbreaks. Factors determining whether an outbreak destroys a reef or not include the initial number of starfish, their movement rates, and the degree of aggregation shown by the starfish. Behaviour of individual starfish may be less critical in determining the course of very large outbreaks compared with smaller outbreaks when a combination of reef shape, coral patchiness and food limitation causes aggregation of starfish regardless of the degree of attraction between individuals.

INTRODUCTION

In many ways, an outbreak of starfish spreading across a coral reef is analogous to a fire spreading across a landscape. Like a bushfire, an outbreak is a disturbance, but more significantly, both are *spatial* processes. Learning how the distribution of starfish changes during an outbreak is as important to our understanding of the outbreak phenomenon as knowing how a fire front expands is to understanding fire behaviour. Therefore models of starfish outbreaks need to deal with spatial pattern explicitly. The importance of this point should not be underestimated. Spatial pattern has rarely been used explicitly in ecological models, yet the few modelling studies that have been carried out on terrestrial (Green *et al.*, 1985; Green, *in press*) and coral reef ecosystems (Reichelt *et al.*, 1986, 1988) all emphasize the importance of "spatial processes" such as disturbance and dispersal in determining the overall dynamics of community change.

Whether for research or for management, models of any process must be *valid* and they must be *useful*. In trying to develop valid models of crown-of-thorns outbreaks a fundamental difficulty is that much has yet to be learnt about crucial aspects of the phenomenon. It is therefore important to develop models that are readily testable. At the same time there is an urgent need for models that are useful, either for testing different hypotheses about outbreaks or else in trying to manage the phenomenon.

In this account I outline an approach to studying crown-of-thorns outbreaks based, by analogy with fire, on the representation of space as a cellular automaton. My aims are, first, to describe a general approach to the problem of modelling outbreaks and, second, to summarize some of the results and their implications. Both of these aspects are important in trying to study, and control, such a poorly understood phenomenon as an outbreak of starfish. I begin by describing the prototype for a flexible, interactive modelling system ("COT") that I have developed in collaboration with Russell Reichelt of AIMS. We have been using this model to study the behaviour of starfish outbreaks in order to develop management strategies in areas important to the tourist industry and to predict the future impact of newly developing starfish outbreaks.

The model has three specific goals: (1) to be a *flexible, generic* system, within which a wide variety of different models and problems can be explored; (2) to be *interactive*, so that the consequences of changes to assumptions or parameter values can be seen immediately; and (3) to provide the most easily interpretable output possible by *displaying outbreaks* visually on maps.

AUTOMATA AND CELLULAR AUTOMATA

The idea of an automaton (Hopcroft & Ullman, 1969), or idealized computer, has been of fundamental importance both in computing and in simulation. With the advent of parallel computing in recent years, the idea of a "cellular automaton" (Wolfram, 1984), a refinement of the automaton concept, has become central to computing science. Equally as important is the vast range of potential applications of the idea (Wilson, 1988).

A *cellular automaton* is an array of identically programmed automata, which can interact with one another. A cellular automaton model of a landscape consists of a fixed array in which each cell represents an area of the land surface (Green, 1987, *in press*; Green *et al.*, 1985, 1986). Associated with each cell are various "states" that correspond to particular environmental features, such as coral cover or topography. This approach is highly appropriate to environmental modelling as it is compatible with both pixel-based satellite imagery and with quadrat-based field observations. Also it enables processes that involve movement through space to be modelled easily. Other applications call for different topologies in the structure of the cell population. For example growth models (Herman & Rozenberg, 1975) call for cellular automata in which the cell population grows and the arrangement of cells mimics the organism's structure, whereas behavioural models (Westman, 1977) call for cellular automata consisting of cells that correspond to individual animals, with the spatial relationships variable and dependent on each animal's location.

Despite their conceptual simplicity, cellular automata are capable of an astonishing variety of behaviour. An important property is that they tend to be "self-organizing". That is, starting from complex, random cell configurations, the rules governing the system cause patterns to emerge from initial chaos. Building appropriate rules into a cellular automaton allows one to simulate any sort of complex behaviour, ranging from the motion of fluids governed by the Navier-Stokes equations (Wilson, 1988) to outbreaks of starfish on a coral reef.

A PROTOTYPE MODEL OF CROWN-OF-THORNS OUTBREAKS

The system (called simply "COT") described here is based upon a series of models developed over the past few years in collaborations with Russell Reichelt, Roger Bradbury and others. Its main features are a pair of cellular automata, one for the reef and one for the starfish. These are structured as follows.

The reef map

The representation of a reef as a cellular automaton is derived from one that was first used in modelling the effects of storms and competition on coral abundances (Reichelt et al., 1986). Each cell in the model grid represents a fixed area of the reef surface. The scale of the map is determined by the satellite images used. The work described here used Landsat images with a pixel resolution of about 70 m. Other satellites (e.g. Spot) provide resolutions down to 20 m per pixel, or even less.

Each pixel in the reef map has the following "states" (attributes) associated with it: the reef "zone" within which the cell lies (a constant defined by the reef map), its maximum percent coral cover (a constant predefined for each zone), its current coral cover (variable), and the number of starfish currently present (variable). In the prototype model coral is assumed to grow logistically, but different types of coral are not differentiated. Values for the parameters used in results reported here are based on figures given by Moran (1986) and Reichelt (*pers. comm.*).

Starfish behaviour

The COT model treats starfish as a separate cellular automaton that interacts with the reef grid. The states (all variable) for each starfish are its position (a pair of real numbers), heading (speed and direction), and behavioural status (e.g. moving, feeding). All work so far has treated this population as constant, but modules for reproduction and predation are planned for future versions.

A good starting point for modelling starfish behaviour is Seymour Papert's "turtle geometry" (Papert, 1973), which deals with the patterns produced by the path of a hypothetical "turtle", whose movement is governed by a fixed program. In the prototype model, the organization of starfish behaviour is treated in very simplistic fashion. It is assumed that the starfish eat coral continually, and at a uniform rate. Furthermore, it is assumed that starfish behaviour is context-sensitive, and dependant on various conditions (chiefly coral cover and numbers of other starfish). These conditions may be *local* (i.e. within the pixel where the individual is currently located), *neighbourhood* (i.e. surrounding pixels) or *global* (i.e. defined for a whole zone or for the whole reef).

The simplest assumption is that starfish movements form random walks. That is

<movement> → Move(r, θ) <movement>   Rule (1)

where Move(r, θ) denotes a random movement of distance r in direction θ degrees from north, with

r ~ Uniform(0, MaxDistance), and θ ~ Uniform(0, 360).

Context-sensitive behaviour is modelled by adding rules to modify the above random movement. For example, to assume that starfish seek areas of high coral cover, we can modify Rule (1) as follows:

<movement> → <Search_for_coral>

IF coral(r, θ) > coral(0, 0) **THEN**
<Search_for_coral> → Move(r, θ) <movement>.   Rule (2)

where coral(r, θ) denotes the coral abundance at a distance r in the direction θ relative to the starfish's present location.

Boundary conditions

As far as possible, the boundary conditions used in outbreak models emulate conditions on real coral reefs. Edge effects are avoided by surrounding the simulated reef with deep water. Normally it is assumed that starfish avoid deep water, so they remain on the simulated reef. Initial starfish numbers and coral abundances can be varied for particular runs. A variety of other initial conditions can be changed as well. Models can be run for arbitrary time periods (mostly 6 months in results reported here) and the spatial scale can be set by the user. Most of our work with the COT model assumes that water currents deposit the bulk of the starfish larvae at one end of the reef and that the size of the starfish population remains constant during the period considered. Also we assume that each coral zone has a fixed upper limit to its coral cover and that below a certain lower limit (usually set at 1%) the starfish could not detect any coral within a pixel.

A common source of problems in simulation models is round-off errors accumulating because too large a time step is used. In this case the critical rates involved are starfish movement (relative to the grid scale), and the relative rates at which corals grow and starfish eat them. The COT model allows all of these parameters to be changed interactively for sensitivity checking. In all of the runs reported below the time step was 1 day, which meant that individual starfish would take several days to traverse a single cell. Some round-off effects are inevitable when the model has large concentrations of starfish within a single cell (see below), but testing indicates that these effects do not bias the results significantly.

THE MODELLING SYSTEM

Graphics

The graphic driver and system of menus used by COT are based on those developed for the interactive bushfire model IGNITE (Green, 1987, *in press*). The graphic driver presents the reef as a map on the computer screen (Figure 1). It uses a "graphics palette" to provide the pattern and colour attributes that represent each symbol in the map. Thus the appearance of a map can be totally changed simply by pressing a key to alter these attributes. An advantage of using an existing graphic driver was that we could use the graphic editor PIXED, which I developed for the IGNITE fire modelling system. PIXED is a graphic editor for pixel-based maps. It has three functions: (1) to create or edit maps, (2) to manage map information (a Geographic Information System), and (3) to generate graphics palettes that define how a map will appear on the screen. For the COT model, PIXED is used to transform classified Landsat images of reefs (Figure 1) into maps showing different reef zones (e.g. slope, crest, flat, and lagoon).

Model flexibility

The COT system is a generic model. That is, it provides a simulation "environment", within which many different models may be defined and used. This flexibility is achieved in two ways. First a system of menus (Figure 2) allows the user to change model parameters interactively while the program is running. Secondly, more complex processes (e.g. the organization of starfish behaviour) can be defined by a sequence of

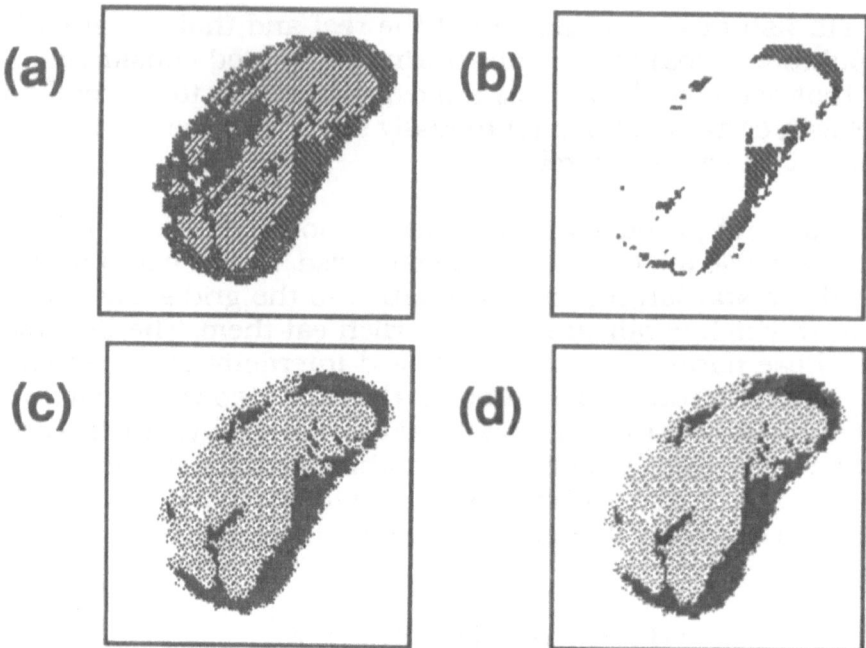

Figure 1. Using the graphic editor PIXED to prepare a reef map for use by the COT model. (a) Landsat image of Davies Reef (classified by depth of penetration) after defining a graphics palette with PIXED, (b) a subset of the image extracted by the editor, (c) final map of Davies Reef after converting subsets to zone identifiers, (d) the map as it appears while running the COT model-infested areas are shown in white.

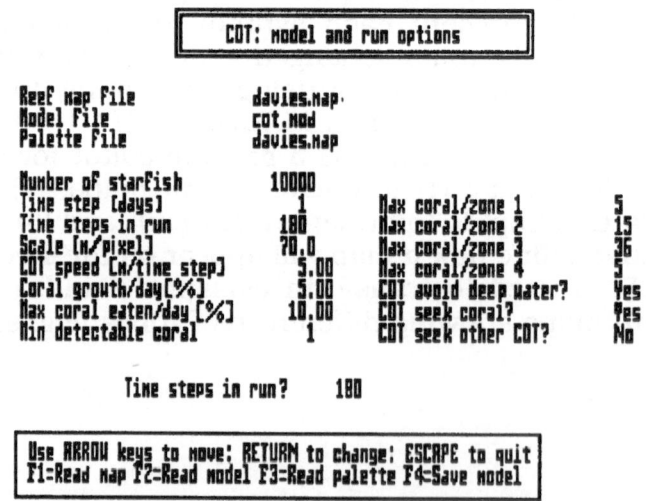

Figure 2. Use of an "ecosystem language" to define a cellular automaton model of growth and competition within a small patch of coral reef (Reichelt *et al.*, 1986). The abbreviations denote types of coral (CM=massive, CB=branching, CE=encrusting, SC=soft, and TA=Turf Algae). The symbol ">>" means that the left hand taxon will overgrow taxa listed on the right. Comments (in lower case) are ignored. Key words are given in boldface type, with system commands in capitals and user options in lowercase.

statements in a file by using a simple application language (Bradbury *et al.*, 1986). This approach, first developed for the MOSAIC vegetation modelling system (Green *et al.*, 1985, 1988) was previously used for defining reef competition models (Figure 3). General "ecosystem languages", currently under development (Green *et al.*, 1988; Green & Bradbury, *in prep.*), will eventually supercede those described here.

SOME RESULTS FROM THE PROTOTYPE MODELS

Effects of starfish behaviour

A question that has been much discussed is whether or not Acanthasters tend to congregate. If they do, then the exact way in which they congregate is important. For example, the simple model assumption that starfish actively seek out the company of other starfish leads to the absurd case of an entire starfish population forming a stationary cluster at a single point (not shown).

The COT model shows that under certain conditions, aggregations of starfish can occur even if the starfish have no gregarious tendency (Reichelt *et al.*, 1988). If we first assume that the starfish move completely at random (Figure 4a), then in a model run they rapidly disperse from their starting point. The rate of spread of the starfish distribution is initially equal to their maximum rate of movement, but rapidly decreases as the population disperses. If we next assume that the starfish can sense areas of high coral cover and seek them out, then they will tend avoid the lagoon and to disperse around the edge of the reef, where coral cover is highest (Figure 4b, 4c). For a small outbreak (<<100,000 starfish, say), the starfish do not aggregate, but spread out more or less uniformly across the reef. However, for large outbreaks, the starfish are initially so concentrated that they eat coral at a much faster rate than it can grow back. Thus all of the starfish move away from their starting point as fast as they can go. Because the rim of the reef is both narrow and roughly constant in width a traffic jam effect occurs, with each new area of coral being completely eaten out as soon as the advancing starfish reach it.

We can weaken the assumption about the starfish behaviour by assuming that they prefer areas of high coral cover, but cannot sense them from any distance. Under these conditions, small outbreaks still spread out around the rim of the reef, but large outbreaks do not produce crowding of starfish. Instead large outbreaks spill out into the lagoon (Figure 4d) and dissipate. Most of the starfish eventually drift back to the rim of the reef, but without producing any marked aggregations.

```
MODEL Reef;                                        (Start model definition              )
        TAXA      CM    CB    CE    SC    TA  ;     (Names for coral taxa               )
        GROWTH   0.005 0.005 0.005 0.005 0.010 ;    (Coral growth rates                 )
        DEATH    0.01  0.01  0.01  0.01  0.01  ;    (Death rate of corals               )
        SETTLE   0.05  0.05  0.05  0.05  0.05  ;    (Recruitment rate                   )
        INTERACTION is by competition               (Assumption about interaction       )
            SC > TA                                 (   between different taxa.          )
            CE > SC TA                              (   Taxa named on the left of        )
            CB > CE SC TA                           (      the ">" replace taxa          )
            CM > CB CE SC TA                         (   on the right.                    )
        STORM 0.1 cyclones/year, average 50 % damage ;  (Define storm regime             )
PROBLEM for this run ;                              (Start problem definition           )
        TIME  20 years at 1 month intervals;        (Time and yime step                 )
RUN                                                 (Run the model                      )
END.                                                (End this model run                 )
```

Figure 3. A menu from the prototype COT model. The user can change any parameter or assumption interactively by selecting the appropriate item from the menu and giving the new value. The values used in model runs were based on currently available information (Moran, 1986, *pers. comm.*).

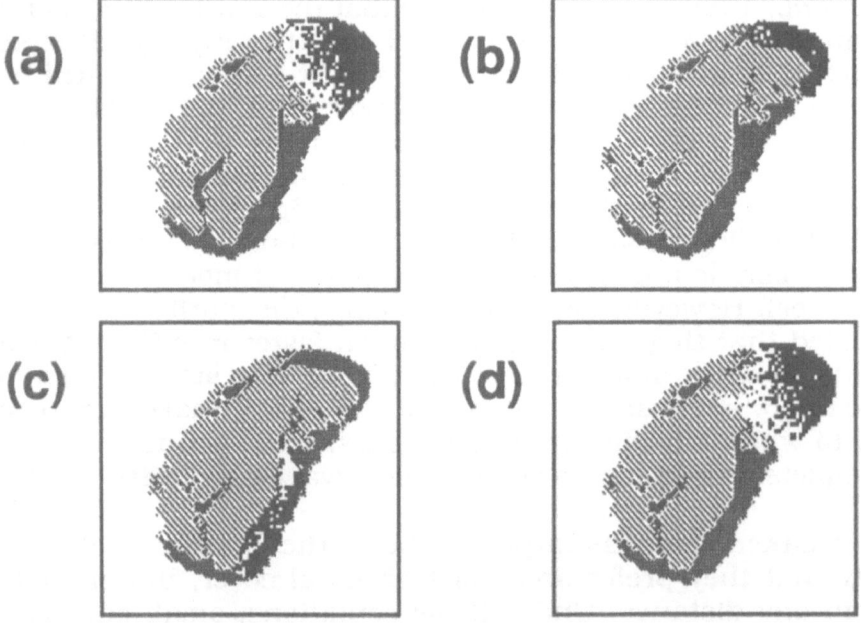

Figure 4. Runs of the COT model for several scenarios at Davies Reef. Infested areas are marked in solid black: (a) random movement by starfish; (b) the starfish prefer areas of highest coral cover; (c) large outbreak; (d) an outbreak starting from southern part of the reef. See the text for further explanation.

Effects of starfish numbers

A series of model runs with different numbers of starfish (not shown) implied that the proportion of reef destroyed in one year by varying numbers of starfish increases linearly for small outbreaks (up to 50,000 starfish), but only neg-exponentially for larger outbreaks (Reichelt et al., 1988). This result illustrates both the importance of initial starfish population size on the course of an outbreak and the limiting effect of speed of movement on the rate of coral destruction by large outbreaks.

Sensitivity of model to size of time step

In a series of model runs the time step length was varied to test its effect on the rate of decline of coral cover on the reef (not shown). The rates of feeding and growth of coral were adjusted accordingly so that the main effect in changing the step length is to change the apparent rate of movement of starfish across the reef. The effect is not proportional, yet all the runs were within the bounds set by field observation of individuals moving in the field (Moran, 1986). This result shows that rate of movement strongly affects the rate of decrease of the reef's coral.

Percolation effects

Starfish outbreaks, like fires, are akin to epidemics. A critical factor in the spread of any epidemic process through a landscape, or seascape, is the availability of sites that can be "infected". Spread of an epidemic across a connected set of sites is termed "percolation" (Stauffer, 1985). If the epidemic starts from a single "seed", then it can spread only to sites connected to its starting point. An important property of random, patchy landscapes is that if more than a critical proportion (~59%) of sites are available, then the epidemic can spread throughout the entire landscape (Turner et al., 1989). At this critical level, the path of the epidemic approximates a fractal curve (Orbach, 1986) and below it the landscape tends to consist of disconnected patches.

Percolation properties apply to starfish outbreaks provided we assume that starfish actively seek out coral: coral reefs often consist of patches of coral, separated by bare sand, so starfish would travel around a reef by moving from one patch of coral to another. In all of the above simulation runs, it was assumed that every cell in the reef map contained coral. In a series of model runs (not shown), coral cover was set to zero in 41% of cells (the critical density), and it was assumed that starfish seek out areas of high coral cover. Two features of these runs are worth noting. First, the tendency to crowd together was greatly increased, because the starfish tended to congregate in cells with coral cover. Secondly, the rate at which the outbreak spread was greatly reduced. The result was that if the model run started with a cluster of starfish, then the cluster tended to persist, simply because the spreading "front" moved so slowly.

CONCLUSIONS

The main lesson to be learnt from the above results is the importance of the initial starfish distribution, and of starfish behaviour, in determining the course that an outbreak takes. As we have seen, apparently minor changes in assumptions about starfish behaviour can produce gross changes in the course that an outbreak takes. Fortunately, these changes are so clear that field testing should be possible, even over relatively short monitoring periods.

On the other hand, the behavioural assumptions made above were very crude. More work needs to be done to construct plausible models for the organization of starfish behaviour. We are currently designing syntax that will enable more complex assumptions about starfish behaviour to be incorporated into the COT system (Green & Bradbury, in prep.). This work builds on the ideas of Westman (1977), who has argued that behavioural models should be based on an animal's "logical structure", with the main elements being syntax for describing the organization and interaction of environmental inputs, the animal's internal state, and its behavioural outputs. Some of this work might best be done at finer local scales than those used here (cf Hogeweg & Hesper, this volume).

The COT system provides a flexible framework within which a variety of outbreak models and questions can be addressed within the one system. For example it would be easy to monitor what happens within different portions of the cell grid and hence compare the relative damage to different reef zones during an outbreak. Likewise management questions can be addressed. For instance the explicit representation of space in COT means that one could examine the effects of starfish movement on (say) the cost-effectiveness of removing starfish from an infested reef. Scientific questions can also be approached this way. For example, by adding modules for fish predation and starfish reproduction, it will be possible to address the effects that spatial distributions have on these processes and on the larger scale dynamics of outbreaks.

The use of satellite data to generate the reef maps used by COT raises the possibility of testing outbreak scenarios for virtually any reef. A fruitful exercise would be to integrate COT with larger-scale models, particularly hydrodynamic models of currents and larval transport between reefs.

The pattern of real outbreaks has been observed to vary greatly. At John Brewer Reef in 1983/4 tens to hundreds of thousands of starfish consumed the bulk of the hard coral cover over a period of about 2 years. The starfish were initially clumped but later dispersed along most of the reef perimeter (Moran et al., 1985). At Davies reef (central GBR) about 150 starfish are currently spread along about 2km of reef slope and the starfish are not consuming large amounts of coral (Moran, pers. comm.).

A third type of outbreak has been observed at Sanctuary Reef (southern GBR) where 75 starfish were counted in 150m strip of reef slope with very little coral damage observed, yet these starfish had moved (as a group) some 500m over a 1 year period (Moran, *pers. comm.*).

The model results presented here suggest that observed aggregations of starfish are not necessarily the result of any "herding" behaviour by the starfish. Several mechanisms produced aggregating behaviour in the simulation runs. These mechanisms include ways in which coral distributions and active coral-seeking behaviour combine, such as the "traffic jam" effect and percolation through patchy reefs.

The traffic jam effect arises with large numbers of starfish because the high rate of decline in coral cover combined with the reef shape (high coral cover occurs in a thin line around the perimeter of the reef) constricts the possible directions that the starfish can move. The aggregation effect seems to break down when food becomes scarce in the late stages of a large outbreak. The observations from John Brewer Reef (see above) support this suggestion.

The percolation results suggest that patchy reefs may slow down outbreaks. The case in which the proportion of coral cover is initially around the critical level leads to the interesting case in which the outbreaking starfish may be able to spread throughout the entire reef, but must follow a long, convoluted fractal path. This mechanism could account for what appear to be chronic infestations of *Acanthaster planci* on some coral reefs (Bradbury, *pers. comm.*) and across whole sets of reefs.

REFERENCES

Bradbury, R.H., Green, D.G. & Reichelt, R.E. Qualitative patterns and processes in coral reef ecology: a conceptual programme. Marine Ecol. Prog. Ser. 29, 299-304 (1986).

Green, D.G. Spatial simulation of fire in plant communities. In Proceedings of National Symposium on Computer Modelling and Remote Sensing in Bushfire Prevention, pp. 36-41 (Ed. P. Wise), National Mapping, Canberra (1987).

Green, D.G. (*in press*) Modelling forest mosaics. In: S.Ikawa & K.Yajima (Editors), System Modelling and Optimization. Springer-Verlag, Heidelberg.

Green, D.G., Bradbury, R.H. & Bainbridge, S. Embodiment of formal languages. Mathematics and Computers in Simulation 30, 39-44 (1988).

Green, D.G., Bradbury, R.H. & Reichelt, R.E. Formal languages and biological pattern. J. Infer. Deduct. Biol, 5, 47-66 (1986).

Green, D.G., House, A.P.N. & House, S.M. Simulating spatial patterns in forest ecosystems. Math. Comput. Simulation 27, 191-198 (1985).

Herman, G.T. & Rozenberg, G. Developmental Systems and Formal Languages. North-Holland, Amsterdam (1975).

Hogeweg, P. & Hesper, B. (this volume) Crowns crowding: an individual oriented model of the *Acanthaster* phenomenon.

Hopcroft, J.E. & Ullman, J.D. Formal Languages and their Relation to Automata. Addison-Wesley, Reading, MA. (1969).

Moran, P.J. The *Acanthaster* phenomenon. Oceanogr. Mar. Biol. Ann. Rev. 24, 379-480 (1986).

Moran, P.J., Bradbury, R.H. & Reichelt, R.E. Mesoscale studies of the crown-of-thorns/coral interaction: a case history from the Great Barrier Reef. Proc. Vth Int. Coral Reef Cong. 5, 321-326 (1985).

Orbach, R. Dynamics of fractal networks. Science 132, 814-819 (1986).

Papert, S.. Uses of technology to enhance education. LOGO Memo no. 8, M.I.T. Artificial Intelligence Laboratory, Boston, (1973).

Reichelt, R.E., Green, D.G. & Bradbury, R.H. Discrete simulation of cyclone effects on the spatial patterns and community structure of a coral reef, Proceedings of the Vth International Coral Reef Congress Vol. 3, pp. 337-342 (1986).

Reichelt, R.E., Bainbridge, S. & Green, D.G. Crown-of-thorns dispersal in the Great Barrier Reef - a simulation study. Mathematics and Computers in Simulation 30, 145-150 (1988).

Stauffer, D. Introduction to percolation theory. Taylor & Francis, London, (1985).

Turner, M.G., Gardner, R.H., Dale, V.H. & O'Neill, R.V. Predicting the spread of disturbance across heterogeneous landscapes. Oikos 55, to appear (1989).

Westman, R.S. Environmental languages and the functional basis of behaviour. In: B.A. Hazlett (Editor), Quantitative Methods in the Study of Animal Behaviour. Academic Press, NY, pp. 145-201 (1977).

Wilson, G. The life and times of cellular automata. New Scientist 120, 44-49 (1988).

Wolfram, S. Cellular automata as models of complexity. Nature 311, 419-424 (1984).

CROWNS CROWDING:
AN INDIVIDUAL ORIENTED MODEL
OF THE *ACANTHASTER* PHENOMENON

P. HOGEWEG AND B. HESPER

Bioinformatica, Padualaan 8, Utrecht, the Netherlands

Abstract. Notwithstanding the fact that the *Acanthaster* phenomenon manifests itself at the scale of the entire Great Barrier Reef, we show in this paper that a model of the behaviour of individual starfish, at a maximum resolution of an order of magnitude smaller than the size of individual starfish (i.e. at a resolution of 1 cm^2) can shed light on the instabilities at the micro-, meso-, and macro-scale of the system. In particular we show that:

1.   Basic data on feeding rate and movement of *Acanthaster*, combined with regrowth rates of coral, are sufficient to explain the outbreak densities reported by Endean & Stablum (1975); the reported outbreak densities are near to equilibrium densities.

2.   Outbreaks can occur with relatively small population fluctuations due to crowding of the starfish; crowding can switch an equilibrium system to a system in which coral is depleted in a relatively short time.

3.   *Acanthaster* is prone to outbreak because of its extremely fast movement as compared to its minimum 'living range' in equilibrium conditions. Therefore *Acanthaster* is virtually independent of the density of its prey (coral) over periods which suffice for complete exhaustion.

4.   If crowding indeed plays an important role in outbreaks, the fact that outbreaks always consist of large individuals is easily understood: it takes more than a year for randomly dispersed individuals to crowd sufficiently to cause coral depletion.

Most important however, we show that the state variable most often used in ecology, namely the number (density) of a population, is insufficient to determine a unique next state function (i.e. the number/density of the population at the next time step) in such locally defined models: the spatial pattern of the population is a crucial additional state.

1. INTRODUCTION: INDIVIDUAL ORIENTED MODELLING IN ECOLOGY

The state variables most often used in ecology are population numbers/densities and/or amounts/concentrations of chemicals, and the ecosystem is described in terms of changes thereof in time and/or space. This is true for both experimental and theoretical work. In contrast individual oriented models take as basic units individuals, and the system is primarily described in terms of the information processing of the individuals (Hogeweg & Hesper, 1979,1981,1985,1988a; Hogeweg & Richter, 1982; Hogeweg et al., 1985; Hogeweg, 1988a; Huston et al., 1988). The obvious disadvantage of an individual oriented approach is that there are many individuals in an ecosystem. Minimisation of the number of state variables thus leads to lumping them into populations or other units. We think however that the advantages of individual oriented modelling, now that the handling of many variables has become feasible through present day computer technology (or that of the near future), far outweigh the disadvantages. These advantages include:

1. By concentrating on the local information processing of individuals, no interactions which are beyond the scope of the organisms studied are introduced into the model. In lumped models one is often tempted to include in the model interactions which implicitly assume that individuals posses more (global) knowledge than they can reasonably have. This is because only global variables are present in the model, and because such global knowledge permits simpler interactions at the lumped level. In contrast, individual oriented models exploit the fact that simple information processing at an individual level often gives rise to fascinating and complex behaviour at the macro level.

2. Because individuals are the basic units for observation and measurement, the model parameters are usually obtainable by model independent observations, whereas in lumped models many parameters are only obtainable from fitting the model to the same (type of) data as the model tries to represent.

3. In individual oriented models one usually studies primarily properties of the model/system which are not included in the basic model formulation. For example population level, community level and/or ecosystem level are studied in individual oriented models. Thus, an integration of formerly separate disciplines is achieved in one model structure.

4.   Individual oriented models are 'variable structure' models: who interacts with whom is not *a priori* given (although the potential for interaction is), but is the result of other interactions. Thus an important observable of the model is who, in fact, influences whom.

5.   Individual oriented models can be 'non-goal oriented models', i.e. one does not have to determine in advance what features will be studied, or are likely to be interesting: specialised entities within the modelling system may detect interesting observables.

In this paper we apply the individual oriented modelling strategy to the *Acanthaster* phenomenon. We break,in an extreme form, with the usual concerns of (*Acanthaster*) ecology, by disregarding population fluctuations of crown-of-thorns entirely, focussing instead on the impact of the behaviour of the individual starfish at fixed population sizes. Nevertheless, we feel that this approach enables us to pinpoint some of the important aspects of the outbreak behaviour of *Acanthaster*.

2. METHODS: MIRROR MODELLING

2.1   Introduction

MIRROR modelling is the modelling methodology we have developed over the past number of years to investigate the concepts of (multi- level) individual oriented modelling, to explore its potential and to exploit its advantages. The main concepts have been recently reviewed (Hogeweg & Hesper, 1988a; Hogeweg, 1988b). Here we will only mention those aspects which are directly needed to understand the here presented model of the *Acanthaster* phenomenon. In MIRROR modelling we set up an 'artificial universe', in the present case an artificial CORAL REEF, and study its behaviour. Such a universe consists of a set of locally defined objects, which are active 'once in a while', i.e. the model is event based, not time step based. More particularly, such a universe consists of (at least) one SPACE in which DWELLERs dwell.

The SPACE is subdivided into (possibly a hierarchy of) PATCHes, i.e. homogeneous parts. It can be (and, in the model presented here, is) an 'active SPACE': each PATCH has in that case its own local behaviour, dependent on its own state (possibly including the DWELLERs which inhabit it) and the state of neighbouring PATCHes, i.e. the SPACE is a generalization of a cellular automaton. The SPACE represents those parts of the universe which have a fixed (or ancestry dependent), local relation to each other; in the model presented here it represents the sedentary coral.

The DWELLERs inhabit the SPACE and move about in it. The behaviour of a DWELLER is typically dependent on other DWELLERs nearby and on the state of the PATCH(es) in which it dwells. Because its behaviour includes movement, its interactions are variable and shaped by previous interactions (which shape its behaviour). They represent the variable relation structure of the model, and in the model presented here represent crown-of-thorns. DWELLERs can, of course, also influence the (state of the) PATCHes and so pull the SPACE into the overall variable structure of the universe.

2.2 Active Spaces and Cellular Automata Planes

Like all other entities of MIRROR worlds, SPACEs and PATCHes are normally 'objects', i.e. a combination of data structure and procedure(s), which are 'active' (i.e. perform (part of) a procedure) 'once in a while'. Activity is caused by explicit activation by some other entity, by the occurrence of some event or at a certain time.

An important function of PATCHes which subdivide the SPACE, is that DWELLERs inhabit them (are stored locally in a list of a PATCH). Their behaviour is often triggered by the (movements of) DWELLERs. In a dynamic hierarchical PATCH structure PATCHes are created and destroyed to fit the detail needed by the DWELLERs (Hogeweg & Hesper, 1981).

As a special case the activity of all such PATCHes of a SPACE can be synchronised; if so, a SPACE can be a cellular automaton (see ,for example, Toffoli & Margolus, 1987). Synchronisation causes a globalisation of the model (Hogeweg & Hesper, 1988; Hogeweg, 1980,1988a) and is therefore undesirable and counter to the basic philosophy of MIRROR modelling. Nevertheless, because such a synchronisation is used so often in other (local) models (i.e. cellular automata) and because it is not always harmful, we provide in MIRSYS (our INTERLISP implementation for MIRROR modelling) the special case of a synchronised Cellular Automaton PLane (CAPL) and exploit the advantages in the potential efficiency provided by the globalisation. Because of the globalisation, 'Patches' do not need to be objects with their own behaviour; they are only local states, in CAPLs they are bits of a bitmap. CAPL uses the basic graphics routines to construct arbitrary bitmap based cellular automata. To this end N(orth), S, E, W, NW, etc neighbourhood planes are constructed by a shift copy from the C(enter) plane, and the cellular automaton rule is composed in terms of logical operations on these planes, which are executed as bitmap-copies, -erases, -paints, etc. The power and generality of this approach is derived from the possibility of having an arbitrary number of CAPLs whose rules use each other('s neighbourhood) (Toffoli & Margolus 1987). By using the bitmap operations, time requirements are relatively small and fairly independent of the size of the cellular automaton (about the square root of number of pixels). This enables us to structure the environment in great detail.

MIRSYS allows the combination of both 'autonomous PATCHes' and 'Pixel PATCHes' in one SPACE. The DWELLERs deal with both: they inhabit the autonomous PATCHes, and interact through them with each other. They can, however also sense the pixel PATCHes in their neighbourhood and change their state. The autonomous PATCHes can activate and be activated by DWELLERs and usually contain several DWELLERs, whereas the pixel PATCHes are subject to a global CAPL updating as well as the local state change by the DWELLERs and may be much smaller than each individual DWELLER.

2.3 DWELLERs

DWELLERs are moving, locally defined, 'selfsufficient' entities, which are 'active' 'once in a while'. The activity is time based, DEMON based on the occurrence of some event, or by explicit activation by some other entity. DEMON based activation includes the activation by the approach of some other DWELLER. The SPACE, as used by DWELLERs, is continuous: their position is given in coordinates as real numbers and they can move any distance into any direction. They have a number of basic actions (e.g.(FORWARD distance), (RIGHT angle)) and senses (e.g. (NEARBY dweller type distance)). Thus, their perception is not influenced by the prevalent size of PATCHes. In this way different DWELLERs may have a different perception and action radius. With respect to their perception of the state of PATCHes, the minimum resolution is of course equal to the PATCH size. As mentioned, this minimum resolution can be quite small when CAPLs are used.

2.4 RECORDERs, REPORTERs and OBSERVERs

A MIRROR universe also contains self-sufficient locally defined entities to observe the universe. Dependent on their sophistication we distinguish RECORDERs (which just record the state of a model variable), REPORTERs (which report on some predetermined property of the universe which is not defined as model variable), and OBSERVERs (who seek interesting features of the universe) (Hogeweg & Hesper 1986). Only simple RECORDERs and REPORTERs were used in the MIRROR universe presented here.

2.5 MICMAC modelling

MICMAC modelling is a strategy for converting a MICro scale universe to MACro scale universes (Hogeweg & Hesper, 1981,1988b) This strategy involves using partial 'empirical' models of the behaviour of the micro universe to confine the events of the macro universe to interesting events, i.e. those events not 'predictable' by these partial models. Because of the local nature of MIRROR worlds and because the partial models are made by inhabitants of the universe (i.e. by DWARFs) the conversion can be a gradual and ongoing process during the life of the MIRROR universe. In the present study, the strategy is used to study the *Acanthaster* phenomenon at the scale interpretable in terms of one entire reef in a separate step after obtaining the results on the impact of feeding behaviour of individual starfish in a reef universe representing a small area (i.e. 7.5 x 7.5 m).

3. THE REEF UNIVERSE

3.1 CROWNs

The DWELLERs inhabiting our MIRROR reef are CROWNs. CROWNs have inherited a number of basic properties of crown-of-thorns; we nickname these entities CROWNs and do not use the more usual abbreviation COT to emphasise that they are entities in their own right, worth studying even if they do not resemble COTs in all particulars. Nevertheless, an almost full definition of a complete 'behavioural entity' can be constructed from known data of *Acanthaster*. These data are given in Table 1, and are all taken from Moran (1988). We have left out data on growth and reproduction of *Acanthaster*: only feeding behaviour of adults is taken into account, and we experiment with various fixed population sizes. CROWNs have a size, a mouth size, a maximum speed of movement, and an amount of food eaten per day. In addition we have used the qualitative statement that they have a 'heading' over periods of days, and that they are attracted to other CROWNs when the latter are feeding; these statements are quantised as seems fit (see below). Missing data are mainly with respect to the selection of a feeding location (e.g. how far do they move between feeding sessions when there is plenty of coral, and whether or not they 'look ahead' to find where the best coral is). Because the mouth size of *Acanthaster* corresponds fairly closely to the area of coral it eats daily, and because it feeds externally by engulfing, our CROWNs eat exactly one mouth size per day if there is 100% CORALcover at the place of feeding; otherwise digestion is assumed to take proportionally less time. Because movement is extremely fast, i.e. takes virtually no time, we have assumed that 'remote sensing' of CORAL is not a strong requirement for CROWNs and therefore is not present. We further assume that CROWNs move away 1.5 times their diameter before feeding again; larger values of this parameter (e.g. 3 times their diameter) change the results only quantitatively not qualitatively.

Table 1. Parameter values in models

Property	cf. Moran (1988)	Values used
size (diameter)	25 - 35 cm	20 -25 cm
mouth size	----	12 - 17 cm
feeding	engulfment	engulfment
amount eaten/day	116 - 378 cm^2	120 - 260 cm^2 = mouth size
movement speed	3 - 20 m/hr	20 m/hr
heading	persistent	-60° + 60° each step
interaction	towards feeding COTs	maximum perception distance = 1m

Thus, the behaviour cycle of a CROWN is:

1.   IF no other CROWN is sensed THEN Move 1.5 times its own diameter in a direction uniformly drawn between -60° and 60° relative to previous direction; OTHERWISE move towards one of the CROWNs 'in front of itself' (-90° to 90° relative to heading). If there is no CROWN in that direction, it moves in the direction of the heading of one of the nearby CROWNs (the latter is an unrealistic assumption; it ensures crowding over a large range of CORALcovers, without trapping a bundle of CROWNs in one place circling each other; it should be considered a simplification for the occurrence of other 'mental processing' of the CROWNs, e.g. noticing nonproductive circling, or noticing behaviour of other CROWNs while feeding). Note that this represents an arbitrary quantification of the properties of the crown-of-thorns discussed above. The results do not seem to depend on the precise values used.

2.   Wait period in which this movement takes place, i.e. DISTANCE-MOVED/20 METER hours (i.e. very short!).

3.   Feed by removing all CORAL under mouth;

4.   Stay put for a period of AMOUNT-EATEN/MOUTHSIZE days.

5.   Iterate.

By itself this behaviour cycle is not very interesting, it is only in interaction with CORAL that a CROWN 'comes to life'.

3.2 CORAL

CORAL is modelled as a cellular automaton (CAPL): because of the fixed spatial structure of coral, this seems a suitable formalism. Moreover, as is stated in Moran (1988), coral regrowth is relatively fast because it is mainly derived from local regrowth from remnants of corals remaining after an outbreak. Such a local regrowth can be modelled easily in the cellular automaton formalism. Synchronicity seems not to be too great a problem for CORAL growth, in particular when CORAL death occurs asynchronously by CROWN feeding. The main choice to be made is with respect to scale: what size does each cellular automaton represent?. We have chosen a scale of 1 cm^2 per automaton (pixel) because:

1. This can represent 1 coral polyp;

2. It is an order of magnitude smaller than the diameter of a CROWN.

Thus the impact of CROWNs can be studied as a spatial process.

The total extent is 7.5 x 7.5 meters (i.e. 750 x 750 pixels) because this is the minimum size which can support 2 CROWNs and the maximum size which fits on the screen.

Three CAPLs are used to model the CORAL regrowth; one representing the colonisable areas (about 73%), one representing the presence or absence of CORAL and one generating (pseudo-)random bits with density 0.5; this provides us with probabilities (as powers of 0.5) by ANDing C(enter)- and neighbouring bitmaps.

The CORAL regrowth rule assumes that the probability of colonisation of a patch depends on the number of its neighbours containing CORAL, reaching its maximum for 4 neighbours (out of 8). This maximum probability is doubled if a randomly drawn pixel in a range of -10 to +10 pixels contains CORAL. This assumption relaxes the localness of regrowth slightly; it is incorporated in order to be able to fit the regrowth data (see section 4.1). Global recruitment is incorporated by seeding 100 randomly drawn (possibly already filled) pixels once a year.

Given these simple assumptions, we have just one parameter left to specify the growth function, i.e. the above mentioned maximum probability (or equivalently, if this is given, the update frequency of the cellular automaton). This one parameter is fitted to the coral regrowth data given by Moran (1988).

4. RESULTS

4.1 CORAL regrowth after an outbreak

The one parameter of the model which was not derived from independent measurements, or by reasonable thumb-based assumptions, is the growth rate of the CORAL. Data are available for regrowth after an outbreak (Moran 1988) and the regrowth rate is fitted thereupon. In order to fit this situation we have to start with a depleted CORAL population, not with a random initialisation of CORAL because spatial patterning is important in cellular automata (compare GAME OF LIFE, the well known cellular automaton rule for which it was proven that the final outcome (survival or extinction) depends so intricately on the initial configuration that no prediction is possible (i.e. it is undecidable)).

It turns out that, starting with a depleted population with a cover of 1%, a maximum probability of regrowth in a patch of 25% with an update frequency of once in 40 days fits the observed data closely. Two snapshots of the CORALreef 4 and 6 years after a CROWN outbreak are shown in Figure 1. The cellular automaton regrowth expressed as percentage cover is compared with the experimental data in Figure 2a. As expected the regrowth is sigmoid shaped; however it does not fit a classical logistic curve: regrowth is relatively fast early, and remains about constant over a wide intermediate range. A generalised logistic curve

$$\frac{dN}{dt} = aN - bN^c$$

is (hand)fitted to it in Figure 2a. It turns out that the power of the self-limiting term (c) is much smaller than the usually assumed value of c=2; it is: c=1.3. Alternatively we can fit a curve which assumes density independent recruitment.

$$\frac{dN}{dt} = I + aN - bN^2$$

A good fit is obtained for I=0.21 per 40 days, i.e. much larger than the I=0.0004 per year global recruitment which is used in the cellular automaton. Figure 2b compares the regrowth of the depleted population (cover 1% clumped) with the growth which occurs if the initial state is a randomly distributed population of 1 %: in the latter case regrowth is much faster. It also resembles a logistic curve more closely: the power in the generalised selflimiting term is c=1.7.

Thus we conclude that a simple local recruitment modelled with a cellular automaton, although resulting in an S shaped regrowth curve, deviates significantly from the classical logistic growth used in most ecological models, including those from *Acanthaster*. Moreover, we conclude that the rate of growth and the shape of the regrowth curve is very sensitive to the spatial pattern of the CORAL. Cover alone is not sufficient to predict the cover at some future time even in the model.

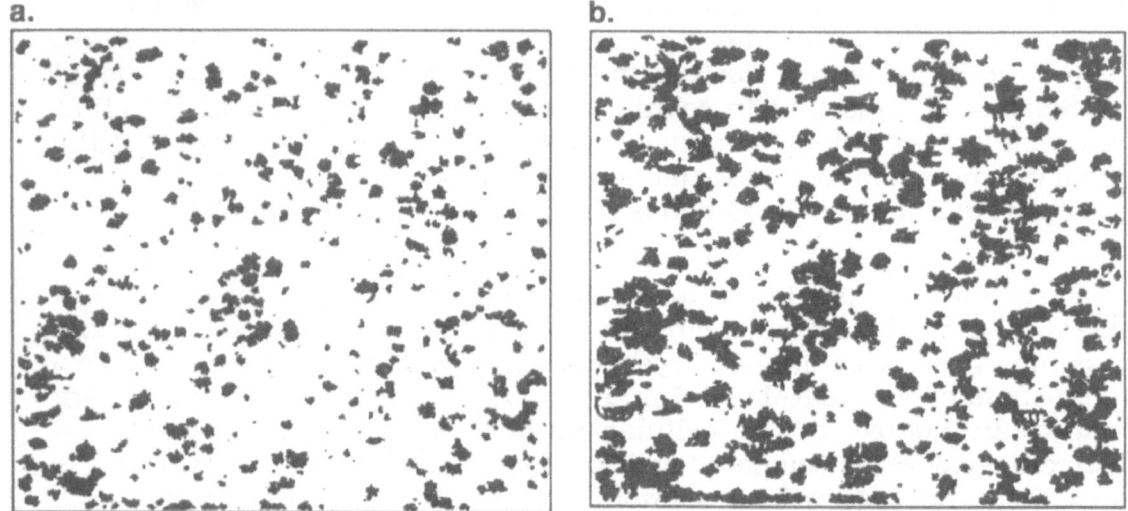

Figure 1. CORAL regrowth after depletion to 1% by CROWNs: spatial representation. **1a.** After 4 years (14% cover).   **1b.** After 6 years (32% cover).

4.2 Impact of CROWNs on the CORALreef

Figure 3 shows the impact of two CROWNs on the reef of 7.5 x 7.5 meter when a stable situation is reached (after 3-4 years). Figure 3a in case of 'noncrowding' CROWNs, Figure 3b for crowding CROWNs. The impact of crowding CROWNs is significantly larger than that of the non- crowding CROWNs: the equilibrium CORALcover is 45% and 58% respectively and, in the case of crowding, certain areas become quite barren. As shown in Figure 4a the amount eaten per week is very constant and independent of either CORALcover or crowding. However, the number of steps taken clearly increases for a decrease of CORALcover and is larger for crowding than for non-crowding CROWNs (Figure 4b). The distance travelled over larger periods increases dramatically as CORALcover decreases further. (Figure 4c). This is measured as the time it takes to get 2.5 meters from the original location, but is expressed as the distance/day.

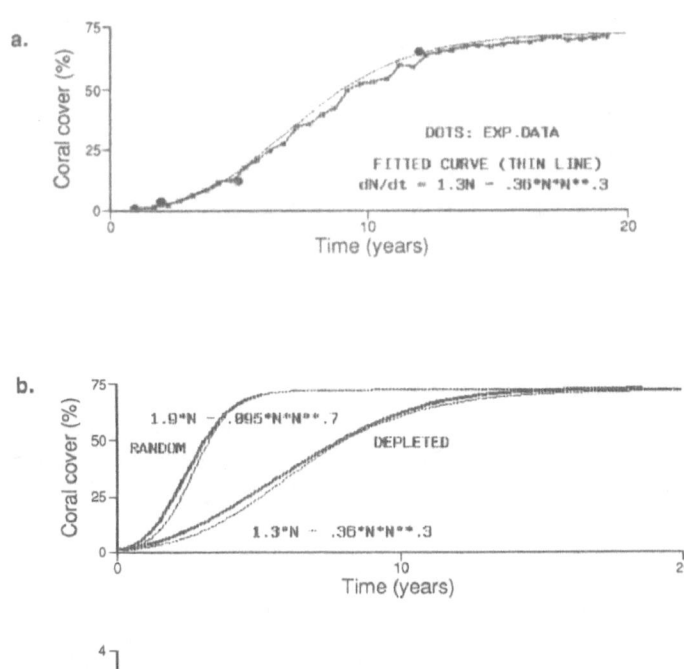

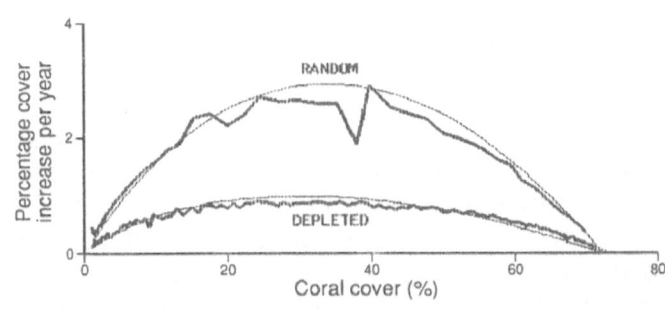

Figure 2. CORAL regrowth after depletion to 1% cover by CROWNs: Percentage cover.
2a. Percentage cover vs time compared to experimental data, and fitted curve (dt=40days).
Initial state depleted CORAL (1% cover).
2b. Percentage cover increase per year vs percentage cover for regrowth after depletion by
CROWNs and after random initialisation with 1% CORALcover.

Two CROWNs of a slightly larger size (mouth diameter = 13 cm, area
eaten per day = 150 cm^2) lead to slow exhaustion of CORAL in 7 years.
Likewise, three standard CROWNs slowly exhaust the CORAL. Thus, two
small CROWNs are the maximum that an area of 56 m^2 can support. The
CROWNs used in our experiment are rather smaller than the crown-of-
thorns in an outbreaking population. Our results can be used for larger
CROWNs: if we change the scale of our experiment slightly, i.e. if we let 1
CORALpixel represent 1.3 x 1.3 cm, our area is 100 m^2 and contains 2
larger CROWNs (mouth diameter: 17 cm for these larger CROWNs). We
can thus conclude that the CORALreef cannot contain 20 CROWNs per
1000 square meter. This compares to 14 COTs cited as the outbreak
number by Endean & Stablum (1975). These figures suggest that
outbreaking populations are of the same order of magnitude as the
'carrying capacity' of the reef, but nevertheless devastate the reef in a
short time (see section 4.3).

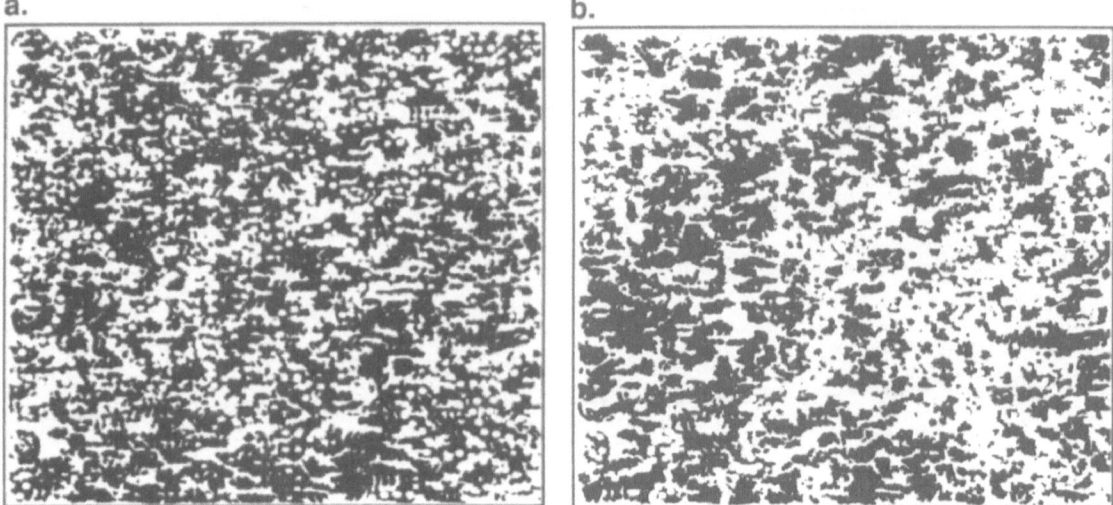

Figure 3. The impact of 2 CROWNs on a reef of 7.5 * 7.5 meter: spatial representation.
3a. 'noncrowding' CROWNs, i.e. CROWNs that do not react in any way to each other. CORAL density in equilibrium 58%.
3b. crowding CROWNs, i.e. CROWNs that move towards feeding CROWNs. CORAL density in equilibrium 45%.

The reason for the larger impact of crowding vs. non-crowding CROWNs is that the regrowth of CORAL is smaller in the case of crowding, because of the more clumped pattern of the CORAL. Figure 5 shows the rate of CORAL regrowth at various CORAL densities for CORAL regrowth after an outbreak, and for CORAL regrowth while 4 crowding or 4 non-crowding CROWNs (introduced in a 'virgin' reef) are feeding (we need a number that exhausts the CORAL to study CORAL regrowth for all CORALcovers). The differences are large: maximum CORAL growth is 7.5%, 15% and 20% per year in the three cases.

In these experiments the CORALreef is shaped as a torus. A torus (i.e. lower and upper as well as left and right borders joined) simulates an infinite area which is everywhere similar to the area represented explicitly, i.e. no larger scale pattern than the pattern in the simulated area is supposed to occur. The introduced CROWNs remain inside the area. In contrast, outbreaking populations of COTs are known to assemble in huge crowds and pass as such over the reef. The effect of such crowds was studied by introducing 100 CROWNs on the south edge of our standard 'virgin' CORALreef, and by letting them pass once, i.e. killing a CROWN as soon as it hits one of the borders of the area.
The impact of such a pass of 100 CROWNs is shown in Figure 6: they cross the field diagonally NE and a small group splits off to the NW. After 47 days all CROWNs have left the area (mean residence time 14.74 days). CORALcover is down to 40% but half of the area is untouched. Maximum regrowth rate of the CORAL is in this case only 10%.

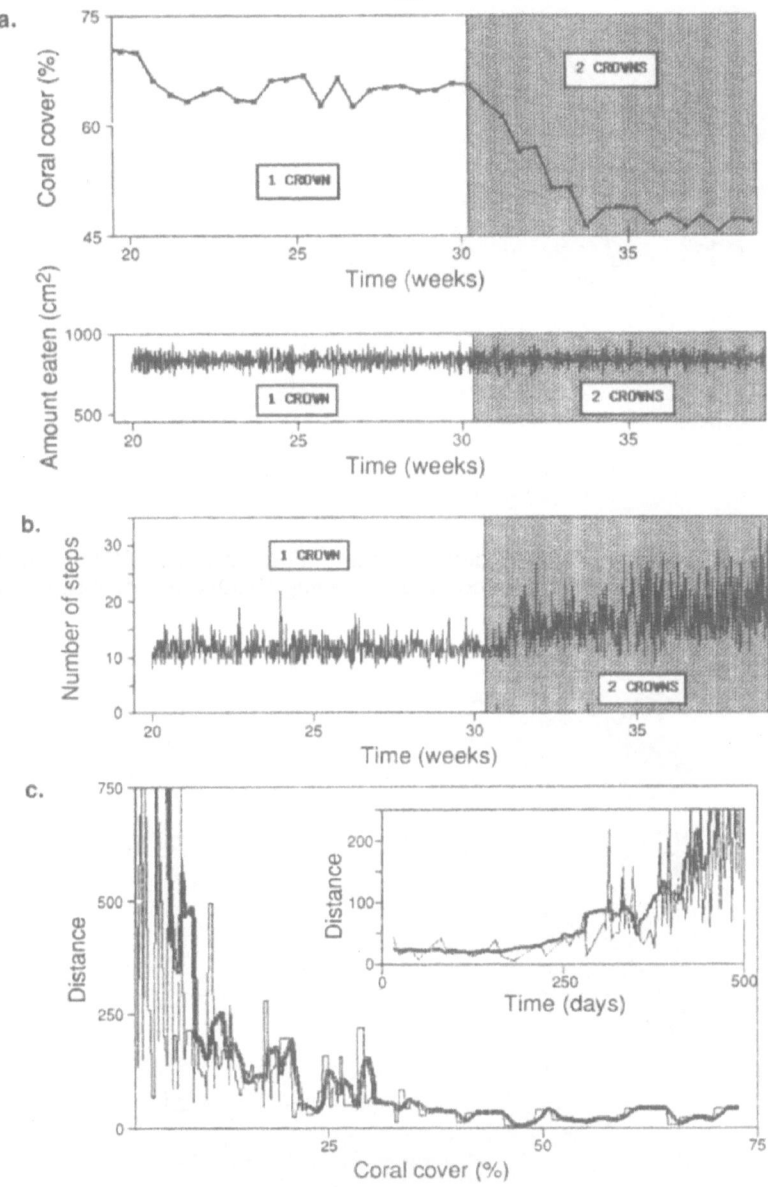

Figure 4. Impact of CORALcover on CROWN behaviour.
4a. amount eaten per week vs time in reefs with 1 and 2; CROWNs respectively; (density of CORAL varies from 73% to 45%). The amount eaten remains constant to a density of 1% CORAL.
4b. Number of steps per week vs time in reefs with 1 and 2 CROWNs respectively; (density of CORAL varies from 73% to 45%).
4c. Distance travelled per day at different CORAL densities. This is measured as the time it takes to travel 2.5 meter from the original location. Distance increases dramatically when CORALcover becomes less then 40%.

These results show that solitary CROWNs structure their environment in such a way that CORAL regrowth is fast. Crowding behaviour decreases this seemingly advantageous effect significantly. Note that coral exhaustion is not necessarily disadvantageous when recruitment is not to the home reef (compare James *et al.*, this volume).

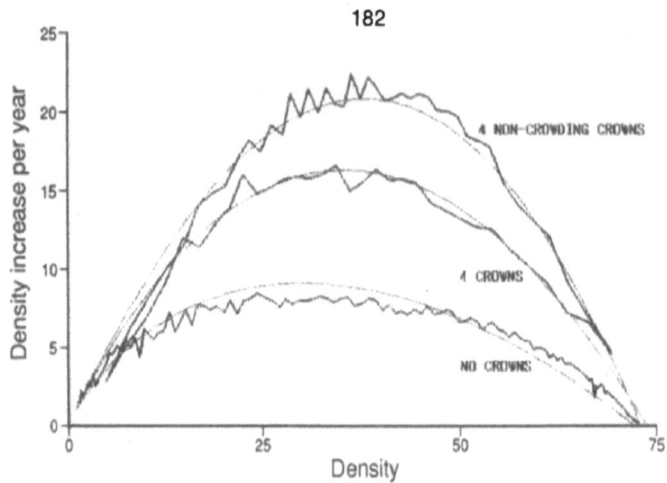

Figure 5. CORAL regrowth vs CORAL density under influence of CROWNs: Fastest regrowth for non-crowding CROWNs, second for crowding CROWNs and slowest in absence of CROWNs.

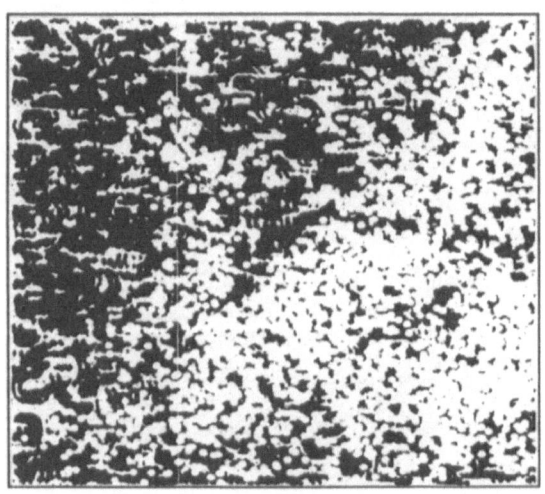

Figure 6. Impact of one pass of 100 CROWNs in reef of 7.5*7.5 m, spatial representation.

4.3 From Micro- to Meso-scale dynamics of CROWNs

In the small area studied so far we demonstrated the impact of crowding of CROWNs on the depletion of CORAL: the CROWNs eat the same amount but CORAL regrowth is slower. We could, however, not study the formation of large crowds as only few CROWNs can live permanently in the area. We therefore used the results (as gathered by RECORDERs and REPORTERs) of the MICro universe to define a MACro model of CROWNs and CORAL.

The PATCHes of this MACro universe represent the SPACE of the MICro universe, i.e. represent an area of 7.5 x 7.5 meters, the SPACE is a ribbon of 25 PATCHes long and 2 PATCHes wide which represent a section of a reef, 15 meters wide. Accordingly, the far edges are joined to represent continuing reef but, at the other sides, the SPACE is bounded and the DWELLERs (macroCROWNs) move back into the reef when they reach these edges. The following features of the CROWNs are used to define the macroCROWNs:

1.   Feeding rate is independent of CORALcover (until CORALcover < 1%)

2.   The distance travelled per week remains approximately constant from 70% to 40% CORALcover and then rapidly increases (see Figure 4c). The macroCROWN uses a set of actual data on distances moved by CROWNs to determine its rate of movement in the prevailing CORAL density, not some arbitrary approximation; this way the large (and density dependent) variance in the data is included.

3.   With respect to crowding two extremes were studied:

3a. Crowding is assumed to be permanent once two CROWNs meet (are closer then 1m from each other). This is an assumption which is not true for the MICro universe (see e.g. Figure 6), but represents an extreme to study maximal crowding.

3b. No explicit crowding mechanism. The longer residence, in areas with larger CORALcover, which occurs automatically by the density dependent movement of (2), may, however cause crowding. This is studied by giving the macroCROWN a bias with general direction (e.g. West) and without such a bias.

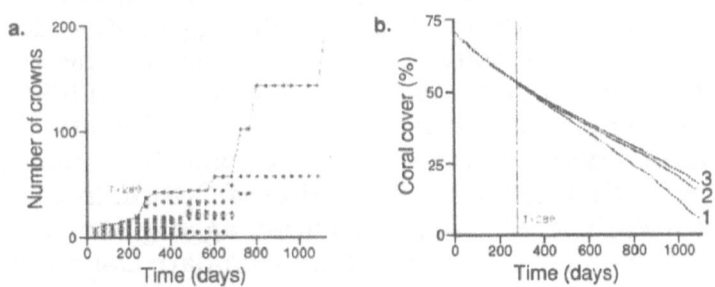

Figure 7. Effect of crowding in the macro reef.
7a. Size of aggregates vs time. The first sizable aggregates appear at t = 280. After exactly 3 years all CROWNs form one big aggregate.
7b. CORAL depletion in the macro reef. (1): explicitly crowding macro CROWNS (fig 7a); (2): non-crowding macroCROWNs with directional bias (fig 8); (3): non crowding macroCROWNs without directional bias and 4 crowding CROWNs in the micro reef.
Note that CORAL regrowth is of 4 crowding CROWNs in all cases until CORAL density

CORAL regrowth is modelled in terms of the approximations given in Figure 4: it follows the curve given for 4 crowding CROWNs until depletion to less than 10% is reached; if so, it switches to regrowth according to the depleted curve. This is a very crude approximation of the growth dynamics observed in the microreef but it should do for our purposes (see below).

Two hundred macroCROWNs of a standard size are initialised at random in this SPACE, i.e. twice the equilibrium population as found in the MICro universe. Small scale movements are of no importance in this larger scale universe. We therefore use as 'interesting events' movements of around 1.5m. Initially (i.e. at large CORALcovers) this step takes about a week, but later (at low CORALcovers) only a fraction of a day (see Figure 4c).

Figure 7a shows that the first sizable aggregates begin to appear at t=280 days. After that CORAL depletion is faster for crowding CROWNs than for a nonexplicit crowding mechanism (Figure 7b). Note that his effect of crowding is in addition to the effect of crowding at the micro scale described above: the PATCHes regrow according to small aggregates of 4 CROWNs at the micro scale, whatever the number of CROWNs in the PATCH, until CORALcover is locally almost exhausted. This extra effect is caused by the fact that CORALcover becomes clumped at the scale of the macro-PATCHes, and the regrowth rate is smaller at high and at low CORAL densities as compared to intermediate densities. Thus, the effect of crowding repeats itself at various scales in a fractal-like manner.

This additional effect of crowding even occurs when no explicit crowding mechanisms are introduced except for a directional bias. In that case sizable (and permanent) crowds appear after about two years due to slower rate of movement in areas of larger CORALcover (see Figure 8); after that CORAL depletion is faster (Figure 7a) than for non- crowding CROWNs without a directional bias and/or in the micro-reef with 4 CROWNs. Without a directional bias small aggregates do form at locations of higher CORAL density, but, after equalising the local higher cover to that of the surrounding areas, they disperse again (like in Figure 8 at high CORAL densities). Thus, in that case, density dependent crowding has overall a density smoothing effect, and is indistinguishable from the micro-reef with 4 CROWNs. Depletion of CORAL in certain areas by external processes (e.g. storms) will cause aggregation of CROWNs and thus enhance the process described here (compare Dana et al., 1972).

We conclude that the movement pattern of CROWNs allow for the formation of large crowds without any additional assumptions on initial distribution. Larger crowds form only after a certain amount of time; this could explain why outbreaking populations are always composed of fairly large COTs.They appear, in fact, just in time for an age-class of CROWNs of twice the 'equilibrium' population to deplete the reef before they are 'due to die'.

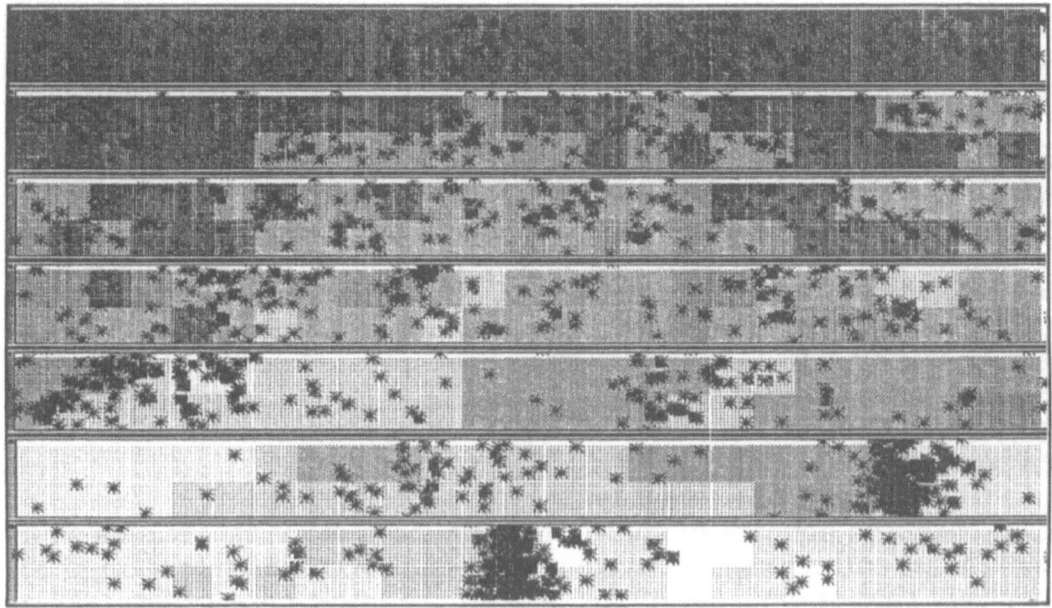

Figure 8. Crowd formation due to differences in CORAL density. CROWNs do not react directly to each other, but have a bias to move 'west'. Note that the CROWNs are drawn much too large (they should be less than 1 pixel). Top: t = 120 days; bottom: t = 1080 days (3 years); interval 160 days.

5. DISCUSSION AND CONCLUSIONS

The picture about the *Acanthaster* phenomenon which emerges from these experiments is as follows:

Acanthaster is prone to outbreak because of its extreme locomotory capabilities: whereas the area which it needs to maintain itself indefinitely is around 5 x 5m it can walk about 500 m a day! This fast movement causes survival of the adult stages to be virtually independent of coral cover. Moreover the consumption of coral does not diminish when coral is scarce, nor does it increase when coral is abundant (rather the opposite is true, see below).

It is well known also from (population level) ecological theory that if the amount of prey consumed is less sensitive to prey density than the growth rate of the prey, limit cycles ensue and such limit cycles, if 'large' enough, can lead to (local) extinction of predator or predator and prey (or only prey if the predator has an alternative food source), notwithstanding their mathematical global stability (Rosenzweig, 1971,1972). These models usually deal, however, with phenomena which take many generations of prey and predator. Our experiments show that the movement and feeding properties of *Acanthaster* are such that one year of a little above normal recruitment can exhaust the entire coral population in one generation and makes the environment unsuitable for the next generation, because the juvenile stages do not yet have such large movement capabilities.

Thus we conclude that the extreme oscillations of *Acanthaster* do not need other more hypothetical explanations than their locomotory capabilities combined with a normally abundant food source (like extra recruitment due to aggregation, or density dependent predation, see this volume). Rather, we need an explanation why it (or coral) has not become extinct (yet), if we consider outbreaks as a frequently recurring phenomenon. We also conclude that any nonspecific factor which causes *Acanthaster* populations to be larger now than before (e.g. by overfishing of their predators) should manifest itself in the form of outbreaks as have been observed.

Crown-of-thorns appears to share the large locomotory capabilities with other famous outbreaking species e.g. locusts. Like these, they seem to be either very rare or very abundant, and in the latter case to form large aggregates which quickly devastate large areas.

We have shown that the role of aggregation in such systems is twofold: 'active' aggregation diminishes regrowth of the food source (if it regrows locally, or if the habitat is locally deteriorated when barren) and (local) depletion of the food source will easily cause aggregation. The positive feedback of these two roles of aggregation seem to be sufficient to let relatively small population fluctuations cause a switch from relatively rare and cryptic to extremely over-abundant. Coincident with this rare/abundant, solitary/crowding switch there seems often to be a physiological and/or behavioural switch. This may be the case also for *Acanthaster*: they seem to feed also diurnally when abundant and only nocturnally when rare (Moran 1988); the propensity to approach each other also seems to be density dependent (Moore, personal communication). We left such switches purposely out of the model; the switches which do occur (see e.g. Figure 8) are 'emergent' properties. However the behavioural switches would only enhance the effects discussed in this paper.

Acknowledgements. We thank Roger Bradbury for urging us to study the *Acanthaster* phenomenon; it was a nice challenge to apply individual oriented modelling to such a large scale problem, and it was interesting to see that it could best be tackled by using an extremely small scale model. We thank Rupert Ormond for communicating useful 'eye ball' observations on real reefs as compared to our artificial reef. One important discrepancy remains: our CORALcover is fragmented at a too small scale in the 'virgin reef'; this does, however, not seem to be of influence to the reported results.

REFERENCES

Dana, T.F., Newman, W.A. & Fager, E.W. (1972) *Acanthaster* aggregations: interpreted as primarily responses to natural phenomena. Pac. Sci. 26, 355-372.

Endean, R. & Stablum, W. (1975) Population explosions of *Acanthaster planci* and associated destruction of the hard coral cover of reefs of the Great Barrier Reef, Australia. Environ. Conserv. 2, 247-256.

Hogeweg, P. (1980) Locally synchronised developmental systems, conceptual advantages of the discrete event formalism. Internat. J. of General Systems 6, 75-73.

Hogeweg, P.(1988a) Cellular automata as a paradigm for ecological modelling. Applied Math. & Computation 27, 81-100.

Hogeweg, P. (1988b) MIRROR beyond MIRROR, Puddles of Life. In: C.G. Langton (ed.) Artificial Life, SFI Studies in the Sciences of Complexity, Addison Wesley Publ Co.

Hogeweg P. & Hesper, B. (1979) Heterarchical self-structuring simulation systems: Concepts and applications in biology. *In:* Methodology of system modelling and simulation (eds. B.P. Zeigler, M.S. Elzas, G.J. Klir, T.I. Oren) North Holland, pp. 221-232.

Hogeweg, P. & Hesper, B. (1981) Two predators and one prey in a patchy environment: An application of MICMAC modelling. J. Theor. Biol. 93, 411-432.

Hogeweg, P. & Hesper, B. (1985) Socioinformatic processes, MIRROR modelling methodology, J. Theor. Biol. 113, 311-330.

Hogeweg, P. & Hesper, B. (1986) Knowledge seeking in variable structure models. In: Simulation in the Artificial Intelligence Era, (Eds M.S. Elzas. T.I. Oren and B.P. Zeigler) pp. 227-243, North Holland.

Hogeweg, P. & Hesper, B. (1988a) Structure oriented-non goal directed modelling.In B.P. Zeigler, TI Oren and M. Elzas. (eds) Modelling and simulation methodology: Knowledge System Paradigms. North Holland Publ. Co.

Hogeweg, P. & Hesper, B. (1988b) Simulation modelling formalisms: Heterarchical. In: Encyclopaedia of systems and control, Pergamon Press pp 4350-4353.

Hogeweg P., Hesper, B. , van Schaik, C.P. & Beeftink, W.G. (1985) The population structure of vegetation, In: The population structure of vegetation. (ed. J. White) pp. 637-666 Dr. W. Junk Publ.

Hogeweg P. & Richter, A.F. (1982) INSTAR, a discrete event model for simulating zooplankton population dynamics. Hydrobiologia 95, 275-285.

Huston, M., DeAngelis D., & Post, W. From individuals to ecosystems: A new approach to ecological theory. BioScience. (in press).

James, M.K., Dight, I.J. & Bode, L. (this volume) Great Barrier Reef hydrodynamics, reef connectivity and *Acanthaster* population dynamics. In: R.H. Bradbury (ed.) The *Acanthaster* phenomenon: a modelling approach. Springer-Verlag, Berlin.

Moran. P.J. (1988) The *Acanthaster* phenomenon. Australian Institute of Marine Science Monograph series, Vol 7, Townsville.

Rosenzweig, M.L. (1971) The paradox of enrichment. Science 171, 385-387.

Rosenzweig, M.L. (1972) Stability of enriched aquatic ecosystems. Science 175, 564-565.

Toffoli, T. & Margolus, N. (1987) Cellular automata machines, a new environment for modeling. MIT press, 259 pp.

TEST OF A MODEL OF REGULATION
OF CROWN-OF-THORNS STARFISH
BY FISH PREDATORS

RUPERT ORMOND[1], ROGER BRADBURY[2], SCOTT BAINBRIDGE[2],
KATARINA FABRICIUS[2&3], JOHN KEESING[2&4], LYNDON DE VANTIER[2&5],
PAUL MEDLAY[1&6], AND ANDREW STEVEN[4]

[1] Tropical Marine Research Unit, Biology Department,
University of York, York YO1 5DD
[2] Australian Institute of Marine Science, Townsville, Queensland 4810
[3] Zoologisches Institut der Universität München, D-8000 Munich 2
[4] Zoology Department, James Cook University of North Queensland,
Townsville, Queensland 4811
[5] Zoology Department, University of Queensland, St. Lucia, Queensland 4067
[6] Marine Resources Assessment Group, Imperial College,
University of London, London

Abstract. Both a population dynamic model and simplified calculations of the numbers of fish predators that would be required to control heavy recruitment of the coral-eating crown-of-thorns starfish (*Acanthaster planci* L.) predict threshold densities of fish predators below which starfish population outbreaks may occur. This prediction has been tested through comparable surveys of putative fish predators in the Red Sea, where no major outbreaks of *A. planci* were found or are known to have occurred, and on the Great Barrier Reef, where two major series of outbreaks have occurred during the last 25 years. In the Red Sea densities of presumed fish predators were found to be well above the predicted threshold. By contrast on the Great Barrier Reef mean densities both of proposed fish predators of juvenile and subadult *A. planci,* and of know fish predators of adult *A. planci,* were below the predicted threshold. Moreover the densities of fish predators on four mid-shelf reefs that are believed to have escaped major outbreaks of *A. planci* were found to be significantly higher than those on otherwise similar but impacted reefs. These data are compatible with the hypothesis that current and recent outbreaks of the crown-of-thorns starfish on the Great Barrier Reef have been facilitated by the presence of only relatively low numbers of fish predators, the numbers of which may have been decreased as a result of the intensification of fishing pressure that has occurred from the 1960s onwards.

INTRODUCTION

It has frequently been argued (Chesher, 1969; Endean, 1977, 1982; Cameron & Endean, 1982; Moran *et al.*, 1986) that the series of population outbreaks of crown-of-thorns starfish (*Acanthaster planci*) (CoTs) that have occurred on the Great Barrier Reef (GBR) and elsewhere in the West Pacific in the last 25 years are effectively a new phenomenon, initiated either by unusual climatic or oceanographic events (Dana *et al*, 1972; Pearson, 1975), or else by anthropogenic factors such as nutrient or pesticide pollution (Birkeland, 1982; Randall,1972), or by the removal of predators (Endean, 1969, 1976). On the other hand, in recent years the predominant scientific view has tended to one that CoT population outbreaks are most likely a natural phenomenon that probably occurred periodically throughout the past (e.g. Dana, 1970; Vine, 1970; Frankel, 1977; Sale, 1980). This view, while given some support by the finding of CoT spines at depth in reef sediments (Frankel 1977; Walbran *et al.*, 1989), has largely been given credit by the lack of clear evidence supporting any of the mechanisms which could account for an increase in frequency and extent of outbreaks (see Moran, 1986; Potts, 1981; Olson, 1987; McCloskey & Deubert, 1972).

Here, however, we both report on a simple population dynamic model that indicates that fish predators could in theory help regulate CoT population levels, and we provide evidence compatible with the hypothesis that some fish species may indeed help regulate CoT populations in this way.

FISH PREDATORS OF THE CROWN-OF-THORNS

The first animal to be clearly established as an active predator of adult CoT was the Giant Triton Shellfish (*Charonia tritonis*), and it was initially proposed that collection of this species by amateur and commercial shell-collectors might have reduced the population level of the shellfish and hence triggered CoT population outbreaks (Endean, 1969). This view did not however gain wide acceptance, both because the pre-fishing population levels of the shellfish were thought to have been relatively low, and also because it seemed unlikely that such an invertebrate could provide the density -dependent mortality necessary to regulate the population numbers of its prey.

However several fish predators of adult CoT have also been described (Endean, 1977, 1982; Ormond *et al.*, 1973; Ormond & Campbell, 1974; Wilson *et al.*, 1974; Owens, 1971). In particular one of us (RFGO) led work from which it appeared that an aggregation of CoT on reefs in the Sudanese Red Sea was eventually dispersed as a result of predation on large adult CoT by the large triggerfishes (Balistidae) *Balistoides viridescens* and *Pseudobalistes flavimarginatus*, and the large pufferfishes (fam. Tetraodontidae) *Arothron hispidus* and *A. stellatus* (Ormond *et al.*, 1973; Ormond & Campbell, 1974)..

In addition, feeding experiments carried out in the Red Sea (Ormond & Sanders, unpubl. data) showed that several abundant species of Emperor Bream (fam. Lethrinidae) feed readily on small echinoids and asteroids, suggesting that these fish might be potentially significant predators of juvenile or subadult CoT. Gut content analyses carried out by Toor (1964), Walker (1978), Birdsey (1988) and ourselves have confirmed that lethrinids, including several common on the GBR (*Lethrinus chrysostomus*, *L. mahsena*, and *L. nebulosus*), are apparently opportunistic predators of medium-sized (eg. 2.5 - 15 cm. diam.) invertebrates, principally crustaceans, molluscs and also echinoderms.

Our study (undertaken during the course of survey work described below) also showed that individual lethrinid guts commonly contain up to 3 or 4 invertebrates (mostly crustaceans and echinoids) in the size range 2.5 - 15 cm. diam. (similar in biomass to a 1 - 2 year old CoT), presumably captured during a single 24 hour period. Prey items included asteroids of the genus *Astropecten* of up to 12 cm. diameter and the echinoid *Echinothrix diadema*, both with spines comparable in size to those of similarly sized CoT. Also, although no CoT were found within the relatively small number of lethrinids sampled, enquiries made during the course of our study provided three recent observations of lethrinids feeding on subadult CoT. J. Paterson (*pers. com.*) offered approx. 30 1 - 2 yr. old CoT to fish feeding on items thrown from a recreational pontoon on John Brewer reef, and observed that most were taken by *L. nebulosus*; R. Bell (*pers. com.*) observed a lethrinid take a small CoT that had just been exposed by a diver; and Birdsey (1988) found the remains of a single CoT (est. diam. 40 cm.) in the stomach of a *L. chrysostomus* collected as part of a recent study for the Great Barrier Reef Marine Park Authority (GBRMPA).

MODELLING THE CORAL-STARFISH-PREDATOR FOOD-CHAIN

To check what numbers of fish predators and rates of predation would be required to limit population levels of CoT to those characteristic of most reefs in the Red Sea and Western Indian Ocean (5 -20 per km. of reef front), a computer-based predator-prey model was developed (Ormond & Medlay, unpubl. data) based on the standard logistic population equations introduced by Lotka (1925) and Volterra (1926) (see May, 1981). The model was based on the equations shown in Table I. Values of the different variables used in the model were either obtained or estimated from the literature, or, where no real values were available, a range of feasible values were tested.

Table I. Formulae used In development of computer model

$$\frac{dN_C}{dt} = r_C(\frac{2N_C}{\pi} + r_C)\,(1 - \frac{N_C}{K_C}) - \frac{a_A\,N_A\,N_C}{D_A + N_C}$$

$$\frac{dN_A}{dt} = r_A N_A\,(1 - \frac{J_A N_A}{N_C}) - a_f N_f\,\frac{V_A\,N_A^2}{D_f + V_A N_A^2}\,E_f$$

$$E_f = \frac{SN_A^2}{SN_A^2 + N_0^2}$$

$$V_A = \frac{N_A^2}{N_A^2 + R^2}$$

$$\frac{dN_f}{dt} = r_f N_f\,(1 - \frac{(J_A + J_0)}{N_A + N_0}\,N_f)$$

Examples of the type of formulae used to model each stage of the coral-CoT-fish predator food chain within the computer model referred to in the text (after Ormond & Medlay, unpublished data). Significance of variables and subscripts is as follows:

variables

N	=	population density (for corals is proportion of substrate covered by coral)
r	=	intrinsic rate of population increase
K	=	carrying capacity
a	=	attack (ie. maximum predation) rate
J	=	equilibrium constant related to intake required to maintain individual in equilibrium
D	=	prey density at which predator feeding rate is half saturated
S	=	switching coefficient of predator between prey (equals relative preference of fish for CoT in relation to other prey)
V	=	vulnerability of CoT to predation
R	=	density of refuges giving CoT protection from fish

subscripts

c	=	coral
A	=	*Acanthaster planci* (CoT)
f	=	fish predators
o	=	other invertebrate prey of fish predators

Runs of the model produced results of the type that might be expected given that the model is essentially a specific case of the more general predator-prey models whose behaviour has been well explored (Hassell & May, 1973; Beddington *et al.*, 1975; May, 1975). In the absence of any predators of CoT a cyclic interaction between CoT and corals (limit cycles) can readily be established. CoT numbers increase either rapidly or more slowly, depending on how oceanographic conditions are presumed to affect the intrinsic rate of increase of the population. The amount of surviving coral declines to the point where in the absence of sufficient food the CoT population itself collapses. Then coral cover slowly recovers to the point where another CoT population outbreak can be sustained.

If numbers of predators of CoT are increased they can first slow and then prevent any large increase in CoT population level. Efficient predators will drive the CoT population to extinction unless refuges in which CoT can avoid predation are introduced into the model. Then, depending on the values selected for critical variables predation of CoT may either tend to return the CoT population density to a relatively low population level, or may generate low-level limit cycles in the numbers of CoT and CoT predators. Thus, as described by Southwood and Comins, 1977) a CoT like species may have two relatively stable population levels, a lower one where its abundance is limited by predation, and an upper one where, when predator pressure is reduced, it becomes limited by its own food supply.

The model confirmed that the observed densities and rates of predation for the large balistids and *Arothron* spp. in the Red Sea would be sufficient to reduce a CoT aggregation of the size observed in Sudan (approximately 2000 adult animals) to the density typical of most Red Sea reef areas (5 - 20 CoT per km. reef face). But the model also indicated that higher predator densities or predation rates would be necessary to control heavy CoT recruitment such as is observed during CoT outbreaks on the GBR. Such heavy CoT recruitment might however be controlled if fish such as lethrinids were capable of feeding on juvenile or subadult CoT and were present in high enough numbers.

The conclusions of the model can be confirmed by simplified calculations of the threshold density of fish predators that will be required to control CoT populations. In view of the importance and controversial nature of this topic these calculations are presented here in order to allow a reader without access to the computer model to assess the probable validity of the conclusions.

The maximum density of CoT recruits that lethrinid type predators can control is determined both by the abundance of these fish, and by the maximum number of prey that individual lethrinids can take per unit time. If it is assumed, as suggested by our examination of lethrinid gut contents, that these fish are capable, in the presence of very high densities of prey, of consuming up to 4 juvenile or sub-adult CoT per day, then both the computer model and the simplified calculations (see Table II) indicate that very heavy recruitment of juvenile CoT (up to 10,000 - 100,000 recruits per 100 m. of reef front) could be controlled by densities of 5 - 20 lethrinids per 100 m. of reef front, ie. most recruits would be consumed by the predators. However, below a density of about 5 lethrinids per 100 m. of reef front, significantly fewer CoT would be consumed than are present; thus over successive breeding cycles CoT numbers could continue to increase and escape predator control.

Table II

	Density of lethrinids per 100m. of reef front				
	1	5	10	20	100
Density of juvenile/ subadult *A. planci* per 100 x 20 m reef	Estimated total number of juvenile/subadult *A.planci* taken by lethrinids				
1	.03	.15	.3	.6	3
10	.3	1.5	3	6	30
100	6	30	60	120	600
1000	100	500	1000	2000	10000
10000	1000	5000	10000	20000	100000
100000	1000	5000	10000	20000	100000

The body of the table shows how the total number of juvenile / subadult crown-of-thorns starfish (CoT) that could be consumed by Emperor bream (Lethrinidae) would vary with both the density of lethrinids and with the density of crown-of-thorns, assuming that the numbers consumed by an individual lethrinid with increasing density of prey (the functional response) would be as shown in the second column (headed 1 lethrinid per 100 m.), and assuming a sustained supply of CoT. Where the figure in the body of the column is equal to or greater than the corresponding density of CoT (to the right of the dashed line) then predation by lethrinids would reduce or control the CoT population. Thus a threshold density of 5 - 10 lethrinids is required to control recruitment of 1000 - 10000 CoT per 100 m. reef front. Note that very high densities of CoT recruitment (eg. 100000 per 100 m.) will swamp even moderately high densities of lethrinids (ie. 20 per 100 m.). Also when juvenile /subadult CoT densities are low (ie. < 10 -100 per 100 m.) even moderately high densities of lethrinid (eg. 20 per 100 m. reef front) do not take all available CoT, ie. very small numbers of CoT escape to grow to maturity, because when small CoT are scarce the fish tend to ignore or not to notice them. As a consequence predation by larger fish (ie. large balistids etc) on adults will also have a role in population regulation.

Both the model and the calculations assume that these fish predators show a type III functional response, ie. not feeding significantly on low densities of a prey, but only 'switching' to feed on them as prey density increases (see Holling, 1965; Murdoch, 1969; Allen, 1972; Greenwood & Elton, 1979). Such feeding behaviour has been found to be characteristic of many significant vertebrate predators, and is widely agreed to be particularly capable of generating density dependent predation leading to prey population control (Holling, 1965; Murdoch, 1969; May, 1981). To date only a few fresh-water species appear to have been studied for evidence of 'switching' and type III feeding behaviour among fish (Elton &

Greenwood, 1970; Werner & Hall, 1974; Maskell *et al.*, 1977). It is probable however that most larger predatory marine fish will also be found to show such behaviour, a suggestion supported by field tests in progress involving these putative CoT predators (Ormond *et al.* in prep.). Switching behaviour of a predator in respect of juvenile CoT is also made more likely by the extreme difficulty of finding juvenile CoT (<10 cm. diam.), except where they are present in high numbers, reported by workers who have attempted to locate them (Moran *et al.*, 1985; Doherty & Davidson, 1989).

A further significant consequence shown by the calculations in Table II, of presuming a type III functional response by the predators, is that if the fish predator population level is below the threshold value of about 5 fish per 100 m. of reef front, then the CoT population will escape predator control even at quite low recruitment densities (1 - 100 juveniles per 100 m. reef front). This is because when fewer CoT are present the fish are expected to take correspondingly fewer of them. This conclusion is robust across a broad range of the exact forms of the functional response.

Calculations comparable to those of Table II can also be made for the larger fish species (large balistids and *Arothron* spp.) thought to be potential predators of adult CoT. These assume, based on feeding experiments carried out in the Red Sea (Ormond *et al.*, 1973; Ormond & Sanders, unpublished data) that these fish are able to consume up to two adult CoT per week. These calculations predict a threshold density of between 1 and 2 large predators per 100 m. reef front.

SURVEYS OF FISH PREDATOR ABUNDANCES

The predictions of the model of threshold levels of fish predators were tested by comparing densities of lethrinids and of larger fish predators of adult CoT between the Red Sea, where no major outbreaks of CoT were found or are known to have occurred, and the Great Barrier Reef, where two cycles of large scale CoT outbreaks have occurred during the last 25 years. The opportunity to determine densities of fish predators on reefs along the length of the Red Sea arose during a marine habitats and resources survey undertaken during 1982-85 (Ormond *et al.*, 1984,1985). On the GBR densities of fish predators were estimated in July 1988 by survey work carried out on 10 different reefs within the central and northern sectors.

The GBR reefs were selected as follows. Firstly the survey was limited to mid-shelf reefs, as opposed to edge-of-shelf or near-shore reefs, since most individual reefs that have escaped significant outbreaks are of the latter two types, and these differ in other ecological or oceanographic respects from mid-shelf reefs. Secondly four reefs (Marx, Michaelmas, Noggin and Wardle) were selected from the relatively small number of mid-shelf reefs between the latitudes of Lizard Island and Townsville that had not been reported or found during recent surveys (Australian Institute of Marine Science, 1986-7) to have experienced a significant CoT outbreak. Four further reefs (Flora, Hastings, Potter and Startle) were then selected as 'control reefs' that had experienced CoT outbreaks, one each close to and apparently similar in general character to each of the first four reefs. The two final reefs (Green Island and John Brewer Reef) were among those best-known to have been seriously impacted during both series of CoT outbreaks.

The estimates of fish predator abundance were carried out using visual census techniques (Chave & Eckert, 1974; Brock, 1982). Divers usually in pairs swam parallel with and a few metres away from the reef edge. SCUBA gear was used on all GBR transects and some Red Sea transects, in which case divers maintained a depth of 3 - 5 m. In either case fish predators were counted both within an approximately 20 m. wide x 100 m. long band transect centred on the diver and running adjacent to the reef crest. In the Red Sea a total of 320 separate 100 m. long transects were completed distributed along the eastern coast from near Haql, not far from Aqaba (in the north) to near Oreste point, at the border between Saudi Arabia and North Yemen (in the south). On the GBR 161 transects were completed with (normally) 8 contiguous transects at each of the northern and southern ends of each reef.

The results of the surveys are summarised in Table III. In the Red Sea 9 species of lethrinid were recorded with an overall combined mean density of 27.0 per 100 m. reef front. The data were also considered in relation to 14 sectors into which the region had previously been subdivided. The three most heavily fished sectors around the three major fishing ports of Jeddah, Haql and Gizan had a mean sector density of 7.4 per 100 m. reef front. The remaining sectors, largely covering as yet lightly fished areas, had (even when one sector with an anomolously high value was excluded) a mean sector density of 21.4 per 100 m. reef front. Thus all sectors were found to have mean lethrinid densities above the predicted threshold, compatible with the absence of CoT outbreaks; but the sectors including the three largest fishing ports were found to have lethrinid densities much lower than the others, and only a little above predicted threshold value.

On the GBR 9 species of lethrinid were also recorded. The mean reef density (i.e. the mean of the mean density calculated for each reef) for all lethrinids on the 4 reefs not previously recorded to have suffered a major outbreak was 9.25 per 100 m. reef front, but both for the 4 paired reefs that had suffered major outbreaks, and also for all 6 impacted reefs taken together, was 2.8 per 100 m. reef front (both differences significant at $p < 0.05$). All 6 impacted reefs had a mean lethrinid density of less than the predicted threshold of 5 per 100 m. reef front (range 1.6 - 4.0 per 100 m.)

Table III. Mean values of densities of fish (per 100 m of reef front) recorded on each reef (GBR) or in each sector (Red Sea)

	GBR major CoT impact reefs	GBR minor CoT impact reefs	GBR all reefs	Red Sea eastern seaboard
Number of reefs (GBR) or Sectors (Red Sea)	6	4	10	14
Total number of transects censused	96	49	161	320
Lethrinidae (all species)	2.82 (0.46)	9.25 (1.91)	5.39 (1.29)	27.01 (10.95)
Balistidae (Large spp.)	0.08 (0.01)	0.19 (0.04)	0.13 (0.03)	1.82 (0.38)
Arothron (2 species)	0.04 (0.02)	0.13 (0.01)	0.08 (0.02)	0.26 (0.07)
Cheilinus undulatus	0.05 (0.02)	0.32 (0.05)	0.16 (0.05)	0.35 (0.05)

The table shows for each fish group the mean values of their mean densities on mid-shelf reefs of the Great Barrier Reef (GBR) and within sectors of the reef-dense eastern seaboard of the Red Sea (Saudi Arabia). Mean densities for each reef on the GBR or for each sector in the Red Sea were based on counts along many (typically 15-30) 100 m. x approx. 20 m. band transects running parallel to the reef face. Mean densities are also given separately for 6 GBR reefs that according to previous survey data had experienced major outbreaks of crown-of-thorns starfish (CoT) (Brewer, Potter, Flora, Green Is. Hastings and Startle) and for four GBR reefs that had not (Wardle, Noggin, Michaelmas and Marx). The figures in brackets are standard errors.

while all 4 relatively unimpacted reefs had a mean lethrinid density of 5 per 100 m. reef front or greater (range 5.0 - 11.2). Thus the total mean reef density of lethrinids on the GBR (5.39 per 100 m.) was less than a fifth as great as their total mean (sector) density in the Red Sea (27.01 per 100 m.).

The mean transect density (i.e the mean count over all transects) of lethrinids was 4.6 per 100 m. reef front. The mean transect density of lethrinids on the 6 reefs known to have experienced major CoT outbreaks was 2.64 whereas the mean transect density of lethrinids on the other 4 reefs was 7.85 (t = 7.03, p < 0.0001; after log+1 transformation). It was also found that lethrinid density appears to be inversely related to the intensity of recent CoT outbreak as estimated independently during the present survey from the proportion of old dead coral showing features diagnostic of CoT attack (see Figure 1).

The mean densities of large triggerfish, large pufferfish (*Arothron spp.*) and *Cheilinus undulatus* are also shown in Table III. In the Red Sea the mean density of these species taken together (2.1 predators per 100 m. reef front) was found to be above the threshold predicted by the calculations described above. By contrast their mean reef density on the GBR (0.36 predators per 100 m. reef front) was (like that for lethrinids) both below the predicted threshold, and also below the mean sector density for these species recorded in the Red Sea. Further the mean overall reef density of these predators on non-outbreak GBR reefs (0.64 per 100 m.) was found to be nearly four times greater than that recorded on outbreak reefs (0.17 per 100 m.). Thus the data indicate that the densities of putative fish predators both of subadult and of adult CoT are lower on the GBR than in the Red Sea, and within the GBR lower on outbreak reefs than on non-outbreak reefs.

CAUSES OF DIFFERING PREDATOR DENSITIES

Several possible explanations suggest themselves for the lower predator abundances on the GBR than in the Red Sea. Most critically they could be the result of greater fishing pressure in the GBR region having reduced stock abundance, especially of lethrinids, which are a significant commercial species. This interpretation is compatible with the available fisheries data which were analysed by one of us (Steven, 1988) on behalf of GBRMPA during the period of this study. The data (collected by the Queensland Fisheries Board and Queensland Fisheries Department) are not especially reliable since, for example, they omit significant recreational and black-market catches; nevertheless, they are still likely to reflect true catch levels and stock trends.

The total catch of reef fish in the central and northern sectors increased approximately tenfold during the early- to mid-1960s to a plateau sustained during the 1970s and 1980s. Separate records were maintained for only one individual species of lethrinid, the Redthroat, *L. chrysostomus*. Analysis of this data shows that the catch of this species expressed as a percentage of total reef fish catch appears to have decreased rapidly during the 1960s from a high proportion of total catch to a low one, then to have oscillated about a lower level. Given that increased exploitation of fish generally leads to lower stock abundance (see e.g. Ricker, 1975; Silliman & Gutsell, 1958) the above data are compatible with a reduction in total lethrinid abundance, including in particular the early targeting and overexploitation of *L. chrysostomus*, an early preferred target of reef anglers and spearfishermen.

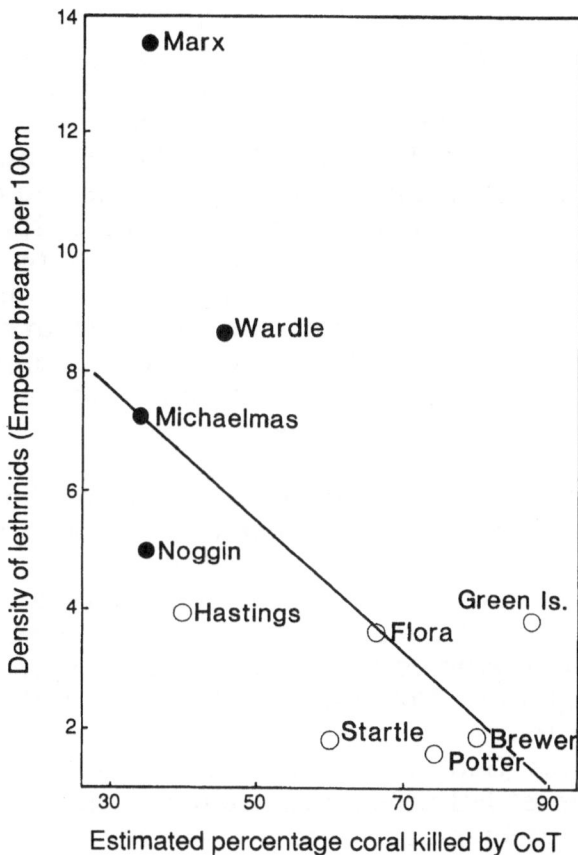

Figure 1. The relationship between mean density of lethrinids on different reefs and the estimated extent of recent damage due to crown-of-thorns starfish. Estimated damage due to CoT was estimated separately 'by eye' for each 100 m. transect, as the proportion of dead and live coral combined that appeared to have died in the recent past (several years) due to attack by CoT as indicated by characteristic features such as old feeding scars etc. The points on the graph are the mean values for the named reefs, the open circles indicating reefs recorded as having experienced a major CoT outbreak since 1980. The line is that obtained from a regression of the separate data from all 161 transects, yielding an equation: *percentage coral loss = 100 - 9.1 lethrinid density* (r = 0.44; p < 0.001).

While it is possible that a decline in lethrinid stocks could have been caused by line-fishing, such a decline could also have been brought about by recruitment failure due to the bycatch of juvenile lethrinids taken by prawn trawlers. Unfortunately studies to examine the impact of prawn trawlers on bycatch species (Jones & Goeden, 1984) have failed to investigate near-reef habitats where trawling for Red-spot king-prawn takes place, but communication with trawler fishermen has revealed that sometimes juvenile emperors can constitute a significant portion of the bycatch (LDV *pers. obs.*).

Contrasting degrees of fisheries exploitation are also a feasible explanation for the lower abundance on the GBR of the larger fish predators of adult CoT. *C. undulatus* is highly prized by line- and spear-fishermen everywhere. Large balistids are not commercial species but are caught unintentionally by line fishermen both on the GBR and in the Red Sea. Grant (1982) describes them on the GBR as 'notorious bait stealers...much disliked by anglers', an observation that not only confirms that these fish are (or were) frequently captured by anglers and line-fishermen, but one that also supports the view that balistids are highly opportunistic foragers such as are especially likely to show type III feeding behaviour. *Arothron* species have become popular tourist souvenirs in many reef areas and are also, as slow swimmers, especially vulnerable to incidental capture by trawling.

There are however alternative possible explanations for the lower abundances of fish predators on CoT impacted reefs or in the GBR by comparison with the Red Sea, and these should be considered. One important alternative is that the lower predator numbers on impacted reefs are a consequence of CoT damage to coral communities, rather than a cause. However several studies have looked for changes in reef fish communities resulting from CoT damage to reefs, but found that differences appear not to be very large except in the case of one or two coral dependent families (Sano *et al.*, 1984; Williams, 1986). It is also relevant to note that most lethrinids and *Arothron spp.*, and also some large balistids, while also foraging over coral dominated slopes, are more characteristic of sandy sea-bed and lagoonal areas. Thus one might anticipate they they would be among the fish families least affected by the fall in coral cover resulting from a CoT outbreak.

The differences in fish numbers between the Red Sea and GBR could also reflect more fundamental differences in community structure or ecology. For zoogeographical reasons the GBR region has a greater species and family diversity than does the Red Sea (Cohen, 1973; Briggs, 1974). It could therefore be that on the GBR there are other fish predators of invertebrates that are not present in the Red Sea. These, by competing for food with lethrinids, large balistids and *Arothron* spp. could reduce their numbers. Alternatively it could be that between the Red Sea and GBR there are major quantitative differences in energy flow through different trophic compartments, such as have been proposed to account for large cross-shelf variation in the abundance of different fish guilds on the GBR (Williams & Hatcher, 1983; Russ, 1984). However neither of these explanations can account for the differences in predator densities observed between impacted and unimpacted GBR reefs, all of which were selected as otherwise comparable mid-shelf reefs.

Thus while none of the above explanations for the observed differences in abundance of CoT fish predators can yet be fully discounted, the simplest explanation for the exact pattern of differences observed is that lower fish abundances are due to greater fishing intensity on the GBR than in the Red Sea, the result of an earlier mechanisation of the fishing fleet and a greater intensification of both recreational and commercial fisheries.

IMPLICATIONS OF THE MODEL

The above models, calculations and data have related to possible roles both for fish predators of juvenile and subadult CoT (e.g. lethrinids), and for fish predators of adult CoT (e.g. the large balistids and *Arothron* species). It is possible that both groups of predators may contribute to CoT population regulation although each would play a somewhat different role. Lethrinid-like predators are anticipated to feed on only juvenile and sub-adult CoT,; but since they have a much higher overall density than do the fish predators of adult CoT, they may, when present in high numbers, bring about the control of high densities of juvenile and sub-adult CoT, such as might result from dense settlement of swarms of CoT larvae. Not only are large balistids and *Arothron* not abundant enough to control very heavy recruitment of CoT, but it seems likely, given current understanding of optimum foraging theory (see Werner & Hall, 1974; Krebs, 1978), that it would not be so profitable for such large predators to invest time in searching for smaller prey.

By contrast the principal significance of the large balistids and *Arothron* spp. may arise as follows: if lethrinids do show type III foraging behaviour, then, as can be seen from Table II, small numbers of juvenile CoT will usually avoid predation and grow to a size too large for lethrinids to handle. This is expected to happen even with low densities of juvenile CoT, since then type III foragers are not expected to switch to feeding on them. Thus it is anticipated that a continual trickle of CoT will continue to grow to maturity, and it is in the control of their numbers that the larger predators are anticipated to be significant.

The calculations predict that different types of CoT population increase would result from differential depletion either of lethrinid-like or of large balistid-like predators. In the absence of lethrinid-like predators there would not be effective control of any heavy settlement of CoT larvae; this would be expected to result in the relatively sudden appearance, two or so years after settlement, of a large single age-group population of CoT. By contrast, loss of predators of adult CoT, in the presence of adequate numbers of predators of the juveniles, would result in the small numbers of CoT that escape predation as juveniles surviving as adults. Consequently, provided there is significant settlement of CoT larvae in successive years, a slow accumulation of a multi-year group population of large CoT would be expected. Such contrasting types of outbreak have indeed been noted (see Moran, 1986).

It is important to emphasise that a number of observations that have in the past been used to reject any role for predators in regulating CoT populations are not only compatible with the present model but are in fact predicted by it.

Firstly, our model presumes that critical fish predators show type III predation, feeding opportunistically on a wide variety of mainly invertebrate prey, and only switching to feed heavily on any one prey type, such as CoT, when that prey population increases. Thus the frequency of predation of CoT by fish among typical low-level CoT populations will be low, and occurrence of CoT remnants among gut contents correspondingly rare. Thus the absence of CoT among gut contents of suspected predators does **not** provide evidence that these predators do not play a critical role in CoT population regulation.

Secondly, the model accepts that predator absence or removal (for example on an experimental basis) will not necessarily or even usually be associated with CoT outbreaks unless, as implied by hydrographic studies (Dight *et al.*, 1989; Black, 1989), water conditions are appropriate, and hydrographic connectivity results in currents from a reef harbouring a significant population of breeding CoT bringing sufficient CoT larvae to the area concerned.

Thirdly, the presence of above threshold numbers of predators on individual reefs will not always prevent outbreaks of CoT on those reefs. Table II indicates that exceptionally high levels of recruitment can still result in more juvenile and subadult CoT than even high densities of lethrinids can consume. Thus once outbreaks become endemic in a region, the extent of larval recruitment may easily be so high that it will swamp predation by normal populations of CoT predators.

Fourthly, in the same way, the model allows that even in the presence of the higher numbers of fish predators characteristic of unfished reef areas, CoT larval recruitment may, as a result of favourable hydrographic conditions or other probabilistic events, be sufficiently heavy to swamp normal predation and give rise to a local CoT population outbreak. Thus demonstration that CoT outbreaks have occurred in earlier historical or geological periods (see Frankel, 1977; Moran *et al.*, 1986; Walbran *et al.*, 1989), prior to significant impact by humans, does not discount the possibility that fish predation of CoT may help regulate CoT outbreaks. On the contrary, the present model suggests that even with pre-impact numbers of fish predators occasional local CoT outbreaks would be expected.

This series of counter-intuitive predictions may excuse previous confusion in this field.

The above evidence provides strong support for a critical role for predators of adults and sub-adults in influencing at least some CoT populations. However we wish to emphasise that predation can only be one of a series of factors which interact to determine observed CoT population levels. Unfortunately the debate regarding the genesis of recent CoT outbreaks in the GBR and Western Pacific has been hindered by advocacy of single factor hypotheses. To contrast our model with oversimplistic 'larval recruitment' or 'predator removal' hypotheses we wish to propose the term 'Recruitment Initiated Predation' (RIP) to describe our more sophisticated concept. The term is intended to emphasise that our model anticipates that significant predation of juvenile and sub-adult CoT will only occur following heavy local recruitment of CoT larvae. Settlement of CoT larvae will in turn depend on a series of semi-stochastic oceanographic and other factors of the type that are more often emphasised. The RIP acronym is also intended as a reminder of the fact that unless a more open-minded attitude to understanding the CoT phenomenon can prevail, continuing degradation of the GBR and other reef areas may lead to their irreversible decline.

REFERENCES

Allen JA (1972) Evidence for stabilising and apostatic selection in wild blackbirds. Nature 237:348-349

Australian Institute of Marine Science (1986-7) Crown-of-thorns study, 1985: an Assessment of the Distribution and Effects of Acanthaster planci (L.) on the Great Barrier Reef. 13 vols. Australian Institute of Marine Science, Townsville, Queensland

Beddington JR, Free CA, Lawton JH (1976) Concepts of stability and resilience in predator-prey models. J Anim Ecol 45:791-816

Birdsey R (1988) Large Reef Fishes as Potential Predators of Acanthaster planci. Report to Great Barrier Reef Marine Park Authority

Birkeland C (1982) Terrestrial runoff as a cause of outbreaks of Acanthaster planci (Echinodermata: Asteroidea). Mar Biol 69:175-185

Black KP (1989) The relationship of reef hydrodynamics to variations in number of planktonic larvae on and around coral reefs. Proc 6th Intnl Coral Reef Symp 2:125-130

Briggs JC (1974) Marine Zoogeography. McGraw-Hill, New York, 475 pp

Brock RE (1982) A critique of the visual census method for assessing coral reef fish populations. Bull mar Sci 32:269-276

Chave EH, Eckert DB (1974) Ecological aspects of the distributions of fishes at Fanning Island. Pac Sci 28:297-317

Cameron AM, Endean R (1982) Renewed population outbreaks of a rare a specialised carnivore (the starfish *Acanthaster planci* in a complex high-diversity system (the Great Barrier Reef). Proc 4th Intnl Coral Reef Symp 2:593-596

Chesher RH (1969) Destruction of Pacific corals by the sea star *Acanthaster planci.* Science 165:280-283

Cohen DM (1973) Zoogeography of the fishes of the Indian Ocean. In: Zeitschel B, Gerlach SA (eds.) The Biology of the Indian Ocean, Ecological Studies 3, pp 451-463

Dana TF (1970) *Acanthaster:* a rarity in the past? Science 169:894

Dana TF, Newman WA, Fager EW (1972) *Acanthaster* aggregations interpreted as primarily a response to natural phenomena. Pacif. Sci 26: 355- 372

Dight IJ, James MK, Bode, L. (1989) Models of larval dispersal within the central Great Barrier Reef: patterns of connectivity and their implications for species distributions. Proc 6th Intnl Coral Reef Symp 3:217-224

Doherty PJ, Davidson J (1989) Monitoring the distribution and abundance of juvenile *Acanthaster planci* in the central Great Barrier Reef. Proc 6th Intnl Coral Reef Symp 2:131-136

Elton RA, Greenwood JDD (1970) Exploring apostatic selection. Heredity 25:629-633

Endean R (1969) Report on investigations made into aspects of the current *Acanthaster planci* (crown-of-thorns) infestations of certain reefs of the Great Barrier Reef. Qld Dept Primary Ind Fish Branch, Brisbane, 30pp

Endean R (1976) Destruction and recovery of coral communities. In: Jones OA, Endean R (eds) Biology and Ecology of Coral Reefs, III. Academic, New York, pp 215-254

Endean R (1977) *Acanthaster planci* infestations of reefs of the Great Barrier Reefs. Proc 3rd Intnl Coral Reef Symp 1:185-191

Endean R (1982) Crown-of-thorns starfish on the Great Barrier Reef. Endeavour 6:10-14

Frankel E (1977) Previous *Acanthaster* aggregations in the Great Barrier Reef. Proc 3rd Intnl Coral Reef Symp 1:201-208

Grant EM (1982) Guide to Fishes, 5th edn. Dept Harbours and Marine, Brisbane, 896 pp

Greenwood JJD, Elton RA (1979) Analysing experiments on frequency- dependent selection by predators. J Anim Ecol 48:721-737

Hassell MP, May RM (1973) Stability in insect host-parasite models. J Anim Ecol 49:693-726

Holling CS (1965) The functional response of predators to prey density and its role in mimicry and population regulation. Mem Ent Soc Can 45:3-60

Jones C, Goeden, GB (1984) The incidence of juvenile redfish species in commercial prawn trawl bycatch. Unpublished report to the Great Barrier Reef Marine Park Authority, Townsville, Queensland

Krebs JR (1978) Optimal foraging: decision rules for predators. In: Krebs JR, Davies NB (eds) Behavioural Ecology: an Evolutionary Approach. Blackwell Scientific, Oxford, pp 22-63

Lotka AJ (1925) Elements of Physical Biology. Williams and Wilkins, Baltimore

McCloskey LS, Deubert KH (1972) Organochlorines in the seastar *Acanthaster planci*. Bull Env Contam Toxicol 8:251-256

Maskell M, Parkin DT, Vespoor E (1977) Apostatic selection by sticklebacks upon dimorphic prey. Heredity 39:83-89

May RM (1975) Stability and Complexity in Model Ecosystems. 2nd edn. Princeton University Press, Princeton.

May RM (1981) Models for two interacting populations. In: May RM (ed) Theoretical Ecology; Principles and Applications. 2nd edn. Blackwell Scientific, Oxford, pp 78-104

Moran PJ (1986) The *Acanthaster* phenomenon. Oceanogr Mar BiolAnn Rev 24:379-480

Moran PJ, Bradbury RE, Reichelt, RE (1985) Mesoscale studies of the crown-of-thorns / coral interaction: a case history from the Great Barrier Reef. Proc 5th Intnl Coral Reef Symp 5:321-326

Moran PJ, Reichelt RE, Bradbury RH (1986) Assessment of the geological evidence for previous *Acanthaster* outbreaks. Coral Reefs 4:235-238

Murdoch WW (1969) Switching in general predators: experiments on predator specificity and stability of prey populations. Ecol Monogr 39:335-354

Olson RR, (1987) In situ culturing as a test of the larval starvation hypothesis for the crown-of-thorns starfish, *Acanthaster planci*. Limnol Oceanogr 32:895-904

Ormond RFG, Campbell AC (1974) Formation and breakdown of *Acanthaster planci* aggregations in the Red Sea. Proc 2nd Intnl Coral Reef Symp 1:595-619

Ormond RFG, Campbell AC, Head SM, Moore RJ, Rainbow PR, Sanders AP (1973) Formation and breakdown of aggregations of the crown-of- thorns starfish, *Acanthaster planci* (L.) Nature 246:167-169

Ormond RFG, Dawson Shepherd AR, Price ARG, Pitts RJ (1984) Report on the Distribution of Habitats and Species in the Saudi Arabian Red Sea. International Union for the Conservation of Nature and Natural Resources, Gland / Meteorological and Environmental Protection Administration, Jeddah.

Ormond RFG, Dawson Shepherd AR, Price ARG, Sheppard C (1985) Distribution of Habitats and Species along the Southern Red Sea Coast of Saudi Arabia. International Union for the Conservation of Nature and Natural Resources, Gland / Meteorological and Environmental Protection Administration, Jeddah.

Owens, D. (1971) *Acanthaster planci* starfish in Fiji: survey of incidence and biological studies. Fiji Agric J 33:15-23

Pearson RG (1975) Coral reefs, unpredictable climatic factors and *Acanthaster*. In: Proc. crown-of-thorns starfish seminar, Brisbane, 6 Sept 1974. Austral Gvnt Publ Serv, Canberra, 131-134

Potts DC (1981) Crown-of-thorns starfish: man-induced pest or natural phenomenon. In: Kitching RL, Jones RE (eds) The Ecology of Pests. CSIRO, Melbourne, pp 55-86

Randall JE (1972) Chemical pollution in the sea and the crown-of-thorns starfish (*Acanthaster planci*). Biotropica 4:132-144

Ricker WE (1975) Computation and interpretation of biological statistics of fish populations. Bull Fish Res Board Can 191, 382 pp

Russ G (1984) Distribution and abundance of herbivorous grazing fishes in the central Great Barrier Reef I. levels of variability across the entire continental shelf. Mar Ecol Prog Ser 20:23-34

Sano M, Shimizu M, Nose Y (1984) Changes in structure of coral reef fish communities by destruction of hermatypic corals: observational and experimental views. Pac Sci 38:51-79

Sale PF (1980) The ecology of fishes on coral reefs. Oceanogr Mar Biol Ann Rev 18:367-421

Silliman RP, Gutsell JS (1958) Experimental exploitation of fish populations. Fish Bull (US) 58:215-252

Southwood TRE, Comins HN (1976) A synoptic population model. J Anim Ecol 45:949-965

Steven AD (1988) An analysis of fishing activities on possible predators of crown-of-thorns starfish *Acanthaster planci* on the Great Barrier Reef. Report to the Great Barrier Reef Marine Park Authority, Townsville, Queensland, 131 pp

Toor HS (1964) Biology and fishery of the Pigface Bream, *Lethrinus lentjan* (Lacepede). I. Food and feeding habits. Indian J Fish 11:559-580

Vine PJ (1970) Densities of *Acanthaster planci* in the Pacific Ocean. Nature 228:341-342

Volterra V (1926) Variations and fluctuations of the numbers of individuals in animal species living together . J Cons perm int Ent Mer 3:3-51

Walbran PD, Henderson RA, Jull AJT, Head MJ (1989) Evidence from sediments of long-term *Acanthaster planci* predation on corals of the Great Barrier Reef. Science 245:847-850

Walker MH (1978) Food and feeding habits of *Lethrinus chrysostomus* Richardson (Pisces: Perciformes) and other lethrinids on the Great Barrier Reef. Aust J Mar Freshwater Res 29:623-630

Werner EE, Hall DJ (1974) Optimal foraging and size selection of prey by the blue-gill sunfish *Lepomus macrochirus*. Ecology 55:1042-1052

Williams DMcB (1986) Temporal variation in the structure of reef slope fish communities (central Great Barrier Reef): short term effects of *Acanthaster planci* infestation. Mar Ecol Prog Ser 28:157-164

Williams DMcB, Hatcher AI (1983) Structure of fish communities on outer slopes of inshore, mid-shelf and outer-shelf reefs of the Great Barrier Reef. Mar Ecol Prog Ser 10:239-250

Wilson BR, Marsh LM, Hutchins B (1974) A puffer fish predator of crown- of-thorns in Australia. Search 5:601-602

EFFECTS OF PREDATION ON *ACANTHASTER:*
AGE-STRUCTURED METAPOPULATION MODELS

H.I. MCCALLUM

Department of Zoology, University of Queensland, Queensland 4072.

Abstract. A model of predation on subadult *Acanthaster planci* is developed within a metapopulation framework in which adult starfish and predators are patchily distributed, but reefs are linked by larval dispersal. The model demonstrates that it is possible for starfish to be maintained at low levels by predators with a type II functional response on some patches, provided larval mixing is incomplete and resource limitation exists on at least one reef. Because of the long pre-reproductive post settlement stage in *Acanthaster*, predation on this stage is found to be of particular importance in then population dynamics of the starfish.

INTRODUCTION

Several models of *Acanthaster* have been constructed within the framework of predator-prey models (eg: Antonelli & Kazarinoff, 1984; Bradbury *et al.* 1985). In each of these models, the starfish have been considered as the predator and the coral as prey. Although *Acanthaster* is taxonomically a predator (as it feeds upon an animal), it is in many ways <u>functionally</u> a herbivore, as it feeds on a sedentary food source and does not necessarily consume entire prey individuals (considering a coral colony as an individual). To the extent that ecological generalisations have any validity, it is more reasonable to expect that generalisations concerning herbivores may be more applicable to *Acanthaster* than are generalisations concerning predators. Invertebrate herbivores are frequently limited or regulated by predators rather than by food limitation (Crawley, 1983).

In this paper I describe a simple model in which *Acanthaster* is considered as a prey species, and the possibility that it may be limited by generalist predators is examined. Like many marine organisms, *Acanthaster* reproduces by production of a large number of pelagic larval stages (Moran, 1986) which disperse over large distances. The post settlement stages of the starfish, whilst not sessile, disperse to a much more limited extent than the larvae. The model is therefore constructed within the metapopulation framework developed by Roughgarden and Iwasa (Roughgarden, Iwasa & Baxter, 1985; Roughgarden & Iwasa, 1986; Iwasa & Roughgarden, 1986), in which a "metapopulation" is defined as one made up of a number of local populations of post-larval stages, linked by larval recruitment from a common pool.

The suggestion that overfishing or over-exploitation of predators may be responsible for *Acanthaster* outbreaks was first made by Endean (1969). His "predator removal hypothesis" remains untested, partially because of the logistical difficulties involved in a large scale empirical test. The principal objective of this study is to determine theoretically the conditions under which the "predator removal hypothesis" is plausible and to identify those parameters which must be measured in order to test it.

Information as to the identity of predators of the crown of thorns starfish is limited at present. Adult starfish have well developed toxic spines on their aboral surface and can be attacked only by those predators with specializations enabling them to overcome these defences. The best known of adult predators is the triton *Charonia tritonis* (Endean, 1969) but some species of puffer fish *Arothron hispidus* and trigger fish *Balistoides viridescens* (Ormond & Campbell, 1974) have also been recorded as predators. Potentially, a wider range of more generalist predators may be able to consume juvenile starfish before the toxic spines are fully developed. Unfortunately, juvenile starfish are cryptic, patchily distributed and rarely discovered in large numbers in the field. Available information on their predators is hence anecdotal. It has been suggested (Ormond *et al.*, this volume) that *Lethrinidae*, which are subject to substantial fishing pressure, may be important predators of juvenile starfish.

MODEL DESCRIPTION

Starfish are assumed to be distributed between n reefs and not to move between reefs except as larvae. I assume that the starfish population on a reef i can be divided into three life-history stages, larvae $L_i(t)$, pre-reproductive starfish $A_i(t)$ and reproductive starfish $N_i(t)$. Within each stage, all individuals are assumed to be equivalent. The model concentrates specifically upon predation on pre-reproductive starfish. For mathematical convenience, resource limitation is also assumed to occur only in the pre-reproductive stage.

Each reef is assumed to have a predator population P_i, which has sufficient alternative prey sources other than the species under consideration that the predator population density is unaffected by variations in A_i. The predators are assumed to have a type II functional response of the form:

$$F(A_i, P_i) = a_i A_i / (A_i + a_i / s_i) \qquad (1)$$

This means that when prey are rare, the prey consumption rate per predator is determined by the rate at which predators encounter prey and is approximately s_iA_i where s_i is the searching efficiency of predators in the patch. At high prey densities the prey consumption rate approaches the constant rate a_i per predator.

Pre-reproductive starfish are assumed to die at a constant rate d_i from causes other than predation. As a crude way of taking account of resource limitation in the prey , I include a term $k_iA_i^2$. After a maturation delay of τ after settlement, surviving starfish enter the reproductive stage N_i, in which they suffer a constant death rate δ_i and have a *per capita* larval production rate of λ_i. Larvae enter a population associated with the patch under consideration, L_i, but transfer from the ith to the jth patch at a *per capita* rate β_{ij}. Finally, starfish recruit to the ith patch at a rate r_iL where r_i is the settlement rate and have a constant death rate μ_i per unit time.

The system is described by the following equations where $f_i(A_i)$ is the death rate of pre-reproductive starfish on the ith patch. $Q_i(t)$ is the proportion of individuals entering the pre-reproductive stage τ time units previously which survive to leave the stage at time t and is dependent on $f_i(Ai)$ between $t-\tau$ and t (eqn. 5). (See Gurney, Nisbett & Lawton, 1983 for details of the construction of this type of model.)

$$\frac{dL_i}{dt} = N_i(t)\lambda_i + \sum_{i \neq j} L_j\beta_{ji} - L_i \sum_{j \neq i} \beta_{ij} - r_iL_i - \mu_iL_i \tag{2}$$

$$\frac{dA_i}{dt} = r_iL_i(t) - r_iL_i(t-\tau)Q_i(t) - A_i(t)f_i(A_i(t)) \tag{3}$$

$$\frac{dN_i}{dt} = r_iL_i(t-\tau)Q_i(t) - \delta_iN_i(t) \tag{4}$$

$$\frac{dQ_i}{dt} = Q_i\{f_i(A_i(t-\tau)) - f_i(A_i(t))\} \tag{5}$$

$$f_i(A_i) = d_i + \frac{a_iP_i}{A_i + a_i/s_i} + k_iA_i \tag{6}$$

RESULTS

Rather than attempt to analyse the model described by eqns. (1)-(6) in its full generality, a number of special cases will be described so that an idea of the likely behaviour of the full model can be developed. Simpler models corresponding to limiting cases of eqns. (1)-(6) without time delays have been analysed by McCallum (1987 & 1989).

In all but the simplest case of a single closed patch, even equilibrium solutions of the equations are transcendental in form and cannot be obtained explicitly. Numerical solutions require estimates of the value of each parameter. Table 1 gives the values used as the basis for the results presented in this paper. Unfortunately, most of these are fairly crude estimates, as little data are available, particularly on juvenile post- settlement stages which have proved to be elusive in the field (eg Doherty & Davidson, 1989). There is no information available to estimate the parameters concerned with the predator functional response: even the identity of the most important predators of the subadult stages remains unknown.

Table 1: Parameters of the model

Parameter	Definition	Approx value (day time units)	Source and justification
τ	Time delay from settlement until first reproduction	730	Moran (1986)
λ_i	larval production rate per adult	2.74×10^4	10^7 female eggs/female/year (Moran, 1986)
$\dfrac{r_i}{r_i + \mu_i}$	proportion of eggs settling as recruits	0.01	At maximum of plausible range
β_{ij}	larval transfer rate from patch i to j.	0.0025	No available evidence
d	pre-reproductive death rate	0.0014	Based on a life expectancy of two years
δ_i	post-reproductive death rate	0.0014	Life expectancy of two years post reproduction (Moran, 1986)
a_t	satiation level of predators	0.14	Based on one prey per week (Endean, 1969), but little available evidence
s_t	searching efficiency of predators	0.016	No available evidence
k_t	resource limitation on adults	0.000571	essentially a scaling parameter
P	Predator population	1	arbitrary

Case 1: a single closed patch

This is discussed in McCallum (1989). Non zero equilibrium values of A satisfy:

$$kA + \frac{aP}{A + a/s} + d = \frac{1}{\tau}\left\{ \ln\left(\frac{r\lambda}{\delta(\mu + r)} \right) \right\} \tag{7}$$

The most important characteristic of this equation is that terms related to fecundity, larval survival and post-reproductive death rate are logged, whereas the terms concerned with the pre-reproductive death rate are not. This means that the system is more sensitive to variation in the pre-reproductive death rate (of which predation is a component) than it is to the same amount of variation in larval survival or fecundity.

Case 2: a single patch with external larval immigration

In this case, there may be 3 non zero equilibria, provided:

$$sP + d > \frac{1}{\tau}\left\{ \ln\left(\frac{r\lambda}{\delta(\mu + r)} \right) \right\} \tag{8}$$

and the larval immigration rate is not too large.

Details of the stability analysis of the system using Laplace transforms following the method described in Murdoch *et al.* (1987) are given in the appendix. Equilibria were located and the characteristic equation was solved numerically for a range of parameter values. For all parameter values examined, the characteristic equation had one real root which determined the stability of the equilibrium. Where there were three equilibria, the lower and upper equilibria were locally stable and the intervening equilibrium was locally unstable. Where there was only one equilibrium, it was locally stable.

Equilibrium values of the immature starfish population A are shown as a function of the immigration rate b and the two parameters associated with predation, s and a in Fig 1(a)-(c). Except for the parameter being varied, all other parameters remained as given in table 1. Over the parameter range investigated, the qualitative behaviour of this model did not differ from that of the non time delayed model discussed by McCallum (1989.)

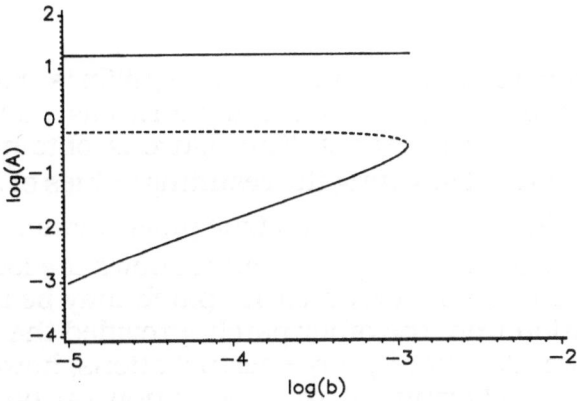

Figure 1. Equilibria of the model with external larval recruitment.

Stable equilibria are shown with solid lines and unstable equilibria are shown dashed. The vertical axis shows the log (base 10) of pre-reproductive starfish numbers. (a) Effect of changing the larval immigration rate b, where $b = (r\beta)/(r+\mu)$ and β is the larval immigration rate per unit time. Other parameters are as given in Table 1.

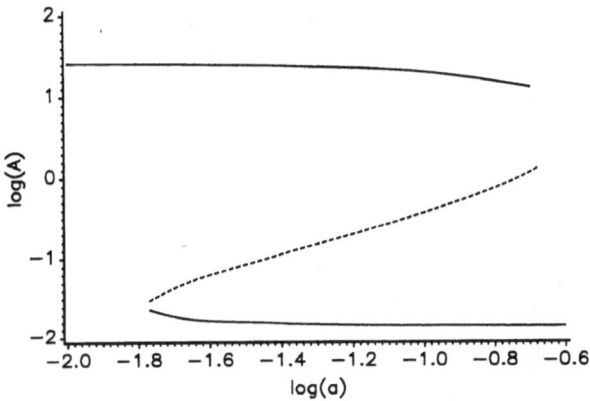

Fig 1 (b) Effect of changing the satiation level of predators a. b=0.0001. All other parameters as in table 1.

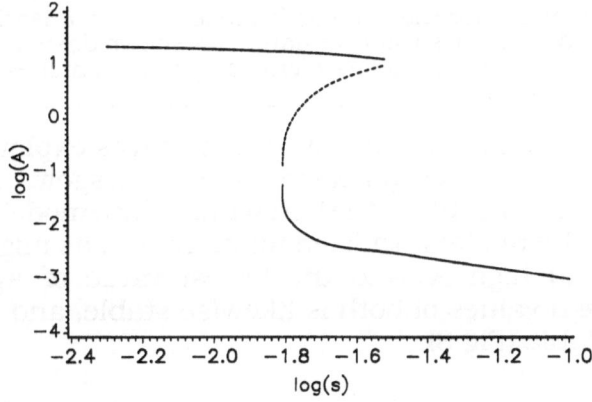

Fig 1 (c) Effect of varying predator searching efficiency s. b=0.0001. All other parameters as in table 1.

Case 3: two patches

In this case there may potentially be 5 non zero equilibria. Figure 2 shows a "pseudo phase plane" for the system in a symmetric case: one in which all parameters are equal for both patches. The figure is obtained by setting $dN_1/dt = 0$ and $dN_2/dt = 0$ and substituting the resulting values of N_1 and N_2 into $dA_1/dt = 0$ and $dA_2/dt = 0$. The diagram is not a true phase plane, as trajectories are not constrained to remain in the plane, but it shows the location of each equilibrium. The starfish population on either patch may be maintained at a low level by immigration from the other patch, provided the second patch is at its high equilibrium. Sufficiently large perturbations, however, will lead to escape to the high equilibrium. If the population on both patches is depressed sufficiently, the entire metapopulation of starfish will become extinct.

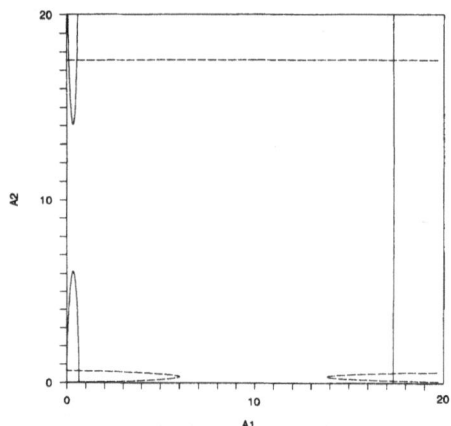

Figure 2. Pseudo phase plane for the two patch model. The zero isocline for patch 1 is shown as a solid line and the zero isocline for patch 2 is shown dashed. Parameter values are as given in table 1, with all parameters identical for both patches.

The behaviour of the model near these equilibria was explored numerically using the SOLVER computer package (APIC, Glasgow, 1988). For the parameter values given in Table 1, the behaviour of the model was essentially the same as that of the model with fixed immigration, as might be expected. The equilibrium with high A_2 and the lowest value of A_1 is stable, the equilibrium with high values of both is likewise stable, and the intervening equilibrium is unstable (Fig 3).

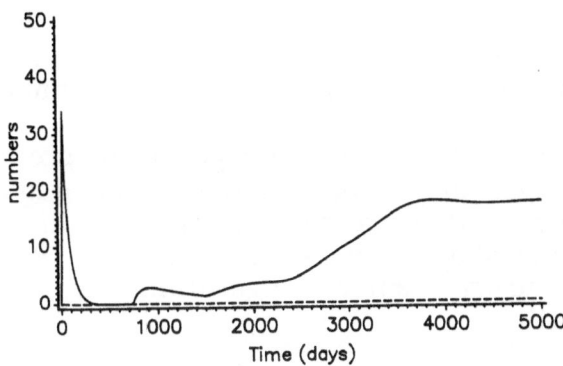

Figure 3. Numerical solutions of the two patch model. The population of immature starfish on patch 1 A_1 is shown as a solid line, and A_2 is indicated with a dashed line. Parameters are as given in table 1. For computational reasons, these solutions were commenced by "inoculating" an empty system with larvae over a short initial time period. The settling of this founding cohort, followed by a period of attrition before further recruitment occurs, generates the spike shown at the commencement of each solution. (a) Inoculation of 5000 larvae to patch 1, 0 to patch 2.

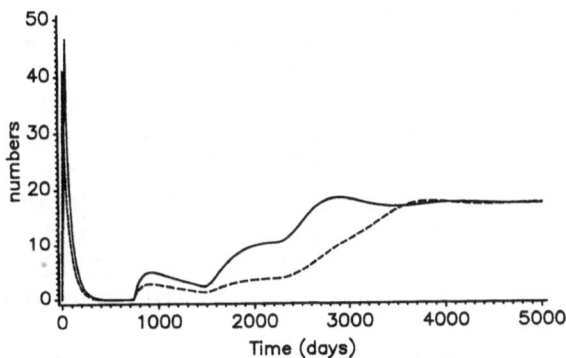

Fig 3 (b) Inoculation of 50000 larvae to patch 1, 5000 to patch 2.

DISCUSSION

These models demonstrate that predators with a type II functional response, feeding on sub-adult stages, are able to maintain starfish populations at sub-outbreak levels on one reef, provided that there is another reef carrying a high starfish population and that larval mixing is incomplete. This conclusion will generalise readily to n reefs: at least one reef at "outbreak" level may maintain a number of others at low starfish levels. Perturbations may transfer reefs to the alternative "outbreak" stable state, which will lead to increased larval immigration rates to all other reefs, decreasing the zone of attraction of the low equilibrium (Fig 1(a)). Eventually, this may lead to a cascade as all remaining reefs transfer to the high equilibrium.

If predation occurs during the extended period of pre-reproductive life, relatively small changes in the predation rate (as might be caused by overfishing) will compound to cause large changes in the proportion of survivors through this stage. This is shown by eqn. 8: recruitment and post-reproductive death enter logarithmically compared with pre-reproductive death.

Despite the importance of pre-reproductive, post settlement factors indicated by the model, there is undoubtedly a large amount of variability in *Acanthaster* recruitment on both temporal and spatial scales. To provide a more accurate quantitative picture of how the system behaves, it would be necessary to integrate this model with details of the water circulation patterns in the reef system under study, along lines developed by Roughgarden, Gaines & Possingham (1988).

Somewhat surprisingly, given the long time delays and high potential fecundity included in the model, no complex oscillatory or chaotic behaviour was detected in this study. This is probably a result of the timing of the density dependence: for mathematical simplicity, resource limitation was assumed to act on the pre-reproductive stages rather than the adult starfish. Biologically, it is probably more reasonable to believe that resource limitation occurs on adult, reproductive starfish because of the greater coral consumption rate of mature starfish compared with juveniles. On the other hand, small starfish are both very patchily distributed and relatively slow moving (Yamaguchi, 1973; Zann et al., 1987) so it may be reasonable to believe that resource limitation does occur within patches of juveniles.

It is well known (Cushing, 1981; Hastings, 1983,1984; Murdoch et al. 1987) that the timing of density dependence and of predation has crucial implications for the stability of age structured models. At this stage, the possibility of density dependence acting on the adults rather than the juveniles was not further explored, because the logistic density dependence is grossly inadequate as a means to provide anything more than a crude caricature of the operation of resource limitation. To provide a valuable description of resource limitation in adults to explore stability properly, it would be necessary to include the dynamics of the coral resource in a three trophic level model. This work is currently in progress, but is beyond the scope of this paper.

Acknowledgements: Computer time for this project was provided by a University of Queensland Innovative Computing Grant.

REFERENCES

Antonelli, P.L. & Kazarinoff, N.D. (1984) Starfish predation of a growing coral reef community. *J. Theor. Biol.* **107** 667-684

Applied Physics Industrial Consultants (1986) *Solver, rev. 2.02* APIC, Strathclyde.

Bellman, R. & Cooke, K.L. (1963) *Differential-Difference Equations.* Academic Press, New York.

Bradbury, R.H., Hammond, L.S., Moran, P.J. & Reichelt, R.E. (1985). Coral reef communities and the Crown-of-thorns starfish: evidence for qualitatively stable cycles. *J. Theor. Biol.* **113,** 69-81.

Crawley, M.J. (1983) *Herbivory.* Blackwell, Oxford.

Cushing, J.M. (1981) Stability and maturation periods in age-structured populations. In: Busenberg, S.N. & Cooke, K.L. (Eds.) *Differential Equations and Applications in Ecology, Epidemics and Population Problems* New York, Academic. 163-181.

Doherty, P.J. & Davidson, J. (1989) Monitoring the distribution and abundance of juvenile *Acanthaster planci* in the central Great Barrier Reef. *Proceedings of the Sixth International Coral Reef Symposium* **2** 131-136.

Endean, R. (1969) Report on investigations made into aspects of the current *Acanthaster planci* (crown-of-thorns) infestations of certain reefs of the Great Barrier Reef. Fisheries Branch, Queensland Department of Primary Industries, Brisbane 38pp.

Gurney, W.S.C., Nisbett, R.M. & Lawton, J.H. (1983) The systematic formulation of tractable single-species population models incorporating age structure. *J. Anim. Ecol.* **52,** 479-95.

Hastings, A. (1983) Age-dependent predation is not a simple process. I. Continuous time models. *Theor. Pop. Biol.* **23,** 347-362.

Hastings, A. (1984) Age-dependent predation is not a simple process. II. Wolves, ungulates and a discrete time model for predation on juveniles with a stabilizing tail. *Theor. Pop. Biol.* **26,** 271-282.

Iwasa, Y. & Roughgarden, J. (1986) Interspecific competition among metapopulations with space-limited subpopulations. *Theor. Pop. Biol.* **30,** 194-214.

McCallum, H.I., (1987) Predator regulation of *Acanthaster planci. J. Theor. Biol.* **127** 207-220.

McCallum, H.I. (1989) Effects of predation on organisms with pelagic larval stages. *Proceedings of the Sixth International Coral Reef Symposium* **2** 101-106.

Moran, P.J. (1986) The *Acanthaster* phenomenon. *Oceanogr. Mar. Biol. Annu. Rev.* **24,** 379-480.

Murdoch, W.W., Nisbett, R.M., Blythe, S.P., Gurney, W.S.C. & Reeve, J.D. (1987) An invulnerable age class and stability in delay-differential parasitoid-host models. *Amer. Natur.* **129,** 263-282.

Ormond, R., Bradbury, R., Bainbridge, S., Fabricus, K., Keesing, J., DeVantier, L., Medley, P. & Steven, A. Test of a model of regulation of crown-of-thorns starfish by fish predators. This volume.

Ormond, R.F.G. & Campbell, A.C. (1974) Formation and breakdown of *Acanthaster planci* aggregations in the Red Sea. *Proceedings of the Second International Coral Reef Symposium* **1:** 595-619.

Roughgarden, J., Gaines, S. & Possingham, H. (1988) Recruitment dynamics in complex life cycles. *Science* **241** 1460-1466.

Roughgarden, J. & Iwasa, Y. (1986) Dynamics of a metapopulation with space-limited subpopulations. *Theor. Pop. Biol.* **29** 235-261.

Roughgarden, J. Iwasa, Y. & Baxter, C. (1985) Demographic theory for an open marine population with space-limited recruitment. *Ecology.* **66** 54-67.

Yamaguchi, M. (1973) Early life histories of coral reef asteroids, with special reference to *Acanthaster planci* (L.). In Jones O.A. & Endean, R. eds. *Biology and Geology of Coral Reefs. Vol II, Biology 1* New York, Academic 369-387.

Zann, L., Brodie, J., Berryman, C. & Naqasima, M. (1987) Recruitment, ecology, growth and behaviour of juvenile *Acanthaster planci* (L.) (Echinodermata: asteroidea) *Bull. Mar. Sci.* **41** 561-575.

APPENDIX

LOCAL STABILITY ANALYSIS OF THE MODEL
WITH CONSTANT LARVAL INFLOW

Assume that the larval stage is short in comparison with the other life-history stages. Equations (2-5) then become:

$$\frac{dA}{dt} = \Lambda N(t) + b - Q(t)\{\Lambda N(t-\tau) + b\} - A(t)f(A(t)) \tag{A1}$$

$$\frac{dN}{dt} = Q(t)\{\Lambda N(t-\tau) + b\} - \delta N(t) \tag{A2}$$

$$\frac{dQ}{dt} = Q(t)\{f(A(t-\tau)) - f(A(t))\} \tag{A3}$$

Here all parameters and variables are as defined in the text, with the exception of the compound parameters $\Lambda = r\lambda/(r+\mu)$ and $b = r\beta/(r+\mu)$, where β is the constant larval immigration rate. The equilibrium value of A, A^* satisfies the transcendental equation:

$$\delta\{A^*f(A^*) - b\} = e^{-\tau f(A^*)}\{\Lambda A^*f(A^*) - \delta\} \tag{A4}$$

and N^* and Q^* can be given in terms of A^* as:

$$N^* = \frac{A^*f(A^*) - b}{\Lambda - \delta} \tag{A5}$$

$$Q^* = e^{-\tau f(A^*)} \tag{A6}$$

Defining:

$$u(t) = A(t) - A^*, \qquad v(t) = N(t) - N^*, \qquad w(t) = Q(t) - Q^* \tag{A7}$$

and neglecting second order and higher order terms, Eqns. (A1)-(A3) become:

$$\frac{du}{dt} = \Lambda v - Q^*\Lambda v(t-\tau) - w(\Lambda N^* + b) - u\left(f(A^*) + A^*\frac{\partial f}{\partial A}\Big|_{A^*}\right) \tag{A8}$$

$$\frac{dv}{dt} = w(\Lambda N^* + b) + Q^* \Lambda v(t - \tau) - \delta v \tag{A9}$$

$$\frac{dw}{dt} = Q^* \frac{\partial f}{\partial A}\Big|_{A^*} \cdot \{u(t-\tau) - u(t)\} \tag{A10}$$

The characteristic equation obtained by Laplace transformation of eqns. (A8-10) is :

$$z^3 + z^2 \left\{ f(A^*) + A^* \frac{\partial f}{\partial A}\Big|_{A^*} + \delta - Q^* \Lambda e^{-z\tau} \right\} \tag{A11}$$

$$+ z \left\{ f(A^*)(\delta - \Lambda Q^* e^{-z\tau}) + \frac{\partial f}{\partial A}\Big|_{A^*} \cdot (A^* \delta - b Q^* (1 - e^{-z\tau}) - A^* \Lambda Q^* e^{-z\tau} - \Lambda N^* Q^* + \Lambda N^* Q^* e^{-z\tau}) \right\}$$

$$+ \frac{\partial f}{\partial A}\Big|_{A^*} \cdot Q^* (\delta - \Lambda)(\Lambda N^* + b)(1 - e^{-z\tau}) = 0$$

If all the roots of this equation have negative real parts, then the equilibrium is stable (Bellman & Cooke, 1963). Eqn. (A11) has a trivial solution at $z=0$. Other solutions were obtained numerically for certain parameter combinations. As equations of this form may have an infinite number of complex roots, stability behaviour was checked using a numerical integration of the original equations.

APPLIED VOLTERRA-HAMILTON SYSTEMS OF FINSLER TYPE: INCREASED SPECIES DIVERSITY AS A NON-CHEMICAL DEFENSE FOR CORAL AGAINST THE CROWN-OF-THORNS

P.L. ANTONELLI

Department of Mathematics, University of Alberta,
Edmonton, Alberta T6G 2G1

Abstract. Local Finsler geometry is used to obtain results on the passive Volterra-Hamilton theory of ecological production. The new Finsler equations are used to model defensive strategies for two species of reef-building corals under attack by starfish, *Acanthaster planci*, on the Great Barrier Reef. These equations are obtained by perturbation of the underlying (formal) cost functional for reef-building so that terms involving species diversity ratios (i.e. polyp number ratios) are included. Such species diversity perturbations result in raising the critical point values at which instability and bifurcation occur, thereby stabilizing the system and its production. Hence, spatial shapes and distributions of hard corals, which result in more efficient growth for the community, may provide a non-chemical defense against *A. planci*.

1. INTRODUCTION TO THE VOLTERRA-HAMILTON METHOD

In 1936 V. Volterra published a paper in French which united his ecological equations with Maupertuis' principle (see [1] for a translation). However, Volterra had been thinking only about animals and because of this his employment of an auxiliary variable to implement the variational technique was unnatural and the paper received little attention. In 1944, the forester J. Kittredge used Huxley's Allometric Law for statistical estimation of leaf biomass from average tree trunk diameters [15]. Today, it is well known that Huxley's Law applies to a wide range of plant species, and can serve as a basis for plant/plant and plant/herbivore interaction theory. Recently, the author and D.F. Rhoades have used the allometric theory of Volterra-Hamilton systems to describe both passive and active plant chemical defenses against herbivory [1, 16, 17]. Rhoades and others have discovered that plants defend themselves from attack by spontaneously increasing amounts of tannins, lignins, etc. in their tissues [16, 17]. The fact that the response time can be as short as a few hours opens up a new area of research using control theory and differential games [11, 12].

While the constant coefficient (passive) theory has been successfully applied to starfish/coral dynamics on the Great Barrier Reef and to the epidemiology of the European rabbit disease called myxomatosis, the variable coefficient (active) theory has been applied to chemical interactions between (poisonous) soft and hard corals on the Great Barrier Reef as well as control and differential game models in plant/herbivore chemical warfare [1, 7, 10, 11, 12]. Furthermore, nonlinear filtering and entropy production are statistical theories which have strongly entered into this field [2, 3, 8].

The first published article on the geometric and stochastic theory of Volterra-Hamilton systems appeared in 1980 [0]. Since then much progress has been made and I believe many of the goals set out there are now being realized. In the present, brief paper, I describe new results on the passive theory derived from Finsler geometry which have application to the crown-of-thorns starfish problem [4, 5, 6, 10]. The latter is given in the last section.

By a *closed eco-development system* $E(P_1, ..., P_n | p_1, ..., p_n)$ we shall mean a set of n kinds of "Producers," $P_1, ..., P_n$ where each set P_i consists of N^i individuals each of whom deposits a product p_i into the environment E. The total amount of product of kind i, due to all N^i individuals in P_i is denoted X^i, and is monotonically increasing through time. The X^i can be thought of as residual accumulation in the case where bioerosive forces are included and so reduce growth accumulation. Producers can interact passively or actively, in general. The latter term means that coefficients G^i_{jk} in the dynamics of *modular units* can depend on $X^1, ..., X^n$ and also on the ratios of the N^i, i.e. on N^i/N^j which we call a *species diversity measure*. This is the major difference with the Riemannian Production Theory which does not allow dependence on these ratios.

By an n-dimensional *Volterra-Hamilton system* we mean a system of ordinary differential equations

$$G: \begin{cases} \dfrac{dX^i}{dt} = k_{(i)}N^i & i, j, k = 1, 2, ..., n \\ \dfrac{dN^i}{dt} = -2G^i_{jk}N^jN^k + \gamma^i_j N^j \end{cases} \qquad (1.1)$$

Use is made of the Einstein summation convention on upper and lower repeated indices except for $k_{(i)}N^i$ where the parenthesis indicates a single term. We assume the growth rates of all the producer populations to be equal to $\lambda > 0$, a constant. Thus, $\gamma_j^i = \lambda\delta_j^i$ for all i, j where δ_j^i is the Kronecker delta. Setting $S = e^{\lambda t}$ defines an *intrinsic time scale, longer than* t, for which (1.1) takes the form

$$\frac{d^2X^i}{dS^2} + 2G_{jk}^i \frac{dX^j}{dS}\frac{dX^k}{dS} = 0, \qquad\qquad (1.2)$$

if $K_{(i)} = 1$ for $i = 1,...,$ n. If $k_{(i)}$ are not so normalized, they enter the G_{jk}^i multiplicatively and in the case n = 2, alter α_1 and α_2 (see 4.1) and thereby effectively alter stability of the ecological interaction and of the production process (see the Theorems of Section 4, ref. (4.8) and discussion). The coefficients are c^∞ functions of X^i, N^i and are homogeneous of degree zero in N^i. We further require that (1.2) has solutions minimizing production time, for small enough S-time intervals. More precisely, (1.2) are assumed to be Euler-Lagrange equations for a convex c^∞-Lagrangian or *cost functional* which is homogeneous of degree one in dX^i/dS. This leads to *Finsler differential geometry* (which includes Riemann geometry as a special case). Requiring G_{jk}^i to be independent of ratios N^i/N^j, we obtain *Finsler spaces of Berwald type* for the background geometry of *production space* (spanned by $X^1,...,X^n$). Specializing to require G_{jk}^i to be n^3 constants, leads to *Berwald spaces of locally constant connection,* which are of primary interest for the starfish problem we consider here. The standard reference for local Finsler geometry is the monograph by H. Rund [18]. I recommend this treatise as well as L. Berwald's fine paper [14].

One good reason for further developing the Volterra-Hamilton theory in the Finsler setting is to obtain descriptive quantification of the influence of the relative proportions of the numbers of production units on these modular units, themselves, and on the production process, itself. But, in this paper we will apply these ideas to a perplexing biological problem, only, leaving the general theory for later publications. The *question* we are concerned with is, *how can a community of reef-building corals defend itself against hordes of devastating crown-of-thorns starfish on the Great Barrier Reef?*

Evidence indicates that such a defense, if it exists at all, is surely not essentially chemical as it is for many terrestrial plant/herbivore interactions [16, 17]. Our method shows that small log-linear perturbations of the reef-building cost functional by terms involving the ratio of coral cover percentages for two species (i.e. species diversity measure), can result in an effective (non-chemical) defense. This is accomplished in Section 4.

2. BERWALD SPACES WITH LOCALLY CONSTANT CONNECTION

Following [1, Chap. 3B], let H^n denote a connected, n-dimensional, c^∞-manifold provided with a global second order differential equation S. This means that S is a c^∞ vector field on the tangent bundle TH and that

$$D\pi \cdot S(\xi) = \xi,$$

for every vector field ξ in the tangent bundle TH, and where $D\pi$ is the differential of the canonical projection map $\pi :$ TH $\to$ H. A global spray is a second order differential equation S for which

$$S(\lambda\xi) = \lambda_* \ \lambda S(\xi),$$

for all $\xi \in$ TH and any real $\lambda \neq 0$. The map λ_* is the bundle automorphism induced by scaler multiplication on each fibre, thusly,

$$\lambda_* : TH \to TH, \quad \lambda_* (\xi) = \lambda\xi.$$

We shall be working locally for the most part, so suppose $u^1,..., u^n$ are coordinates in a trivializing neighbourhood (U, h) of the bundle TH, h being the coordinate diffeomorphism onto an open subset of $\mathfrak{R}^n$.

Then our c^∞-spray has the *local* description

$$\frac{d^2u^i}{dt^2} = -2\underset{2}{G^i}(u \cdot \frac{du}{dt}) \tag{2.1}$$

where the number 2 under the G^i indicates that G^i is 2nd order homogeneous in du^i/dt. Clearly, (1.2) is a special case of (2.1).

Now define the *Berwald* (local) *connection for the spray* by

$$G^i_{kj} = G^i_{jk} = \frac{\partial^2 G^i_2}{\partial \dot{u}^j \partial \dot{u}^k} \qquad (2.2)$$

where $\dot{u}^i$ denotes du^i/dt, (see [14, 18]). These n^3 c^∞-functions of u^i and $\dot{u}^i$ transform according to the usual law of the classical linear affine connection,

$$\frac{\partial \bar{u}^r}{\partial u^j} \frac{\partial \bar{u}^s}{\partial u^k} \bar{G}^i_{rs} = \frac{\partial \bar{u}^i}{\partial u^r} G^r_{jk} - \frac{\partial^2 \bar{u}^i}{\partial u^j \partial n^k}. \qquad (2.3)$$

Note that G^i_{jk} are symmetric in their lower indices and that they are

homogeneous of degree zero in $\dot{u}^i$.

We now suppose (2.1) to be the Euler-Lagrange equations for a c^∞ cost

functional $F(u,\dot{u})$ by the usual calculus of variations technique

$$\delta \int_{t_0}^{t_1} F(u,\dot{u})\, dt = 0, \qquad (2.4)$$

where $F(u,\dot{u}) > 0$ for all $\dot{u} \neq 0$ and $F(u,\lambda\dot{u}) = |\lambda|\, F(u,\dot{u})$ for all real λ with non-vanishing absolute value $|\lambda|$. The quadratic form of degree zero in $\dot{u}$ defined by

$$g_{ij} = \frac{1}{2} \frac{\partial^2 (F \cdot F)}{\partial \dot{u}^i \partial \dot{u}^j} \qquad (2.5)$$

defines the *metric tensor for the Finsler geometry with metric function* F.

By Euler's Theorem on homogeneous functions

$$F^2(u,\dot{u}) = g_{ij}\, \dot{u}^i \dot{u}^j \tag{2.6}$$

Another important tensor is defined

$$C_{ijk}(u,\dot{u}) = \frac{1}{2}\frac{\partial g_{ij}(u,\dot{u})}{\partial \dot{u}^k}. \tag{2.7}$$

The geometry is *Riemannian* if and only if $C_{ijk} \equiv 0$.

For *Berwald spaces* one has a "Conservation Law"

$$\frac{\partial C_{ihj}}{\partial u^k} = \frac{\partial C_{ihj}}{\partial \dot{u}^\ell}\frac{\partial G^\ell}{\partial \dot{u}^k} + C_{ljh}\, G_{ik}^{\,\ell} + C_{i\ell j}\, G_{hk}^{\,\ell} + C_{ih\ell}\, G_{jk}^{\,\ell}. \tag{2.8}$$

This is a restricted class of Finsler spaces. For example, *the condition (2.8) is necessary and sufficient for G^i_{jk} to be independent of $\dot{u}$*, and we have the *Berwald version of Ricci's lemma*:

$$0 = g_{ij(k)} \equiv \frac{\partial g_{ij}}{\partial u^k} - \frac{\partial g_{ij}}{\partial \dot{u}^\ell}\frac{\partial g^\ell}{\partial \dot{u}^k} - G_{ijk} - G_{jik} \tag{2.9}$$

Here, the superscripts on G^i_{jk} are lowered via g_{ij} and the parenthesis indicates that *Berwald's covariant derivative* is being used (see [14, 18]). One *does not* have Ricci's lemma for Berwald's covariant derivative in general Finsler spaces [14, 18].

We are especially interested in Berwald spaces whose connection coefficients are constant in some coordinate system $u^1,...,u^n$. One example of such a space is given by

$$F^2 = e^{2\alpha_i u^i}(du_i \cdot du^i) \tag{2.10}$$

where the (Berwald) Connection is

$$
\left.
\begin{aligned}
2G_{ii}^{i} &= \alpha_i \\
2G_{ij}^{i} &= 2G_{ji}^{i} = \alpha_j \qquad (i \neq j) \\
2G_{jj}^{i} &= {}^{-}\alpha_i \qquad (i \neq j) \\
G_{jk}^{i} &= 0 \qquad (i \neq j \neq k)
\end{aligned}
\right\}
\qquad (2.11)
$$

As a matter of fact this geometry is *Riemannian* because $C_{ijk} \equiv 0$, hence the connection is Levi-Civita. In order to obtain non-Riemannian spaces of locally constant connection we make use of a conformal factor $\sigma = \sigma(u)$ and write

$$
g^{-ij}(u,\dot{u}) \equiv e^{-2\sigma} \cdot g^{ij}(u,\dot{u}) \qquad (2.12)
$$

(see [18, p. 225]).

The *Christoffel symbols of 2nd kind* (see 2.17) are

$$
\overline{\gamma}_{ij}^{h} = \gamma_{ij}^{h} + (\sigma_i \delta_j^h + \sigma_j \delta_i^h - g^{hk} g_{ij} \sigma_k) \qquad (2.13)
$$

where $\sigma_k \equiv \partial\sigma/\partial u^k$.

The *Berwald Connection* is then,

$$
G_{ij}^{h} = \gamma_{ij}^{h} + \frac{\partial \gamma_{rj}^{h}}{\partial \dot{u}^i}\dot{u}^r + \frac{\partial \gamma_{is}^{h}}{\partial \dot{u}^j}\dot{u}^s + \frac{1}{2}\frac{\partial^2 \gamma_{rs}^{h}}{\partial \dot{u}^i \partial \dot{u}^j}\dot{u}^r\dot{u}^s \qquad (2.14)
$$

and the *conformally transformed version* is

$$\bar{G}_{ij}^h = G_{ij}^h - \frac{\partial^2\, B^{hk}}{\partial \dot{u}^i \partial \dot{u}^j}\, \sigma_k \quad \Bigg\}$$

$$B^{hk} = \frac{1}{2}\, F^2\, g^{hk} - \dot{u}^h \dot{u}^k \quad \Bigg\} \tag{2.15}$$

It follows that

$$\bar{G}^h = G^h - B^{hk}\, \sigma_k. \tag{2.16}$$

Now if we choose g_{ij} to be *only a function of* $\dot{u}$ *and not of* u, then the Christoffel symbols of 2nd kind are zero because

$$\gamma_{ij}^h = \frac{1}{2}\, g^{\lambda h}\, [ij,\lambda]$$

$$= \frac{1}{2}\, g^{\lambda h}\Big(\frac{\partial g_{i\lambda}}{\partial u^j} + \frac{\partial g_{j\lambda}}{\partial u^i} - \frac{\partial g_{ij}}{\partial u^\lambda}\Big) \quad \Bigg\}$$

$$\equiv 0. \tag{2.17}$$

From (2.14) we obtain $G_{ij}^h \equiv 0$ so that $G^h \equiv 0$ and from (2.15)

$$\bar{G}_{ij}^h = -\frac{\partial^2\, B^{hk}}{\partial \dot{u}^i \partial \dot{u}^j}\, \sigma_k. \tag{2.18}$$

In the next section we compute the two dimensional Finsler (non-Riemannian) spaces which can arise from the metric (2.10) by the above conformal trick. In our context these non-Riemannian metrics arise by perturbation of the metric (2.10).

3. 2-DIMENSIONAL BERWALD SPACES
OF LOCALLY CONSTANT CONNECTION

We postulate the metric function

$$F^2 = e^{[2\alpha_i x^i(L^2 + 1) + 2L \cdot \mathrm{Tan}^{-1}(\dot{X}^1/\dot{X}^2)]} \cdot ((\dot{X}^1)^2 + (\dot{X}^2)^2) \qquad (3.1)$$

where X^1, X^2 are Cartesian coordinates on $\Re^2$ and L is a (perturbation) parameter. If we set $\alpha_1 = \alpha_2 = 0$ and compute the metric tensor, we arrive at

$$g_{11}(X, \dot{X}) = [1 + \frac{2L\dot{X}^2 (\dot{X}^1 + L\dot{X}^2)}{(\dot{X}^1)^2 + (\dot{X}^2)^2}] \, e^{\, 2L \, \mathrm{Tan}^{-1}(\dot{X}^1/\dot{X}^2)}$$

$$g_{21}(X, \dot{X}) = g_{12}(X, \dot{X}) = L[1 - \frac{2(\dot{X}^1 + L\dot{X}^2)\dot{X}^1}{(\dot{X}^1)^2 + (\dot{X}^2)^2}] \, e^{\, 2L \, \mathrm{Tan}^{-1}(\dot{X}^1/\dot{X}^2)} \qquad (3.2)$$

$$g_{22}(X, \dot{X}) = [1 - \frac{2L(\dot{X}^2 - L\dot{X}^1)\dot{X}^1}{(\dot{X}^1)^2 + (\dot{X}^2)^2}] \, e^{\, 2L \, \mathrm{Tan}^{-1}(\dot{X}^1/\dot{X}^2)}$$

so, g_{ij} is independent of X^i.

If we return to $\alpha_1 \neq 0$, $\alpha_2 \neq 0$ and *define*

$$\bar{g}_{ij}(X,\dot{X}) = e^{\, 2\alpha_i X^i(L^2 + 1)} \, g_{ij}. \qquad (3.3)$$

we see that if $L \equiv 0$, then $\bar{g}_{ij}$ is independent of $\dot{X}$; in fact it is the Kronecker symbol δ_{ij} multiplied by $e^{2\alpha_i X^i}$ and is consequently the metric tensor for (2.10). Returning to the $L \neq 0$ case, we denote by γ^i_{jk}, the Christoffel symbol for g_{ij} and see that it vanishes identically. Consequently, as in Section 2,

$$\bar{G}^h = -B^{hk}\sigma_k,$$

where $\sigma_k = \frac{\partial\sigma}{\partial X^k}$, $\sigma = \alpha_i X^i (L^2 + 1)$, so that the spray equations (or Finsler geodesic equations) are

$$\frac{d^2X^i}{dS^2} - 2B^{hk}\sigma_k = 0 \qquad (3.4)$$

where from (2.15)

$$B^{hk} = \frac{1}{2} F^2 g^{hk} - \dot{X}^k\dot{X}^k.$$

Thus, we can write (3.4) as

$$0 = \frac{d^2X^i}{dS^2} + [\alpha_1 - L\alpha_2] (\dot{X}^1)^2 + [L\alpha_2 - \alpha_1] (\dot{X}^2)^2 + 2[\alpha_2 + \alpha_1 L] \dot{X}^1\dot{X}^2$$

$$(3.5)$$

$$0 \equiv \frac{d^2X^2}{dS^2} + [-\alpha_2 - L\alpha_1] (\dot{X}^1)^2 + [\alpha_2 + L\alpha_1] (\dot{X}^2)^2 + 2[\alpha_1 - \alpha_2 L] \dot{X}^1\dot{X}^2$$

Setting $\alpha_1 = \alpha_2 = 0$ we obtain

$$\frac{d^2X^i}{dS^i} = 0, \quad i = 1,2,$$

while setting $L = 0$ we obtain the geodesic equations for the Riemannian metric (2.10). In previous work, the latter have been used as production equations for a two-species community of scleractinian corals, [4, 5, 6, 13].

We can now state an important uniqueness *Theorem*:

> *The only 2- dimensional Berwald spaces of locally constant connection which arise by a conformal change from the Riemannian*
>
> *ones are those which are perturbations by* $\tan^{-1} (X^1/X^2)$ *as in (3.1). The Riemannian class is obtained by setting* $L = 0$ *in (3.1).*
>
> Interchanging X^1 with X^2 reverses the sign of L. (See last paragraph of Section 4).

The proof of this theorem depends largely on Berwald's classification of

all 2-dimensional Finsler spaces for which G^i_{jk} are independent of X^i [14]. I

do not know if uniqueness would hold in higher dimensions. That is, whether the only Berwald spaces of locally constant connection are those

obtained by perturbation using $\tan^{-1} (X^i/X^j)$, $i \neq j$. But, note that the conformal change technique gives rise to a different class of Berwald spaces of locally constant connection than the above. But, these are not perturbations of the Riemannian ones.

If we modify $\sigma = \sigma_i X^i$, $i = 1,2$ to include quadratic terms in X^i so that

$$\sigma = a_i X^i + \frac{1}{2}\beta_1(x^1)^2 + \beta_{12}X^1X^2 + \frac{1}{2}\beta_2(x^2)^2 \qquad (3.6)$$

then one obtains a Finsler space of Berwald type.

Theorem:

> *The scaler curvature of this Berwald space is*
>
> $$K = -(\beta_1 + \beta_2)e^{-2\sigma}(L^2 + 1) - 2L\,\text{Tan}^{-1} (X^1/X^2) \qquad (3.7)$$

This formula agrees with the known result for $L = 0$, [1, Chapter 2F]. For the ecologically interesting cases α_i are positive and the curvature is negative depending on the signs of β_i.

Recall from [1, 3, 8, 19] that probabilistic properties of Brownian motion in negatively curved Riemannian geometries make it possible to interpret, -R, the negative of the scaler curvature, as *relative community vigour*. Corresponding to the classical result of Jacobi for Riemannian geometries, the more negative K is, the less (Jacobi) stable is the production process (3.5). (The extremal paths for the cost functional F are exponentially divergent and even more so with increasing negativity of K.) This quantity has been shown to enter the solutions of the Zakai equations for a large class of nonlinear filtering problems and also in the entropy production equations of chemical ecology. Although it is possible to extend this interpretation in a formal way to the scaler curvature K of Finsler spaces, a rigorous stochastic theory has been wanting until recently. Our goal is to investigate stochastic canalization in negatively curved Berwald spaces along the lines of the theorem of Y. Takahashi and S. Watanabe [19].

4. APPLICATION TO CORAL REEF ECOLOGY

We wish to consider the following Volterra equations

$$\frac{dN^1}{dt} = -2[\alpha_2 + \alpha_1 L]N^1N^2 - [L\alpha_2 - \alpha_1](N^2)^2 - [\alpha_1 - L\alpha_2](N^1)^2 + \lambda N^1$$

$$(4.1)$$

$$\frac{dN^2}{dt} = -2[\alpha_1 - \alpha_2 L]N^1N^2 - [\alpha_2 + L\alpha_1](N^2)^2 + [\alpha_2 + L\alpha_1](N^1)^2 + \lambda N^2$$

If $L = 0$, (4.1) is the extensively studied 2-coral community model of scleractinian corals [2, 4, 5, 6]. The parameter L measures the influence of the ratio of polyp numbers on the *cost of reef-building* and enters into the interaction coefficients themselves. But, the time- scale for the production process is variable and for large time, much
longer than the ecological timescale, t, of (4.1). This is reflected in the formal passage from (4.1) to (3.5) via

$$\frac{1}{\lambda} dS = e^{\lambda t}.dt. \qquad (4.2)$$

The coefficients of (3.5) are

$$
\begin{aligned}
2\overset{\sim 2}{G}_{11} &= -\alpha_2 - \alpha_1 L, & \overset{\sim 2}{G}_{12} &= \alpha_1 - \alpha_2 L \\[2mm]
2\overset{\sim 2}{G}_{22} &= \alpha_2 + \alpha_1 L, & 2\overset{\sim 1}{G}_{11} &= \alpha_1 - \alpha_2 L \\[2mm]
\overset{\sim 1}{G}_{12} &= \alpha_2 + \alpha_1 L, & 2\overset{\sim 1}{G}_{22} &= -\alpha_1 + \alpha_2 L
\end{aligned}
\right\} \qquad (4.3)
$$

These $\tilde{G}$-coefficients exhibit the "conformal pattern" of (2.11)! This fact enables us to apply a previously proved theorem of Antonelli and Kazarinoff [4] to analyse equilibria of the following *Finsler type* (2-coral/1-starfish)-*system,*

$$
\left.
\begin{aligned}
\frac{dN^1}{dt} &= -2\overset{\sim 1}{\underset{12}{G}}\, N^1N^2 - 2\overset{\sim 1}{\underset{11}{G}}\, (N^1)^2 - 2\overset{\sim 1}{\underset{22}{G}}\, (N^2)^2 - \delta_1 FN^1 + \lambda N^1 \\
\frac{dN^2}{dt} &= -2\overset{\sim 2}{\underset{12}{G}}\, N^1N^2 - 2\overset{\sim 2}{\underset{11}{G}}\, (N^1)^2 - 2\overset{\sim 2}{\underset{22}{G}}\, (N^2)^2 - \delta_2 FN^2 + \lambda N^2 \\
\frac{dF}{dt} &= \beta F(N^1 + N^2) + \gamma F^2 - \varepsilon F
\end{aligned}
\right\}
\qquad (4.4)
$$

Here, F denotes the starfish population density and $\gamma \cdot F^2$ is the *cooperative term* or aggregation term. The coefficient γ is called the *aggregation coefficient*. This parameter serves as Hopf bifurcation parameter in [4, 5, 6]. If we suppose $\delta_1 = \delta_2 = \delta$ and $k\alpha_1 = \alpha_2 = \alpha k$, $k > 0$, then (4.7) has a unique equilibrium in the positive orthant. Namely, we have

$$
\left.
\begin{aligned}
N_0^1 &= \frac{\lambda - \delta F_0}{\tilde{\alpha}_1(1+k^2)}, \quad N_0^2 = kN_0^1 \\[2mm]
F_0 &= \frac{\tilde{\alpha}_1 \varepsilon(1+k^2) - \beta(1 + k)}{\tilde{\alpha}_1 \gamma(1+k^2) - \beta\delta(1 + k)}
\end{aligned}
\right\}
\qquad (4.5)
$$

by applying results of [4] to (4.4). This being allowable because the $\tilde{G}$ -coefficients have the "conformal pattern".

If $k_{(1)}$ and $k_{(2)}$ are very small in the production equation (1.1). The equilibrium (4.5) is strongly affected causing N_0^1 and N_0^2 to be the same

order of smallness (close to zero) while $F_0 \sim \varepsilon/\gamma$. This is a description of degradation of the (4.5) equilibrium to the point of total destruction of the reefal community. Such an effect could for example be caused by extreme algal overgrowth.

Referring again to (4.4) we have *Theorem* (Antonelli and Kazarinoff):

If $\beta\lambda(1 + k) > \tilde{\alpha}_1\varepsilon(1 + k^2)$, then there is, $\gamma_c > 0$, such that, $\gamma > \gamma_c$, implies (N_0^1, N_0^2, F_0) is unstable, while if $\gamma < \gamma_c$, then (N_0^1, N_0^2, F_0) is stable. Furthermore, Hopf bifurcation to a stable limit cycle occurs at

$$\gamma = \frac{\tilde{\alpha}_1\delta\varepsilon}{\beta\lambda(\dfrac{1+k}{1+k^2}) + \tilde{\alpha}_1(\lambda - \varepsilon)},\tag{4.6}$$

(see [4]). *Here, we have set* $k = \dfrac{1+L}{1-L}$ *and* $\tilde{\alpha}_2 = k\tilde{\alpha}_1$ *where* $\tilde{\alpha}_1 = \alpha(1-L) > 0$

Note that $k = \dfrac{1+L}{1-L} > 1$ if $0 < L < 1$, whereas $k < 1$ where $-1 < L < 0$.

It is now easy to see that as L increases in $(-1, +1)$, k increases and because differentiation with respect to k yields

$$\left(\frac{1+k}{1+k^2}\right)' = \frac{1-2k - k^2}{1+2k^2 + k^4},\tag{4.7}$$

we conclude that γ_c increases.

Finally we have the statement: *Relatively higher levels of aggregation by starfish are required to destabilize the equilibrium when L-values are incorporated.* Letting V denote -K, we have the Bianchi identity for Berwald spaces

$$\partial_0 V + \frac{2L}{\sqrt{1+L^2}} V = 0,\tag{4.8}$$

where ∂_0 is derivative with respect to the so-called *Landsberg angle* for this geometry. This angle o is a measure of species diversity of N^1 and N^2 and is computed to be

$$o = \sqrt{1+L^2} \ \text{Tan}^{-1}\frac{N^1}{N^2}.$$

Theorem.

>*For fixed L, increasing o, decreases V. Thus increased diversity implies greater production stability. The larger the L-values the more Jacobi stable the production system.*

A final comment: If $L \cdot \tan^{-1} \dfrac{\dot{x}^2}{\dot{x}^1}$ had been used in (3.1), the results would still be the same. This is because the 2-coral system obtained from setting $\delta = 0$ has invariance under interchange of superscripts 1 and 2. Such will convert $\tilde{\alpha}_1 = k\tilde{\alpha}_2$ into $\tilde{\alpha}_2 = k\tilde{\alpha}_1$ and converts $\dfrac{1+L}{1-L}$ into $\dfrac{1-L}{1+L}$, where $\delta \neq 0$. This is of course the same as replacing L by -L with $L \in (-1, +1)$. But, both these cases are covered in the above statements and arguments. It was only required that L be increasing to secure the increase of γ_c (or $\tilde{\gamma}_c$). Anyway, after an interchange, $\tilde{\alpha}_1$ should be replaced by $\tilde{\alpha}_2$ in (4.8), (4.9) and in the first inequality of the above quoted theorem.

Acknowledgements. I would like to thank Dr Eva Ruhnau for conversations regarding portions of this work. Partially supported by NSERC-A-7667.

References

0. Antonelli, P.L.: Optimal growth of an ideal coral reef, Acta Cient. Venezolana, 31, 521-525 (1980)

1. Antonelli, P.L.: (Ed.) Mathematical Essays on Growth and the Emergence of Form, Univ. of Alberta Press, Edmonton, Canada, 354 pp (1985)

2. Antonelli, P.L. & R.J. Elliott: Nonlinear filtering theory for coral/starfish and plant/herbivore interactions, Stoch. Anal. and Appl., 4, 1-23 (1986)

3. Antonelli, P.L., R.J. Elliott & R.M. Seymour: Nonlinear filtering and Riemannian scaler curvature, K. Adv. Appl. Math. 8, 237-253 (1987)

4. Antonelli, P.L. & N.D. Kazarinoff: Starfish predation of a growing coral reef community, J. Theor. Biol., 107, 667-684 (1984)

5. Antonelli, P.L. & N.D. Kazarinoff: Letter-to-editor, "Comments on starfish/coral cycles after R. Bradbury et al.", J. Theor. Biol., 119, 501-502

6. Antonelli, P.L., K.D. Fuller & N.D. Kazarinoff: A study of large amplitude periodic solutions in a model for starfish predation of coral. IMA J. Math. Appl. Med. Biol. 4, 207-214 (1987)

7.   Antonelli, P.L., P.W. Sammarco & J.C. Coll: A model of allelochemical interactions between soft and scleractinian corals on the Great Barrier Reef. 20 pp., J. inf. deduct. Biol. (in press)

8.   Antonelli, P.L. & R.M. Seymour: Entropy production in Stochastic Riemannian geometries with applications to chemical ecology. Adv. Appl. Math. 8, 254 - 280 (1987)

9.   Antonelli, P.L. & R.M. Seymour: A model of myxomatosis based on hormonal control of rabbit-flea reproduction. IMA J. Math. Appl. Med. Biol. 5, 65 - 80 (1988)

10. Antonelli, P.L. & J.M. Skowronski: Identifications of states and parameters in a model of starfish predation on corals. Math. Comp. Modelling. 10, 17 - 25 (1988)

11. Antonelli, P.L. & J.M. Skowronski: Adaptive identification of environmental stress for the management of plant growth. Math. Comp. Modelling. 10, 27 - 35 (1988)

12. Antonelli, P.L. & J.M. Skowronski: Differential offensive- defensive games between plants and herbivores. IMA J. Math. Appl. Med. Biol. 8, 319 - 340 (1987)

13. Antonelli, P.L. & B.H. Voorhees: Nonlinear growth mechanics - I. Volterra-Hamilton systems, Bull. Math. Biol., 45, 103-116 (1983)

14. Berwald, L.: Two-dimensional Finsler spaces with rectilinear extreamals, Ann. Math., 42, 84-112 (1941)

15. Kittredge, J.: Estimation of the amount of foliage of trees and stands, J. Forestry, 42, 905-912 (1944)

16. Rhoades, D.F.: Evolution of plant chemical defense against herbivores, In: G.A. Rosenthal and D.H. Janzen, eds., Herbivores: their interaction with secondary plant metabolites, Academic Press, 3-54 (1979)

17. Rhoades, D.F.: Offensive-defensive interactions between herbivores and plants: their relevance in herbivore population dynamics and ecological theory, in Amer. Nat. 125, 205 - 238 (1985)

18. Rund, H.: The Differential Geometry of Finsler Spaces, Springer- Verlag, Berlin, 285 pp (1959)

19. Takahashi, Y. & S. Watanabe: The probability functionals (onsager- machlup functions) of diffusion processes, in Stoch. Integrals, Lecture Notes in Math., #851, Proc. LMS. Durham Symp., 433-4563 (1980)

PERSISTENT AND TRANSIENT POPULATIONS OF THE CROWN-OF-THORNS STARFISH, *ACANTHASTER PLANCI*

RICHARD J. MOORE

School of Biological Sciences, Queen Mary College (University of London), Mile End Road, London E1 4NS

Abstract. Current models to explain 'primary outbreaks' of *Acanthaster planci* assume recruitment of larvae to the natal reef. In the light of recent studies and the current awareness of the importance of larval advection in the recruitment of coral reef organisms, two new models are offered; particular attention is drawn to the possible importance of persistent breeding populations of *A. planci* as sources of larvae initiating sequences of secondary outbreaks. Reef habitats harbouring stable *A. planci* populations are characterised by hydrodynamic systems retentive to larvae, and poor coral prey availability, in contrast to the strong flushing and rich coral cover of open-water reefs where outbreaks occur. Other features of stable *A. planci* populations are a significant contribution to diet by alternatives to hard corals; and reduced predator pressure. A simple model of the important processes influencing starfish population stability is developed. The ability to alternate between endemic and epidemic life-styles in different habitats is a common feature of 'outbreaking' organisms.

INTRODUCTION

A decade ago, two papers discussing the life strategy of the crown-of-thorns starfish, *Acanthaster planci*, were published (Cameron 1977, Moore 1978). Both sought to explain the causes of instability of *A. planci* populations in relation to the role of its morphological, reproductive and behavioural characteristics in its population ecology.

We reached contradictory conclusions. Cameron argued that *A. planci* is specialised in its diet, and its large adult size and spiny armature indicate a conservative strategy with respect to predator pressure, i.e. it is a typical 'K-strategist' (Pianka 1970) in a complex, equilibrial and biologically-accommodated ecosystem. It could be expected to have a stable, low density population, and homeostatic mechanisms would - under natural circumstances - maintain this situation, and thus Man was implicated in the occurrence of outbreaks.

While acknowledging the K-selected aspects of *A. planci's* ecology, I argued that in other important respects, *A. planci* appeared to be r-selected: exploiting a super-abundant prey resource (a virtual 'ecological vacuum'); having a high fecundity; and the consequential massive recruitment episodes of irregular occurrence combined with the vulnerability of the larvae to transport by currents to give a fugitive life-style over succeeding generations.

In retrospect, the disparity in our views could be dismissed as the failure of a misconceived attempt to explain life histories as outcomes of single selective pressures (Wilbur *et al.* 1974), and on the basis of inadequate information. However, the issue is an important one because it hinges on the extent to which *A. planci* populations are structured by density-dependent controls or by the influx of recruits from the plankton. This question is the crux of the renewed interest in so-called 'supply-side ecology' (Lewin 1986, Underwood & Fairweather 1989).

Considerably more information on the recruitment and ecology of *A. planci* populations has now become available, sufficient to re-assess the place of *A. planci* within coral reef ecosystems. In this paper, I first present a succinct review of our current knowledge of recruitment processes on coral reefs, and the implications for the known patterns of spatial and temporal distribution of *A. planci* populations.

Two major characteristics of starfish populations on open-water reefs - high hydrodynamic flushing rates and very low endemic starfish population densities - argue against their role as initiators of primary outbreaks. I point to the occurrence of some relatively stable and persistent populations of *A. planci*, and suggest that they are important as sources of larvae seeding outbreaks. A comparison of their ecology suggests that they have several common characteristics which may distinguish them from outbreak populations. A simple model is developed to show how the varying ecological characteristics of different types of reef habitat may influence *A. planci* population behaviour and dynamics. Finally, I reconsider *A. planci's* life strategy within the r-K conceptual framework, and compare it with other animals with markedly fluctuating populations.

MODELS OF RECRUITMENT IN CORAL REEF ORGANISMS

Inevitably, early models of coral reef community structure imitated those prevalent in terrestrial ecology, which were seen as primarily resource-limited and structured by competition and predation. Populations of terrestrial animals are primarily considered to be enclosed within a constant range, and maintained close to carrying capacity by density-dependent controls on recruitment, i.e. they are primarily K-selected (Cody & Diamond 1975). Recruitment is thus limited, except at the edge of a species range where recolonisation becomes important, as a result of catastrophic density-independent mortality factors. The unfavourable environment there makes populations less stable, and the species appears to be more r-selected (Huffaker and Messenger 1964, Whittaker 1971).

Taylor and Taylor (1977) pointed out that this view underestimated, or even ignored, the importance of migration or dispersal, which tended to be relegated to the status of unaccountable mortality. They criticised population models that were primarily concerned with equilibrium conditions, and proposed a model describing the spatial dynamics of populations, of general applicability to a wide range of mobile organisms. They visualised a distribution map with patches constantly changing their geographical position over time, a space-time reticulum which, they suggested, could be visually represented by a drawing of the stelar (vascular) structure of a particular species of fern.

All populations are spatially fluid to some extent, because movement is a fundamental biological response to adversity (Taylor & Taylor 1977). However, most marine organisms have a life history that involves an obligate period of dispersal in the plankton, giving a much greater potential for removal from the parental habitat under the influence of physical transport processes. Curiously, this was rarely evident in the early models of intertidal communities (Connell 1975, Paine 1974), which seemed to match the determinism of terrestrial models.

It now seems clear that the intertidal systems from which such models were developed may not have been typical. They appeared deterministic because space created in them by predation or physical disturbance was reoccupied by a superfluous supply of recruits of the dominant species from the plankton. However, these models do not reflect the recruitment processes occurring at many sites, because such a predictable, saturated supply of larvae is exceptional. If propagules of important species do not arrive in a habitat with great regularity, and in sufficient numbers to fulfil their interactive roles in the system, the patterns of occupancy of resources will not be determined by those species, and the structure of assemblages becomes indeterminate (Connell 1985, Underwood & Fairweather 1989).

The supply of pelagic larvae to many benthic populations has, in fact, been known to be notoriously unpredictable for many years (e.g. Thorson 1950). Much of the temporal and spatial variability in settlement is believed to result from patchiness in larval distribution, which has been attributed to many physical and biological factors (see Gaines et al. 1985 for references.). It is possible that Taylor and Taylor's fern stele model may be appropriate to describe the temporal and spatial distribution of the supply of larvae to vacant habitats.

The archipelagic nature of coral reef ecosystems confers its own patchiness to larvae starting their pelagic life. The potential movement of planktonic larvae in the central region of the GBR has been modelled by Williams *et al.* (1984). They considered that following spawning, larvae of most corals and other invertebrates, and of fishes, with pelagic lives of several weeks, are unlikely to be trapped over or around reefs until settlement; instead, they would be rapidly entrained in the 'mainstream' circulation over the continental shelf, and advected to other reefs.

The conclusions of this model appear to have been challenged by preliminary results of experimental studies of recruitment of corals at varying distances from a reef acting as a source of larvae (Sammarco and Andrews 1988). They found that numbers of corals recruiting to moored settlement plates declined logarithmically with distance from the source reef, decreasing by an order of magnitude more than 2.5 km from the reef; 70% of all recruits settled within 300 m of the reef. They concluded that coral reefs are primarily self-seeded with respect to coral larvae, and suggested that the scale of dispersal allowed only 'highly limited' recruitment to other reefs, although it was sufficient to ensure gene flow over oceanic distances.

This is an important study, but it should be pointed out that a number of factors limit the generality of the authors' conclusions. First, data for brooded planulae with planktonic lives of only a few days were combined with that for species which are externally fertilised and have planktonic lives of several weeks. Clearly the latter are more likely to escape from the reef before settlement. Secondly, the potential for exporting recruits at the 'background level' of recruitment, attained several kilometres from the source reef, is very considerable: 6-7 per 600 cm2, equivalent to approximately 106 ha^{-1}. Thirdly, if we accept that rich reefs are normally space-limited (see below), recruitment success on the natal reef will be very much lower than on the settlement plates; while if exported larvae are able to settle on another reef with a much greater availability of space (e.g. following an outbreak of *A. planci*), their recruitment rates could be relatively much higher.

The high diversity of coral reefs initially appeared to be adequately explained by classical niche theory, i.e. through the evolution of organisms specialising on the large number of spatial and biotic niches, created by environmental predictability in reef microhabitats coupled with strong environmental gradients across them (Grassle 1973). This did not mean, of course, that all species associated with coral reefs are specialists (i.e. K-strategists) constrained by these niches, it merely explained the high diversities observed. However, certainly with respect to the major sessile components, rich reefs appear to be nearly saturated communities: hard corals, soft corals and sponges are clearly K-selected in view of their massive size, long life, adaptations for inter-specific competition for space (e.g. Connell 1973), and a variety of morphological and chemical defences against predation (Bakus & Green 1974, Vermeij 1978, Bakus 1981).

Among the more mobile components of coral reef ecosystems, fish communities have also tended to be treated as persistent and stable assemblages, with homeostatic control over both species richness and patterns of relative abundance. As with the apparently deterministic temperate inter-tidal systems, it was assumed that reproduction produced a surplus of potential recruits, but that the abundance of populations was limited by post-settlement processes (see review by Sale 1980). Sale concluded, however, that coral fish communities are predominantly unstable and suffer from large fluctuations in recruitment.

In a review of more recent studies on the replenishment of tropical reef fishes, Doherty & Williams (1988) found that the available evidence did not support the hypothesis that coral reef fishes are generally limited by the carrying capacity of reef environments. Rather, abundances of at least some species are driven by recruitment dynamics, and as such are not fundamentally different from those of temperate marine fisheries. In accordance with the model of reef flushing of Williams *et al.* (1984), Doherty & Williams describe various pieces of circumstantial evidence suggesting that fish populations at the scales most often studied by coral reef ecologists (i.e. a single reef) do not constitute closed self-recruiting populations ('stocks' in fisheries terminology), but rather local populations are more probably replenished by larvae spawned outside a reef population.

Synoptic surveys of recruitment across replicate units of habitat have shown that much of the stochastic variation in larval replenishment is associated with transient pulses, which are coherent across spatial scales up to tens of kilometres. The coherent pulses, Doherty & Williams suggest, are most likely to be the result of dense patches of pre-settlement fishes, such as have been occasionally observed passing over reefs. The authors do not discuss how such larval patches might arise, but patches up to a few tens of kilometres wide are consistent with the scale of larval clouds generated by a population on a single reef, under the advection-diffusion model of Williams *et al.* (1984), after a period of time corresponding to that of larval development.

PATTERNS OF RECRUITMENT OF *A. PLANCI*:
SECONDARY INFESTATIONS

No such larval clouds have been observed for *A. planci*, indeed *A. planci* larvae have never been identified in plankton surveys. But *A. planci* does provide a single important advantage to its study over that of other coral reef organisms. The uniquely conspicuous impact of its predatory activities on coral colonies has allowed large-scale synoptic surveys, using simple techniques, to describe the wide-scale distribution of dense populations of the starfish.

A. planci populations are patchily distributed in space and time, and occur at a wide range of densities, fluctuating through up to 5-6 orders of magnitude. While it may not be be possible to define an 'outbreak' precisely (Potts 1981), its meaning is generally used to indicate that numbers of the outbreaking species markedly exceed the carrying capacity of its habitat. That this may be occurring in a population of *A. planci* is made conspicuous by the presence of numerous white feeding scars, and synoptic surveys are sufficient to estimate distribution and relative abundance of the starfish at order of magnitude level.

Where such surveys have been repeated over a number of years, it has been possible to interpret, with some confidence, shifting patterns of dense populations of *A. planci* in several distinct areas of the Indo-Pacific. Dense populations identified early in a series are plausible sources of clouds of larvae which may be advected to other reefs by prevailing currents; and thus dense populations later in the series can be construed as subsequent generations of dense starfish populations, connected by larval advection.

The sudden appearance of outbreaks could also be attributed to migration of large numbers of adults from another reef area (Endean 1974). However, while evidence for the role of occasional large recruitment episodes from the larvae, in the occurrence of outbreaks, has accumulated with time, no specific evidence for adult migration has emerged, and any such contribution of migration of adults is likely to be restricted to reefs nearby in relation to the distances that larvae may be transported.

Crown-of-thorns starfish outbreaks often appear rather suddenly as large numbers of previously cryptic juveniles mature and start to feed conspicuously in the open (Zann *et al.* 1987). These can occur as the result of large inputs of larval recruits in single years (Doherty & Davidson, 1989). In Fiji, recruitment of *A. planci,* monitored over 9 years, varied considerably from year to year, but was low or very low in all years except one. In that year, recruitment was so massive that, in spite of the juvenile population suffering >99% mortality, an outbreak still ensued when the survivors matured (Zann *et al.* 1987).

While suggestions that successions of outbreaks of *A. planci* progressed in a certain direction, reflecting transport of larvae by currents, had been made earlier, the first detailed analysis of the geographical distribution and timing of outbreaks on the GBR, attempting to relate them to rates of growth to reproductive maturity, was made by Kenchington (1977). He suggested that the southward spread of dense starfish populations from 1957 to 1972 could be interpreted as several successive waves, arising through recruitment of larvae from mother populations to the north. The direction and distances of larval advection required by his model are consistent with known current directions and velocities (Williams *et al.* 1984). Kenchington's ascription of distinct parent and progeny populations, connected through transport of larvae by currents, is striking; indeed, *A. planci* may be the first marine benthic organism for which such a relationship between geographically separated populations has been claimed.

The resurgence of *A. planci* outbreaks on the GBR since 1979 has prompted more regular synoptic surveys of its distribution, the results of which (Moran *et al.* 1988) have tended to substantiate the broad principle of Kenchington's model, i.e. a general trend of southerly drift, although deviations from this trend may occur due to temporary current reversals.

The pattern of outbreaks in the first documented starfish plague on the GBR, as interpreted by Kenchington (1977), provided the basis for surveying the distribution of cryptic juveniles in a transect of reefs extending 300 km downstream of a region with active outbreaks in the previous year (Doherty & Davidson, 1989). The settlement of *A. planci* in 1985 across a broad geographic area was an order of magnitude greater than that during either 1986 or 1987. While the 1985 settlement was not very high (mean densities of 1+ juveniles of 0.03 m-2 = 300 ha^{-1}), it is 3 orders of magnitude higher than the 'normal' low-densities found on uninfested reefs (Endean 1974).

While the number of reefs surveyed by Doherty & Davidson was insufficient to allow identification of the total effective larval settlement area, it is noteworthy that the distribution of juveniles found is consistent with the occurrence of two distinct larval clouds -one settling on Lodestone, Keeper and Little Broadhurst, the other on reefs between Stanley and Bait inclusive - perhaps originating from two different source reefs.

Starfish outbreaks in other parts of the Indo-Pacific have similarly been attributed to parent populations up to several hundred kilometres distant. Yamaguchi (1986) has interpreted the timing and geographical distribution of *A. planci* outbreaks in the Ryukyu islands with a similar conclusion, larvae being advected to different reefs according to variations in the course of the Kuroshio current.

In the Red Sea, I predicted (Moore 1985) that an outbreak would occur on the Sudanese (western) coast, following observation of a large breeding population on the Saudi Arabian (eastern) coast in 1983, and the inference of a current traversing the 300 km width of the Red Sea. A survey of Sudanese reefs showed that starfish population densities had increased from <10 ha^{-1} in 1984 to >100 ha^{-1} in 1987. The occurrence of a westerly cross-current at c. 19°N has been confirmed following the identification of a cyclonic gyre using CZCS satellite imagery (R.J. Moore, unpublished data).

If these ascriptions of connected parent and progeny populations are correct, they indicate that dense adult populations of *A. planci* can rather regularly generate further dense populations on distant reefs, which, if currents patterns are constant and unidirectional, can somewhat predictably be identified. The simplest interpretation of this is that, following spawning by dense *A. planci* populations, sufficient larvae regularly survive mortality in the plankton, at settlement, and as juveniles to generate another high-density population of adult starfish on any reef that is suitably positioned (with respect to advection velocities and larval development time) down-current.

Like many other coral reef invertebrates, *A. planci* is highly fecund (Kettle & Lucas 1987), but unlike most, breeding in dense populations frequently occurs between starfish that are already aggregated by feeding activity (Ormond & Campbell 1974). Moreover, spawning appears to be synchronised by a pheromone (Beach *et al* 1975), and together these factors will tend to favour high fertilisation success in dense starfish populations. The huge numbers of larvae which can thus be expected to be produced will suffer some mortality through starvation, predation and other factors; and the spatial density of the larval cloud will be further reduced by diffusion by physical processes.

Larvae settling on a reef and developing as juvenile starfish then become subject to predation and other mortality factors associated with the reef. To the extent that this mortality is density-dependent, dilution processes in the plankton are apparently insufficient to reduce the larval settlement density to a level at which the resulting population of juvenile starfish will remain under the control of predators and other mortality factors. Thus the patchiness in distribution of the parent population may be maintained during the pelagic life of the larvae, and reproduced in the next generation of adults.

This interpretation appears to be in conflict with that of the vast majority of population studies on other marine benthic organisms. It is well known in fisheries ecology that recruitment is often independent of stock densities, and Kenchington's (1977) conclusions were challenged by Ebert (1983), who questioned whether variation in numbers of adult *A. planci* causes variation in subsequent recruitment. To illustrate this effect in echinoderms, Ebert (1983) used the example of a population of a temperate asteroid, *Asterias forbesi*, which in 25 years of study had shown little correlation between adult density and settlement density (Loosanoff 1964).

In Loosanoff's study, settlement of the starfish larvae was studied at 10 stations along a c. 25 km stretch of the northern shore of Long Island Sound. Adult abundances were surveyed outside this area, but again in a restricted part of the Sound. Long Island Sound is c. 160 km long, but

Loosanoff provided no data to indicate whether or not there was uniformity of larval settlement or adult population densities over different part of the Sound. Nor did he consider the possibility that the larvae settling may have originated from adults located in another, unsurveyed, part of the Sound - or, indeed, from outside it. He estimated that *A. forbesi* larvae spend c. 21 days in the plankton before settlement, thus providing considerable opportunity for advection.

In the light of the previous discussion on patchiness of larvae, there is clearly an alternative explanation why variation in numbers of adult marine organisms is often not related to variation in subsequent recruitment. Like coral reef organisms, populations of temperate benthic organisms are to some extent patchily distributed, and thus larvae will also start their pelagic lives patchily distributed. This patchiness could be either enhanced or diminished by subsequent physical and biological factors (see above), but some patchiness in settlement seems inevitable.

Using inferred patterns of larval advection and settlement of *A. planci* larvae on the GBR as a model for benthic invertebrates generally, we can see that to an observer investigating settlement at a fixed study site, there will be little relationship between adult densities and settlement densities; whereas to an observer travelling with the larval cloud produced by an *A. planci* population at one site, being carried to settlement at another site, the relationship between adult and settlement densities will be much closer.

THE INVASION STRATEGY

I suggest that the way in which *A. planci* achieves persistence at high density levels on archipelagic reef systems over many generations may be seen as part of a life strategy to invade an ecosystem in which it is under considerable pressure from predators.

Many reef organisms have evolved a variety of morphological and chemical defences against predation (Bakus & Green 1974, Vermeij 1978, Bakus 1981). However, while the spiny armature of the crown-of-thorns starfish may appear formidable, it is only a partial defence against certain generalist fish predators, particularly triggers and puffers (Ormond & Campbell 1974). Most populations of *A. planci* on rich reefs have a high (30-60%) incidence of damaged or regenerating arms (Moran 1986), which is generally attributed to attack by such predators. Nocturnality is selected for by diurnal pressure from visually-oriented predators (Fricke 1974), and mediated through a light-avoidance response. Accordingly, in low population densities *A. planci* are usually primarily nocturnal, remaining concealed in the daytime.

At higher starfish densities, a number of factors may operate to alter this behaviour. A shortage of prey and coral cover, and attraction by mutual feeding stimulation (Ormond & Campbell 1974), overrides the light-avoidance response. At extremely high densities, starfish may aggregate in bands, feeding in the daytime. At such densities, *A. planci* clearly suffers a greater risk from starvation than from predation, and at the same time, grouping reduces the impact of predators. Defensive superiority is the adaptive advantage of social behaviour reported most frequently in field studies, and is the one which occurs in the greatest diversity of organisms (Wilson 1975). The dense groupings formed will tend to reduce the effect of territorial - or otherwise spaced-out-predators, by satiation with an excess of available prey, i.e. 'predator swamping'. This type of relationship between predators and the population density of their prey is potentially destabilising, and can be described mathematically as a consequence of the type of functional response of predators to changes in the abundance of their prey (Holling 1959).

McCallum (1987) has developed a model of this type for *A. planci*, to investigate the properties that its predators would need to possess if they were capable of regulating starfish numbers at a low endemic level. Using either a Type II or Type III functional response (Holling 1959), the model may produce two stable equilibria. At the high (epidemic) level, *A. planci* escapes the control of predators and is limited by availability of its coral prey. However, only the Type III functional response allows starfish to persist at the lower equilibrium. Conclusions from McCallum's model are, however, limited in that it describes only a closed system, i.e. it does not consider recruitment from outside the population on an individual reef. As he recognises, continual immigration of larvae from other reefs would be likely to result in some starfish being present on most reefs. Thus, as argued here, sufficient numbers of larval recruits could immigrate from other reefs to break through the predator 'barrier' and establish the epidemic level.

The defensive advantage of grouping is rendered most effective by synchronisation in place and time of all life stages (Wilson 1975). The synchronisation of spawning of aggregated starfish by a pheromone (Beach *et al.* 1975) has already been mentioned. The increased reproductive effectiveness would tend to amplify the predator-swamping effect, further destabilising population densities through an inverse (positive) dependence of reproductive rates on density.

Once larval development has begun, and larvae which have escaped the influence of the natal reef become advected by wider-scale current systems, there is an advantage in synchronisation of larval development to reduce dispersion of the larvae at settlement. Laboratory studies by Lucas (1982) showed that larval development times can vary considerably. However, with phytoplankton levels giving the highest survival rates (5000-10,000 cells/ml), corresponding most closely to oligotrophic levels of phytoplankton availability found on open-water reefs of the GBR, Lucas found that larval development proceeded very uniformly (pers. comm.). If such uniformity occurs under field conditions, it could have an important effect in tending to concentrate settlement over a relatively small area. This would tend to reduce the impact of territorial predators on juvenile starfish, thus enhancing the 'predator-swamping' effect.

Rich reefs can thus be regarded as 'high risk, high reward' habitats for *A. planci*, which it can only invade temporarily, and inevitably with a destabilising influence. At the epidemic level, reef populations are limited by availability of coral prey, and will decline when it becomes exhausted; but they can be maintained at their high levels in succeeding generations by invading other reefs by larval advection. Thus *A. planci* is a 'fugitive' in its larval stage by virtue of current advection; and in its adult stage, from predator pressure.

THE ORIGINS OF OUTBREAKS OF *A. PLANCI*

MODEL 1: Primary outbreak populations as initiators

We have seen, in the preceding section, that high density *A. planci* populations show a considerable measure of persistence over several generations, given a suitable supply of reefs to which the larvae produced by each generation may be transported and settle. Crown-of-thorns starfish outbreaks are more difficult to explain when no large aggregated populations, which might be responsible for an abnormally large input of settling larvae, are known in the vicinity. This is the case with outbreaks on isolated oceanic reefs, and with the (as yet unidentified) population on the GBR which initiated the sequences of secondary outbreaks (Kenchington 1977). These are usually referred to as 'primary' outbreaks or infestations, following Endean (1973) who wrote:

> 'a distinction should be made between primary infestations of reefs following local increases in starfish numbers on certain reefs owing to the operation of factors peculiar to those reefs or the waters around those reefs, and secondary infestations stemming from carriage by currents of large numbers of larvae from the primary centers of infestation to other reefs, or from migration of large numbers of adult starfish from devastated reefs to other reefs' (Endean 1973).

This concept of the role of a primary outbreak population in generating secondary outbreaks remains intuitively plausible. The factors Endean proposed as peculiar to the central region of the GBR, which he was considering as a primary outbreak area, were an increase in numbers of adult *A. planci* resulting from removal of predators, and a 'closed or semi-closed' system of surface currents causing larvae to settle in the same area.

There has been considerable controversy over whether or not the first of these factors, in particular, has operated, but this has tended to obscure the fact that Endean's distinction between primary and secondary outbreaks has not been substantiated. In most cases there is insufficient knowledge of the processes leading to outbreaks to classify them with much confidence (Moran 1986).

The basic criterion for a population of *A. planci* to have the primary role described above is that it should be self-sustaining, i.e. juveniles from the larvae must be recruited to the parental population with sufficient regularity to maintain a viable breeding population. Most models to explain the occurrence of primary outbreaks implicitly assume this, but focus on occasional factors that might cause a abnormally large settlement of larvae.

Thus Birkeland (1982), developing the 'larval recruitment hypothesis' of Lucas (1973) and Pearson (1975), proposed that terrestrial run-off from occasional heavy rains may provide enough nutrients to stimulate phytoplankton blooms. He hypothesised that, with the resulting super-abundant supply of phytoplankton food, abnormally large numbers of larval *A. planci* would survive, and subsequently appear as an outbreak. The correlate of this hypothesis, that high mortality of *A. planci* larvae occurs under normal levels of food availability, has been refuted by in situ culturing of larvae, which showed that high survival rates occur under the oligotrophic levels of phytoplankton availability typical of the GBR (Olson 1987).

Olson suggested, instead, that the occurrence of primary outbreaks might be explained by abnormally low wastage of larvae by dispersal of currents away from the reef, i.e. by the occasional occurrence of oceanographic conditions favourable to the retention of larvae. Such an explanation is appealing because of its simplicity and generality. The transience of dense adult populations of *A. planci* (outbreaks tend to disappear within a few years) is an indication of the strength of influences on the dispersal of *A. planci* larvae. On isolated oceanic reefs such as in Micronesia, secondary outbreaks rarely occur. Potts (1981) suggests that this is because the islands and atolls are separated by great distances of deep oceanic waters which preclude adult migrations, and cause larvae produced by even the most massive spawning events to disperse. This is particularly so in regions where the trade winds are fairly constant during long periods of the year, and where the topography of the reefs results in their direct exposure to open ocean currents, as on Guam and other islands in the Marianas (Yamaguchi 1975).

As with previous hypotheses, Olson's hypothesis implicitly assumes that the adults pre-existing before a primary outbreak breed successfully enough to maintain the population at that level, and under favourable environmental conditions, or release of predator pressure, to actually increase.

This assumption should be examined more closely: although there is no evidence either way, population densities prior to primary outbreaks are extremely low, and doubts have often been expressed that fertilisation of gametes from such widely spaced individuals can occur (Lucas 1973, Vine 1973, Ormond & Campbell 1971, Potts 1981). Endean estimated a density of 0.06 ha^{-1} for *A. planci* on uninfested reefs of the GBR (Endean 1974). This was dismissed by Potts as 'almost certainly too low', but other reports from the Indo-West Pacific indicate that *A. planci* is usually equally scarce. Vine (1970) saw no *A. planci* on 81% of 83 reefs in the Pacific which he surveyed in 1968-70, and it was 'scarce' (1-5 per 20 min dive) on another 14% of these reefs. The frequency distribution of *A. planci* population densities presented by Dana *et al.* (1972) showed that on 400 of 617 surveys (65%), no *A. planci* was observed, and on another 173 (28%), less than 20 starfish were observed (usually in 20 min swims or tows). Birkeland (1982) cites various reports that *A. planci* was very scarce in American Samoa between 1966 and 1977. He himself observed a total of less than 20 individuals during c. 120 h diving between 1975 and 1979 on Guam, where large outbreaks appeared in 1967 and 1979.

The usual justification for the assumption that such sparse populations are capable of effective breeding is the high fecundity of *A. planci*. Pearson & Endean (1969) calculated that mature females can spawn from 12-24 million eggs, while Conand (1983) found that the largest individuals can produce up to 60 million eggs in a season. However, many other coral reef invertebrates are also highly fecund, yet few are known to exhibit large population fluctuations (Birkeland 1982), so that this reasoning should continue to be treated with reservation. In view of the genetic homogeneity of *A. planci* populations throughout the Pacific Ocean (see below), it seems more plausible that the sparse 'normal' populations on oceanic reefs are only stray survivors of larval recruits exported from dense breeding populations on distant reefs, rather than self-recruiting populations.

MODEL 2: The 'fern stele' model

Outbreaks on isolated oceanic reefs have been considered as primary only because it appeared a remote possibility that larvae or adults could have immigrated across the barriers of long distances and deep water which separate them (Moran 1986). This view may hold for adult migration, but needs to be reconsidered for larvae.

Many echinoderm larvae, including those of *A. planci*, are planktotrophic with long pelagic lives; planktotrophic larvae are generally longer-lived than non-feeding larvae (Grahame & Branch 1985). The larvae of *A. planci* can spend as little as 12 days developing in the plankton (Olson 1987) before settlement, but larval development rates are very variable, and laboratory studies have shown that they can continue for more than 6 weeks (Lucas 1982).

In his review of recruitment in echinoderms, Ebert (1983) gave little consideration to the possibility that planktonic larvae recruited to a population may have been transported considerable distances. However, Yamaguchi has suggested that the geographical distribution of Pacific coral reef asteroids is related to the mode of swimming of their larvae, which determines the extent to which the larvae are exposed to the influence of transport by surface currents. *A. planci* larvae are negatively geotactic and thus swim at the surface: *A. planci* and other asteroid species with this swimming mode have a wider distribution than those with positively geotactic behaviour (Yamaguchi 1977).

Some indication of the amount of gene flow between *A. planci* populations has been obtained in recent electrophoretic studies of enzyme loci (Nishida and Lucas 1988). They have shown remarkable genetic homogeneity in *A. planci* populations throughout the Pacific, so that there now appears to be no basis for considering eastern Pacific *A. planci* populations as a separate species. The major source of this genetic similarity is most likely to be ongoing or recent gene flow though the larvae. Larval development times may exceed 6 weeks on marginal diets (Lucas 1982), but no specific adaptations for teleplanic life are known for *A. planci* larvae. However, they must be capable, occasionally at least, of traversing vast oceanic distances, perhaps through delaying metamorphosis.

The appearance of *A. planci* populations on remote islands at high latitudes where water temperatures may be too low for the starfish to breed is an indication of the possible importance of transport of larvae over oceanic distances. McKnight (1978) suggested that an *A. planci* population in the Kermadec Islands, in New Zealand waters, had arisen through transport of larvae from Tonga (500 km), but more likely from Rarotonga (1200 km), in view of the direction of prevailing currents.

Cool water temperatures could also assist *A. planci* larvae in crossing the eastern Pacific barrier, by prolonging larval life: at temperatures below 25°C, *A. planci* larvae do not advance into the brachiolaria stage, though they continue feeding (Yamaguchi 1973). Development is resumed when water temperature is increased to 28°C (Henderson & Lucas 1971). So larvae can tolerate arrested growth due to low temperature and then develop normally, although the length of period of arrested development which can be tolerated has not been investigated.

The developing view of considerable larval flux both within and between different reef areas is tending to undermine the classification of outbreaks on oceanic reefs as primary, simply because one cannot be confident that the larvae which produce an outbreak are derived from the resident population. Given the doubts over the potential of the pre-existing population to produce sufficient larvae to generate a subsequent outbreak, it is thus worth considering that outbreaks on isolated oceanic reefs might conceivably be seeded by large aggregations on reefs in the order of thousands of kilometres distant. This might require cohesion of the larval cloud, e.g. in mesoscale eddies (Scheltema 1986, Lobel & Robinson 1986) to counteract diffusion processes over such distances.

In the light of the interpretation of sequences of secondary outbreaks on the GBR and elsewhere discussed above, astronomical fecundities alone are scarcely sufficient to account for the generation, by a few individual starfish, of an outbreak population as large as those believed to be produced by a known pre-existing outbreak population on the GBR.

While no field experiments have been reported in the literature, any plausible model for larval production based on fertilisation success would include not only numbers of mature starfish, but an inverse function of inter-starfish distance, perhaps geared under the influence of a pheromone inducing starfish to aggregate and spawn synchronously (Beach *et al.* 1975). I suggest that it is this aspect of *A. planci*'s ecology which distinguishes it from other reef organisms, and allows outbreaks to occur, by focussing the reproductive effort of a reef population into a concentrated cloud which can persist in sufficient concentration to establish further generations on reefs in the order of 1000's of kilometres distant.

Comparing total production of larvae by a dense, partially aggregated population with that of the sparse pre-existing populations postulated to generate primary outbreaks on isolated oceanic reefs: differences in population density of seed populations of, say, 4 orders of magnitude could be expected to result in a greater total production of larvae by the dense population, of several orders more than this.

A large breeding population could conceivably also seed several distant oceanic reefs, themselves separated by large distances, through fragmentation of the larval cloud by physical processes. This might explain the simultaneous appearance of outbreaks on many Pacific islands in 1969 (Endean & Chesher 1973). Another noteworthy coincidence in timing of outbreaks is that between the GBR and Fiji. On the GBR, the first documented series of outbreaks ran from 1962 to the early 1970s, and the second plague started in 1979 (Moran 1986). On Fiji, outbreaks occurred from 1963-70, followed by a gap until 1979 (Zann *et al.* 1987).

On this basis, a model could be proposed of a continuous succession of transient dense populations, i.e. 'secondary' outbreaks, maintained indefinitely over succeeding generations by transport of larvae by large-scale circular systems of currents. This is, in effect, Taylor and Taylor's (1977) 'fern stele model' in the marine environment, the structure of the fern stele representing the space-time reticulum of the distribution of dense breeding populations, which constantly change their geographical position over succeeding generations, connected through larval advection.

MODEL 3: Persistent starfish populations as initiators

Under the above Model 2, however, chains of outbreaks on reefs connected by larval advection are vulnerable to extinction by unfavourable current patterns, or conditions unfavourable to the cohesion of a larval cloud, generating only sparse adult populations with little effective breeding potential. The main purpose of this paper is to point out that some types of reef habitat allow *A. planci* populations to persist at density levels intermediate between the extremes found on rich open-water reefs, at or around the carrying capacity of the reef; and to suggest that such populations may be important sources of larvae initiating the outbreaks that are observed on open-water reefs.

This model has similar hydrodynamic preconditions to that of Endean's described above, in that the prevailing hydrodynamic system must be 'closed or semi-closed' to favour retention of larvae and thus ensure persistence of the population through self-recruitment, but also allowing occasional release of larvae which could seed outbreaks on other reefs. However, *A. planci* populations in such habitats are not 'primary' outbreak populations in the sense used by Endean, as such habitats are unable to support high density outbreak populations; rather, they are permanent breeding populations in habitats which are distinct from those reefs on which outbreaks occur.

Insofar as populations in such 'permanent habitats' do fluctuate, the likelihood of them seeding an outbreak on distant open-water reefs will be increased by higher densities of the population generating the larval cloud, but those could occur as part of a natural cycle of fluctuations through a narrower range of population densities. However, other conditions, particularly hydrodynamic, may need to be favourable for the larval cloud to remain coherent, and a dense settlement of larvae to be effected.

The evidence for the capacity of persistent populations to initiate outbreaks on other reefs is limited to one known population, and is indirect. Nevertheless, if the principle of connection of sequences of secondary outbreaks through transport of larvae by currents over large distances is accepted, then this model is a plausible alternative to other models in the literature, given the lack of evidence for 'primary outbreak' populations sensu Endean. By examining what is known of persistent populations in several reef areas in the Indo-Pacific, I hope to identify common features which may aid the identification of hitherto unknown persistent populations elsewhere which may be important in initiating outbreaks, e.g. on the GBR.

THE IDENTIFICATION OF PERSISTENT POPULATIONS

Unless immigration of adults contributes a significant input to recruitment to a population, the major factor distinguishing persistent populations from transient populations must be the frequency at which significant recruitment of juveniles occurs relative to the longevity of the starfish. However, juveniles up to c. 20 months old are very cryptic (Zann et al. 1987), and thus the literature of studies in which direct data have been obtained, on a regular basis at fixed sites, is minimal.

Inferences from size data are usually treated with suspicion: growth rates of A. planci are very variable, so that it is not possible to directly determine the age of a starfish from its size. In theory, it is possible to resolve size frequency data from a population comprising several age classes into component normal distributions around the mean sizes of individual year classes (Harding 1949, Cassie 1954). The criticism of this method that it depends on a subjective assessment (Grant et al. 1987) is compounded by a wide range of variances of size distributions of year classes reported for A. planci, which thus makes it difficult to check on the validity of the analysis. In studies of growth rates of laboratory-reared A. planci up to 2 years old, Yamaguchi (1974) reported S.D.s in the range 15-30 mm; while Zann et al. (1987) found a similar range (18-35 mm) for a single year class studied on a reef in Fiji.

If variances were always this small, it would be relatively easy to distinguish different age-classes in a mixed population. However, Kenchington's analysis of size-frequency data from GBR populations using the Harding-Cassie method yielded S.D. values ranging from c. 20 mm to more than 50 mm (Kenchington 1977). In part this wide range of values may have resulted from higher growth rates on the GBR, an effect also observed by Zann et al. (1987) in relatively favourable habitats in Fiji. However, the highest variances were observed during the breeding season: spawning is associated with a significant decline in diameter (Branham et al. 1971, Yamaguchi 1974), so that during the breeding season starfish of the same age which have spawned, and those yet to spawn, will tend to produce sub-groups of different size mode.

Such high and seasonally variable variances make it difficult to analyse size-frequency data with confidence, particularly for data collected during the breeding season. Even treating size-frequency data as a whole, i.e. considering the population as unimodally distributed, standard deviations for most field populations rarely exceed 60 mm (see Table 1), on which basis it is rarely possible to be confident that it comprises multiple age-classes, rather than a single one, without prior knowledge of the variances of that population. Migration of starfish of certain size classes may also distort size-frequency distributions (Wilson & Marsh 1974, Moore 1985).

Table 1. Density ranges in some populations of A. planci

Reef area	Pop.density range, ha^{-1}	Years surveyed	Diameter range,mm	S.D. mm	Source
GBR, Australia	0.06-1400 <400-15.10^4	1962-	100-500	<80	P&E 1969, Endean 1974, Potts 1981, Kenchington 1977
Guam	10^4 '<20/120 h'	1967 1975-79	-	-	Chesher 1969, Birkeland 1982
Dampier Archipelago W. Australia	40-240 'common' 45 92	1972-4 1978 1983 1985	140-450 100-400	37-60 66	Wilson & Marsh, 1975, L.M.Marsh, p.comm., Simpson & Grey 1989
E. Pacific: Uva I Panama	45-130 1-27	1970-74 1976-83	-	-	Glynn 1974 Glynn 1985
Red Sea, Wingate and Towartit complexes	3-400	1969-87	150-480	50	O&C 1971, 1974 Moore 1985
Dunganab Bay, Red Sea	8-127	1975-84	120-380	58	Moore 1985

P&E = Pearson & Endean; O&C = Ormond & Campbell.

Thus the only reliable criterion suitable to judge literature data regarding persistence of a population is relative stability in population density over periods of time longer than the longevity of *A. planci*, observed at intervals much shorter than it. However, there are several factors which have contributed to a dearth of studies of populations over long periods of time. The discussion above suggests that hydrodynamics are an important influence on regularity of recruitment to the parental population. We may thus predict that the most persistent populations of *A. planci* will be found in areas where displacement of larvae by currents is inhibited. Land masses are an important impediment to current movements, particularly where the coast is indented, or in the waters of island archipelagoes.

However, reefs near large land masses are likely to be influenced by terrigenous nutrient input, which increases both algal overgrowth of corals, and plankton primary production. The latter both reduces water clarity and increases bioerosion of corals (Highsmith 1980). The resulting reefs are unattractive relative to those in clear oligotrophic waters further offshore, and thus the persistence, or indeed the presence, of unalarming numbers of *A. planci* has tended to excite little interest compared to the spectacular destruction of rich open-water reefs by infestations. Similarly, once the outbreak has died away, interest in the residual sparse population wanes rapidly.

There are thus few long series of estimates of *A. planci* population density to be found in the literature. However, in Table 1, I present data for a number of *A. planci* populations, showing a wide range of population stabilities. The populations which have attracted most attention are, of course, those which have fluctuated most widely: the GBR and reefs of Pacific islands, with fluctuations in population density through 4 or more orders of magnitude.

Towards the other end of the scale, most other populations listed are more stable, fluctuating through less than two orders of magnitude. Two reef areas stand out as having persistent, relatively stable, *A. planci* populations over periods of more than a few years: Dunganab Bay and the Dampier Archipelago. The Dampier Archipelago population has been described as 'normal' (Wilson & Marsh 1975, Potts 1981), with the implication that outbreak populations are 'abnormal'. However, such labels are neither appropriate in respect of the observed frequency distribution of *A. planci* populations (Dana *et al.* 1972), nor do they contribute anything to our understanding of the ecological processes which shape them. Much of the data on these *A. planci* populations has not, as yet, been fully published, and it is therefore worth presenting it here in some detail in order to highlight differences between their ecology and that of the more well-known volatile populations of *A. planci*.

There is one more reef area - the Gulf of California - knowledge of whose *A. planci* populations is, I believe, sufficient to justify inclusion in a list of 'persistent populations'. While only a few underwater surveys of the Gulf have been made in recent years, there is a long series of records of the starfish extending to well before the advent of SCUBA -and together these investigations provide an indication that the *A. planci* population in the Gulf of California is more persistent than most, to the extent that the starfish may be predictably found there in moderate numbers.

Dunganab Bay

Most reefs on the Red Sea coast are rich and diverse, and of three basic types: (i) a coastal fringing reef on the edge of a deep ship channel; (ii) a line of barrier reefs lying 10-20 km off-shore; and (iii) numerous patch reefs in the lee of the barrier reefs (Head 1987). *A. planci* occurs in low densities (<10 ha^{-1}) on most reefs, but transient dense populations sporadically occur on patch reefs, somewhat analogously to their preponderance on mid-shelf reefs of the GBR. For instance, in 1970, aggregations of starfish were found on some patch reefs of the Wingate and Towartit complexes, with population densities reaching 500 ha^{-1} (Ormond & Campbell 1974).

Dunganab Bay is anomalous in having only poorly-developed reefs. It is a shallow, semi-enclosed bay, and although small patch reefs occur within it, scleractinian corals are rarely found at depths greater than 2 m (Crossland 1907, 1911; Vine & Vine 1980). Records of *A. planci* in Dunganab Bay date from early this century, when Crossland described it as 'the only fairly common species [of starfish] .... in the Red Sea' (Crossland 1919). In 1975, the Vines found it 'fairly commonly' in the Bay, apparently subsisting on soft corals, in the absence of sufficient hard coral prey (Vine & Vine 1980).

Their report prompted the present author to make a special investigation of the starfish population in Dunganab Bay, and I subsequently made three surveys: in 1978, 1982, and 1984. I estimated *A. planci* population densities at 10-22 sites with hard substrates in the main basin of Dunganab Bay by swimming for 100 m, scanning a width of 5m for individual starfish, which were exposed to view by virtue of the poor coral cover (Moore 1985).

Mean population densities ranged between 8 ha^{-1} in 1978 and 127 ha^{-1} in 1982, with a decline in 1984. Crude estimates of population densities in 1972 and 1975, based on descriptions given by D.H. Nasr (pers. comm.) and Vine & Vine (1980), indicate that they also fell within this range. The relative regularity of the surveys thus indicates with some certainty that, over the 12-year period 1972-84, population densities of *A. planci* in

Dunganab Bay have ranged over little more than an order of magnitude. This compares with a range of approximately two orders of magnitude for patch reefs in open-water reef complexes near Port Sudan (Table 1) (Ormond & Campbell 1974; Moore 1985).

Dampier Archipelago

Acanthaster populations in the Dampier Archipelago of Western Australia were first studied by Wilson & Marsh (1974, 1975) on the reefs of Kendrew Island, in the west of the archipelago, between 1972 and 1974. *A. planci* was also found 'commonly' on Kendrew Island during a survey of reefs in 1978 (L.M. Marsh, pers.comm.).

In 1983, Simpson & Grey (1989) estimated *A. planci* population densities at another reef in the western Dampier Archipelago, and followed this up in October 1985 (op. cit.) with the first wide-ranging survey of the archipelago. They found *A. planci* only on reefs on the seaward edge of the archipelago, close to the 20 m depth contour. *A. planci* population densities there, estimated on transects laid at c. 3m and 7 m at each site, averaged 58 ha^{-1} (N=20), but higher densities were generally found at six sites spanning a 15-km length of the periphery of the western half of the archipelago (mean 92 ha^{-1}, N=12). In April 1987, Johnson and Stoddart (1988) recorded substantial numbers of *A. planci* at Enderby Island - somewhat closer to shore than the 20 m depth contour - as well as at Kendrew Island.

Gulf of California

Acanthaster in the Gulf of California was originally considered a distinct species *A. ellisii*, but the ranges of morphological characters of *Acanthaster* from eastern and western Pacific populations overlap (Glynn 1974), and genetic studies indicate *A. ellisii* is conspecific with *A. planci* (Nishida and Lucas 1988).

Acanthaster was collected in the the Gulf of California by several expeditions prior to the 1960s (Barham *et al.* 1973), one of the earliest being that of Steinbeck & Ricketts (1941). Surveys between 1966 and 1972 found its distribution to be concentrated on the western shore of the southern half of the Gulf (Barham *et al.* 1973). In 1970, Dana & Wolfson (1970) surveyed 10 sites on the inshore coasts of several islets at or near the entrance to La Paz Bay, and estimated a mean population density of the starfish of 45 ha^{-1}. More recently, Lucas *et al.* (1985) collected 53 specimens in the same area. There are thus reasonable grounds for considering that *A. planci* has been rather predictably found in moderate densities, perhaps in the order of 10-100 ha^{-1}, over many years.

CURRENT SYSTEMS AND RECRUITMENT IN PERMANENT HABITATS

If observations of laboratory-held starfish reflect the longevity of A. *planci* in the wild, its natural life-span may be in the range 5-8 years (Lucas 1984). The persistence of an adult population over a period much greater than this indicates that relatively regular recruitment to the adult population is occurring.

Recruitment of A. *planci* to any reef population can occur in three ways:

1. by immigration of adults from populations on other reefs;

2. by settlement of larvae produced by the resident population;

3. by immigration of larvae produced by populations on other, perhaps more distant, reefs.

In Dunganab Bay, process 1 was discounted as no significant populations of A. *planci* have been detected within 10 km of the entrance to Dunganab Bay (Moore 1985). Processes 2 and 3 are difficult to distinguish: the reason for discounting process 1 might also indicate that process 3 is relatively unimportant, as otherwise some incoming larvae would settle on reefs in the approach to Dunganab Bay. However, post-settlement mortality could possibly be much higher on reefs outside the Bay, and so prevent the establishment of significant adult populations there.

Nevertheless, the persisting adult population of A. *planci* within Dunganab Bay provides a significant continuing source of larvae, and hydrodynamic influences may retain this input within the Bay: the system of water exchange between Dunganab Bay and open waters is typical of semi-enclosed bays in arid regions (Grasshoff 1975), in which evaporation exceeds freshwater input, resulting in a sub-surface outflow of hypersaline water, which is replaced by a surface inflow into the Bay (Moore 1985). This constitutes a powerful system for retention of surface-swimming larvae, such as those of A. *planci* (Yamaguchi 1977), within the Bay. The contribution of larvae from the resident population seems likely, under most scenarios, to be normally greater than that from populations outside Dunganab Bay, so that the Dunganab Bay population appears to be primarily self-maintaining. However, during periods of low numbers of A. *planci* in Dunganab Bay, supply of larvae within the Bay may be significantly supplemented by entrapment of larvae originating outside the Bay.

There is also some indirect evidence that larvae produced within Dunganab Bay may occasionally establish starfish populations on reefs outside it. Under certain meteorological conditions, the retentive circulation described above can break down, allowing surface waters to flow out of the Bay. During one occurrence of this, outflowing waters pursued a meandering track through coastal patch reefs to join the open-water current system. Reefs surveyed along the effluent track had significantly higher *A. planci* population densities than other reefs in the same area which were not in the track of the effluent, suggesting that larvae transported from the Bay may have settled on reefs in the effluent track (Moore 1985).

On the western Australian coast, large populations of *A. planci* appear to be restricted to a band between latitudes 17°S and 22°S. Outside this range, water temperatures are relatively unfavourable for larval development and survival of *A. planci* (Simpson & Grey 1989). During the breeding season, winds are usually from the southwest, probably resulting in a net northerly drift of surface waters (Hearn *et al.* 1986, cited by Simpson & Grey 1989). Thus it is unlikely that *A. planci* populations in the Dampier Archipelago could receive any except occasional larvae from distant *A. planci* populations, although the extent of its genetic isolation from other Indo-Pacific populations must await electrophoretic studies of enzyme loci. Nevertheless, it seems likely that the the Dampier Archipelago population is primarily self- recruiting; presumably the many islands act as barriers to the larvae being flushed away.

The Gulf of California, like Dunganab Bay, is situated in an arid zone, and evaporation exceeds freshwater input from precipitation and the influx of the River Colorado. However, current flows at its entrance are complex, with fronts and gyres, and vary seasonally (Alvarez-Borrego 1983). In summer, though (the breeding season of *A. planci*), there is a surface flow northward into the Gulf. Within the Gulf, La Paz Bay, with a string of islets restricting its entrance, can be expected to exert a considerable retentive influence on the recruitment of *A. planci* larvae to the resident population.

OTHER FACTORS PROMOTING STABILITY

In Table 2, I offer a tentative ranking according to three levels of relative stability of a number of populations, together with data on other ecological parameters of possible relevance, and which have frequently been collected during surveys of *A. planci* populations. Ranked as 'stable' are populations which fluctuate through a density range of only one order of magnitude; 'intermediate' are those fluctuating through two or three orders of magnitude; while the most volatile are classified as 'transient' and fluctuate through up to five or six orders of magnitude.

Table 2. Some ecological characteristics of *A. planci* populations of differing stability, classified according to range of population density.

Reef area	Mean coral cover (range)	Diurnal feeding	Arm injuries	Source
TRANSIENT POPULATIONS: 3+ orders of magnitude.				
GBR, Australia	42% (25-69%)	(agg.) 90% (sol.) 'concealed'	33%	P&E 1969, Endean 1974, Done 1982
Guam	-	(agg.) 'diurnal' (sol.) 'nocturnal'	43%	Chesher 1969, Glynn 1982, Birkeland 1982
INTERMEDIATE: 2-3 orders of magnitude.				
Red Sea, Wingate & Towartit complexes	-	12%	29%	Ormond & Campbell 1971,1974
E. Pacific: Uva I., Panama	45% (10-90%)	"forage day and night"	17%	Glynn 1972, 1973, 1982, 1985
STABLE POPULATIONS: 1 order of magnitude.				
Dunganab Bay (2-60%)	16%	52-74%	1.5%	Moore 1985
Gulf of California	3-10%	72%	2.6%	D&W 1970, Barham *et al.*1973, Lucas 1985
Western Dampier Archipelago	11% (2-64%)	5-66% 65%*	33*-44%	Wilson & Marsh 1975, Simpson & Grey 1989*

D&W = Dana & Wolfson, P&E = Pearson & Endean, agg. = aggregated populations,
sol. = solitary populations.
* Mature starfish only - see text.

The other ecological data presented are, specifically: percentage cover of hard corals; percentage of starfish feeding in the daytime; and percentage of starfish bearing one or more damaged or regenerating arms, presumed to be the result of attack by a predator. There are inevitably inadequacies in the comparability of these data, owing to differences in methodology; and to the fact that they are not constants within a habitat, but will vary over time under the influence of many variables including *A. planci* population density. Nevertheless, a comparison at this preliminary level will be attempted in order to identify possible patterns and common factors with a view to developing testable hypotheses.

Considering first the data for the 'transient' populations, in order to illustrate the ecological and selective pressures on *A. planci* on open-water reefs, these are characterised by a relatively rich cover of hard corals. Coral cover varies according to the type of community or species assemblage, which in turn depends in part on the physical conditions which prevail at the reef zone, and the reef type, under consideration.

In his study of the distribution of coral communities across the central GBR, Done (1982) categorised communities according to their association with wave exposure, classified from maximum exposure (Class I) to most sheltered (Class III). On mid-shelf reefs, where *A. planci* predominates (Moran 1986), all classes of wave exposure occur, but we may exclude coral community types restricted to Class I conditions from our consideration, because they occur on mid-shelf reefs only in narrow zones in which *A. planci* is unlikely to occur, owing to its avoidance of zones subject to extreme wave action. The remaining Class II and III community types occurring on mid-shelf reefs in the central GBR have mean coverage of hard coral ranging between 25% and 69% (Done 1982), with a mean of 40% (N=7).

Such a relatively rich cover, while patchy, provides adequate food and protection from predators at starfish density levels well below the carrying capacity. As mentioned previously, low-density *A. planci* populations on rich reefs are usually primarily nocturnal in activity (although a small percentage may feed diurnally), remaining concealed under coral in the daytime; while at higher population densities, starfish tend to aggregate, feeding on adjacent patches of coral, and with a greater tendency to feed in the daytime. Nevertheless, predator pressure is considerable at all population levels, as indicated by the high frequency of occurrence of arm injuries (>30%).

Turning now to the characteristics of 'permanent habitats', Table 2 indicates that, while coral cover in them is similarly patchy, overall it is markedly lower, at under 20%, than on typical open-water reefs. In part, this may reflect the continuous pressure of predation by *A. planci*, although other factors may also contribute. Crossland (1911) was struck by the chronic paucity of corals in Dunganab Bay, although he was apparently unaware of the possible role of *A. planci* in this regard. Persistent predation pressure by *A. planci* and another corallivorous asteroid, *Culcita coriacea*, undoubtedly make a major contribution to low coverage of corals in Dunganab Bay (Moore 1985).

It was suggested that this, and other factors distinguishing *A. planci* ecology in the Bay from that on open-water reefs outside, might contribute to the relative stability and persistence of *A. planci* populations in Dunganab Bay (Moore 1985):

a) *Non-specialised diet*: The *A. planci* population density cycles around the carrying capacity of scleractinian corals in Dunganab Bay, so during years following periods of high population density, its favoured prey has become virtually exhausted. However, it is capable of persisting on an alternative diet of soft corals and other non- scleractinian Cnidaria. In 1975, the major prey of *A. planci* were alcyonarians *Xenia* spp. (Vine & Vine 1980). In 1982, even with an abundance of hard corals available, 15% (N=20) were feeding on alternative prey; and in 1984, 23% (N=22), on items which included the alcyonarian *Sarcophyton glaucum*, and a zoantharian *Palythoa* sp..

The high level of diurnal activity by *A. planci* may also reflect paucity of hard coral, for two reasons: because starfish must devote more time to foraging for food to meet their energy requirements; and because the lack of availability of cover would tend to produce habituation to the light stimulus the starfish cannot avoid, so that daylight no longer inhibits feeding activity.

b) *Low predator pressure*: Known fish predators of *A. planci* are common in Dunganab Bay, particularly trigger-fishes (Balistidae), yet the incidence of arm injuries attributable to such predators is very low: in 1984, only 1.5% (N=67) of starfish had one or more shortened or regenerating arms. This is the lowest value reported for any population of *A. planci* (Moran 1986), and contrasts with a frequency of 29% in populations on Sudanese reefs in open waters (Ormond & Campbell 1971). The low incidence may reflect the abundance of alternative prey, particularly echinoids, for such predators; additionally, it is possible that *A. planci* acquires some protection against fish predators by consuming soft corals such as *Sarcophyton glaucum* and *Xenia* spp., which are known to be toxic to fish (Ne'man *et al.* 1974, Bakus 1981).

The ecological data in Table 2 for other permanent habitats show considerable similarities to that for Dunganab Bay. In the Gulf of California, coral cover in the *A. planci* habitats around the islets off La Paz Bay studied by Dana & Wolfson (1970) averaged 10%, presumably reflecting both the sub-tropical temperature regime in the Gulf, and its very high plankton primary production (Alvarez-Borrego 1983), as well as grazing pressure by *A. planci*.

As in Dunganab Bay, *A. planci* populations in the Gulf of California show a marked propensity for diurnal feeding, and alternative items to hard corals are occasionally fed upon: Barham *et al.* (1973) reported that coralline algae was the food of 5% of starfish observed feeding, and *A. planci*'s diet in the Gulf of California also includes *Padina* algae and gorgonians (Dana & Wolfson 1970, Barham *et al.* 1973).

The very low incidence of arm injuries in *A. planci* populations in the Gulf of California (Lucas *et al.* 1985) is comparable to that found in Dunganab Bay. No predators of *A. planci* have been identified in the Gulf of California, and it is possible that predator pressure is generally less there than in most other coral reef areas, reflecting its sub-tropical nature (Vermeij 1978).

In the Dampier Archipelago, living hard coral cover on seaward reefs, estimated on the same transects as for *A. planci* by Simpson & Grey (1989), varied between 1.6% and 72.6%, with a mean of 24.0% (N=20). But at the six western sites (where *A. planci* was more abundant), mean coral cover was only 11.1% (N=12), and remaining space was occupied by turf algae, soft corals and zoanthids in varying proportions.

Overall 76% of starfish on seaward reefs remained concealed in the daytime, and only 28% were feeding, primarily on hard corals (Simpson & Grey 1989). However, all starfish less than 20 cm in overall diameter were cryptic, and the overall incidence of diurnal feeding was depressed by the predominance of high densities of juvenile starfish at some sites. It is not possible to separate Simpson & Grey's data for size and feeding, but if we exclude sites 5, 6 and 9, where the majority of stars smaller than 20 cm diameter were concentrated, 24/37 starfish (=64.9%) were feeding in the daytime at the remaining sites. Wilson & Marsh (1975) found a range of incidence of diurnal feeding of 22-66%, depending on season, in their 'Northern Sector', where hard corals were 'scattered'. Somewhat fewer starfish fed diurnally (5-46%) in the 'rich' corals of the shallow 'Western Sector'.

Coralline algae was the only food item of *A. planci*, other than hard corals, observed by Simpson & Grey (1989), at a relative incidence of 15%, and in some cases, adult starfish (diameter >24 cm) were observed feeding on it. Although Wilson & Marsh (1974, 1975) did not report any feeding by *A. planci* on alternatives to hard coral prey at Kendrew Island, significant feeding on soft corals (primarily nephtheids) was observed (L.M. Marsh, pers. comm.) in the Western Sector in mid-summer, after high water temperatures had caused starfish aggregations to disperse to deeper water (c. 20 m), where only scattered hard coral colonies were present.

Predator pressure on *A. planci* appears to be high in the Dampier Archipelago, as evidenced by incidence of arm injuries. Simpson & Grey (1989) found 47% overall but, as with diurnal feeding, they found a strong size effect: of starfish less than 20 cm in diameter, 78% had one or more arms missing or regenerating, compared to 33% of starfish with diameters of 20 cm and over. Incidences of 44% (Wilson & Marsh 1975) and 38% (Wilson, Marsh & Hutchins 1974) were reported for the primarily large adult starfish populations present at Kendrew Island in 1972-4.

The incidence of arm injuries in the Dampier Archipelago is in marked contrast to the low levels in Dunganab Bay and the Gulf of California, and is similar to that observed in rich reef areas with more transient *A. planci* populations (see above). Such injuries may be largely attributable to attacks by known generalist predators which are known to be present in the area, particularly puffer fishes and trigger fishes (Wilson, Marsh & Hutchins 1974). In Dunganab Bay, such predator pressure may be mitigated by the presence of alternative, perhaps preferred, prey, but information regarding this in the Dampier Archipelago has not been published.

Data on populations of intermediate stability has been included to illustrate that there is a continuum between the two extreme habitat types discussed above. In part, the narrower range of fluctuations in density, compared to the most volatile, may reflect relative proximity to a more persistent source of larvae. *A. planci* populations on reef complexes in the Port Sudan area of the Red Sea may be in receipt of larvae from the population in the Farasan Archipelago (Moore 1985), which, though it has only been surveyed once, shares many ecological and hydrographical features with the habitats described above as favouring persistence in *A. planci* populations, viz. a combination of bays and numerous islets to inhibit flushing of larvae, and poor coral cover in a generally productive environment owing to a high allochthonous energy input (Moore in press).

ECOLOGICAL STRATEGIES IN PERMANENT HABITATS

To summarise, I suggest that the persistent populations discussed above are associated with habitats with the following common characteristics:

1. Retentive hydrodynamic systems afforded by bays and archipelagoes, as earlier predicted;

2. Chronically low coral cover: in all three reef areas, authors of studies have commented that coral development appeared to be restricted by predation pressure by *A. planci*. However, high plankton primary production arising through terrestrial and oceanographic nutrient inputs could (as discussed above) also play an important part in creating a habitat with sparse coral cover. Such an influence could be identified by other associated differences in nutrient-enriched reef communities, e.g. the predominance of filter feeders (Moore, in press);

3.   Low availability of coral for food and cover may produce a number of associated behavioural changes in *A. planci*, viz. diurnal feeding and alternative diet. These characteristics may, of course, also occur in the late stages of an outbreak on a previously rich, open-water reef ('temporary habitat'), so these characteristics alone are insufficient to distinguish the two types of habitat;

4.   Low predator pressure; this is not universal, but two of the three 'persistent' populations have the lowest incidences of arm injuries recorded.

The effects of many of the ecological factors which tend to increase reproductive output and reduce mortality in epidemic populations on open-water reefs act in reverse in the permanent habitats to promote stability of *A. planci* populations, and some of these opposing effects are summarised in Figure 1.

Just as a rich coral cover is clearly a prerequisite for the formation of feeding aggregations of starfish, the poor coral cover in permanent habitats will conversely tend to inhibit the formation of aggregations. This effect may be enhanced by *A. planci*'s social behaviour being facultative rather than obligate: in Y-maze experiments performed to test the olfactory attractiveness of feeding conspecifics relative to unattacked coral prey (Ormond *et al.* 1972), a greater attraction to feeding conspecifics was observed only after several days of repulsion (*pers. obs.*). There may thus be a mutual avoidance, or 'spacing-out', effect in low population densities, a suggestion which concurs with some observations of scattered populations in the field, e.g. Goreau's (1964) suggestion that *A. planci* was territorial.

Reduced coral cover in inshore reefs may have several other important effects on *A. planci* ecology and behaviour. Predation of larvae by coral polyps may be reduced (Yamaguchi 1973), but counteracting this, mortality of juvenile starfish may be increased by the unavailability of hard coral prey at the critical stage of transfer from a diet of coralline algae (Lucas 1984). Such mortality may be important in generating population cycles in Dunganab Bay, where availability of hard coral becomes exhausted following periods of relatively high numbers of starfish (Moore 1985).

The suggested importance of poor coral availability in the 'permanent habitats' in promoting stability of *A. planci* populations is reminiscent of Watt's (1965) suggestion regarding forest insect pests, that stability of a population is greater when a lower proportion of the environment is filled with usable food.

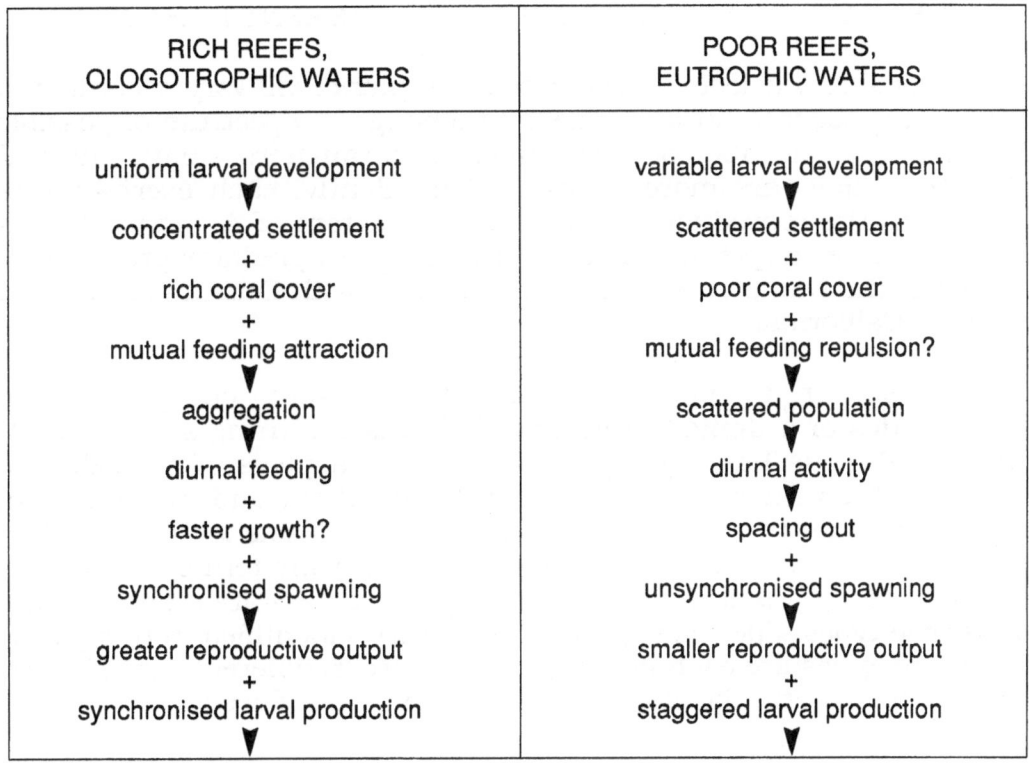

RICH REEFS, OLOGOTROPHIC WATERS	POOR REEFS, EUTROPHIC WATERS
uniform larval development ▼ concentrated settlement + rich coral cover + mutual feeding attraction ▼ aggregation ▼ diurnal feeding + faster growth? + synchronised spawning ▼ greater reproductive output + synchronised larval production ▼	variable larval development ▼ scattered settlement + poor coral cover + mutual feeding repulsion? ▼ scattered population ▼ diurnal activity ▼ spacing out + unsynchronised spawning ▼ smaller reproductive output + staggered larval production ▼

Figure 1. Some ecological influences on social behaviour of *A. planci.*

Other ecological influences in the poor reef areas on *A. planci* development and behaviour would also tend enhance stability by reducing or neutralising the gearing effect of a contagious distribution. Scattered populations of adults are likely to have a smaller reproductive output in terms of production of larvae owing to the greater distance between individuals and thus lower fertilisation success. Larval development may also be more variable in conditions of high phytoplankton availability: some larvae develop rapidly, while others develop more slowly or abnormally (Lucas 1982, pers. comm.). This staggered development, coupled with transport by currents, could lead to more scattered settlement in the productive waters of some coastal reefs than under oligotrophic conditions.

We can thus see that the different ecological pressures in different reef habitats tend to favour different, or even opposing, ecological strategies. In essence, if predator pressure in some poor-reef habitats is less high than on rich, open-water reefs, greater advantage may be gained from a shift towards exploiting scarce prey resources than gaining protection from predators.

ALTERNATING LIFE STRATEGIES IN DIFFERENT HABITATS

The ecological and selective pressures on *A. planci* will vary according to the habitat it occupies at any time, and among the spectrum of possible habitats a large number of environmental parameters - both physical and biotic - may vary more or less independently, each exercising its influence on the reproductive and mortality factors of *A. planci*. This is illustrated, for example, by the difference between predator pressure on *A. planci* in the Dampier Archipelago compared with Dunganab Bay and the Gulf of California.

Nevertheless, it is possible that the present comparison of characteristics of a limited number of habitats favouring stability of *A. planci* populations has highlighted the critical aspects of the selective pressures which have shaped *A. planci*'s evolution, and in the face of which it continues to persist as an extant species. If this is the case, then we can define *A. planci*'s 'life strategy' in the following terms: the species persists by being adapted to life in a range of reef habitats between the two extreme types - 'permanent' and 'temporary - identified in this paper. The extreme types are each hostile and insecure in different ways, but by occupying them both, the chances of extinction are reduced.

This concept may be expressed visually in Figure 2. It attempts to represent schematically the relationship between a shift in ecological pressures in different habitats, and the *A. planci* population density which is sustainable over more than one generation. In a more sophisticated model, resource availability and predator pressure would be considered as independent variables, but at the present level of understanding, we are limited to a simple, two-dimensional model covering the range of habitats in which the gradient of prey resource availability (i.e. coral cover) is positively correlated with a gradient in predator pressure. Thus it is not valid for habitats such as the western Dampier Archipelago, where coral cover is relatively poor, but *A. planci* populations suffer relatively high predator pressure.

The sustainable population density is represented by a space bounded by: a) at high levels, availability of coral prey, which increases with increasing richness of the reef habitat; b) density-independent mortality, which is proposed to be higher in the poor reef areas by virtue of the more extreme physical conditions. I have drawn arbitrarily straight lines to represent these constraints, joining values of sustainable population density which reflect the ranges between which starfish populations are known to fluctuate (at order of magnitude level) in the two extreme habitat types (see Table 1), forming a truncated triangle.

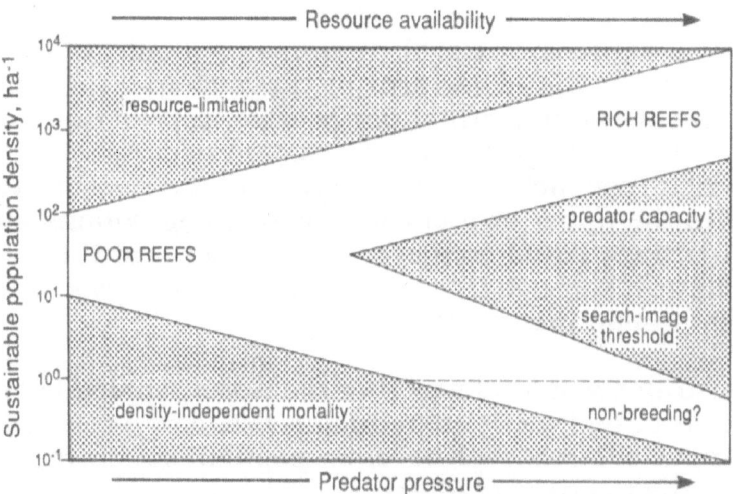

Figure 2. Schematic model suggesting relationships between some ecological factors and *A. planci* population densities observed in a spectrum of habitat types.

Different habitat types are connected, within the triangle, through larval advection. On rich reefs, the schema requires that dense populations ('outbreaks') are sustained by different reefs for each generation, i.e. they are connected through larval advection. If cycles of outbreak occur infrequently enough, allowing recovery of corals in the intervening period, dense populations may persist indefinitely within an area defined by a circular system of currents.

The space bounded by the truncated triangle is diminished by a wedge driving into it from the wide end, producing the shape of a 'Y' on its side. The wedge represents pressure from predators of *A. planci*, selecting either for high density populations, which can 'swamp' or satiate territorial predators; or for populations which are below the density threshold at which visual predators may develop a 'search image' (Tinbergen 1960, Tinbergen *et al.* 1967). The implication of this 'predator wedge' is that populations of a density intermediate between the two extremes are not sustainable over multiple generations, and will be tend to be reduced, sooner or later, by predators to the low-density arm of the Y-space.

The right hand side of this schema reflects the general conclusion of McCallum's (1987) model, regarding the possibility of stable equilibria at epidemic and endemic population densities, bifurcated by density-dependent predation. However, the additional dimension of variable habitat quality provides an insight into the different ways that the 'predator barrier' (in this case 'wedge') may be breached.

McCallum's model does not consider that low density 'resident' populations on rich reefs may be non-breeding, but in this paper I have argued that input of larvae from populations on other reefs may be much more important as sources of larval recruits. To illustrate this, I have placed a dotted line in the lower arm of the Y-shape, arbitrarily at a population density of one individual per hectare, representing a density threshold below which insignificant fertilisation occurs. Thus resident populations on rich reefs below this density level are non-breeding, and cannot connect to the upper arm of the Y-shape. Populations above this level (say 10^0-10^1 ha^{-1}) could connect to the upper arm if the 'predator wedge' was withdrawn somewhat, i.e. predator pressure was reduced, for example through human intervention.

The diagram illustrates a second way that the upper, epidemic arm of the Y-shape may be attained, without reducing predator pressure on rich reefs: through input of larval recruits from populations on reefs towards the poorer end of the richness gradient.

Earlier in this paper, *A. planci* was described as a 'fugitive' on rich reefs, which constitute a habitat that it can only invade temporarily, while *A. planci*'s habitats in poor reef areas in semi-enclosed coastal waters have been here described as 'permanent', and their stability suggests that they are more favourable habitats for *A. planci* than are rich reefs. Aspects of *A. planci*'s ecology such as relatively low reproductive output and territorial feeding behaviour can be seen as a conservative strategy to maintain the population close to the carrying capacity of the habitat, i.e. K-selected.

Note, however, that an important part of this habitat favourability results from reduced predation, i.e. is biotic. The physical environment afforded is relatively harsh, and in some respects would appear to favour r-selection: their vulnerability to extremes of temperature, and of salinity through evaporation and flooding, must render them less predictable environments than coral reefs in open waters, leading to occasionally catastrophic, density-independent mortality, and perhaps even local extinction.

However, *A. planci* is well adapted to cope with at least some of such perturbations to its physical environment. *A. planci* larvae have a high tolerance to low salinity (Henderson 1969). Moreover, reduced salinities, at least down to 30 ppt, enhance survival of the larvae over that in the 35-37 ppt salinity of open-sea water (Lucas 1973). This observation is consistent with the hypothesis that poor-reef habitats in coastal waters, subject to terrestrial run-off, play an important role in *A. planci*'s life strategy.

It may seem paradoxical that the rich coral reefs favoured by Man for their beauty and diversity are not the most favourable habitat for *A. planci*, if favourability of a habitat is reflected by the stability of its populations. However, the favourability of an environment can only be defined with regard to a particular species. Thus although coral reefs are complex, high-diversity ecosystems, this does not allow us to predict that all species found on coral reefs should have stable populations.

While it may be unwise to expect organisms to conform faithfully to a general theoretical scheme (Grahame & Branch 1985), conceptual frameworks are created as tools to aid understanding, and it may be useful to compare the life strategy of *A. planci* (as proposed here) with that of other organisms.

Alternation of life strategies between different habitats is known in a number of terrestrial animals with markedly fluctuating populations. Many species of microtine rodents exhibit population cycles, accompanied by dispersal during the increase phase of the cycle. In *Microtus* populations, genotypes with the highest reproductive potential (r- strategists) tend to disperse, leaving behind genotypes selected for spacing behaviour - K-strategists (Krebs *et al.* 1973). Nichols *et al.* (1976) document temporally dynamic reproductive strategies in a number of rodents, birds and anurans inhabiting arid regions. Individuals of such strategists may shift their position along the r-K continuum by changing their reproductive activity in response to variable environmental cues.

Locusts - the paradigms of pests - have evolved a highly flexible ability to respond to fluctuations in the relative favourability of two alternative habitats: the habitats are distinct, and to survive in them locusts are capable of adopting completely different ways of life, solitary and gregarious (Kennedy 1956). These life styles, or phases, involve changes in colour, morphology, physiology and behaviour which are effected through phenotypic responses to crowding, as well as nongenic and genic inheritance.

The solitary phase is expressed in relatively restricted 'outbreak areas' (or the more diffuse 'recession area' of the desert locust *Schistocerca gregaria*), which are areas of permanent inhabitation and the source of swarms. Following increases in population density, swarms migrate out of the outbreak areas, and during plagues, may persist over many generations and invade huge areas of land from which they are normally absent, or present in only small numbers, during which time they transform to the gregarious phase (Uvarov 1977).

The 'invasion areas' of African locusts constitute millions of square kilometres of grassland and savannah of fairly uniform cover, as well as agricultural land (Uvarov 1977). Locust swarms in the 'invasion habitat' are subject to continuous attack from predators (Moore 1979). Gregarious phase desert locusts produce fewer but larger eggs than solitaria phase, but their synchronised development and faster maturation permit them to have much faster rates of increase, r_{max}, than solitarious populations. This gives gregarious locusts a selective advantage in the invasion area by being exposed to predators for shorter periods in each life cycle, while the slower development of the solitarious phase allows a greater probability of some offspring surviving in the permanent habitat after a protracted period of drought (Cheke 1978).

The permanent habitat varies with species, but is characterised by more diffuse and patchy vegetation, often regarded as 'marginal'. For the desert and brown locusts, this is in diffuse areas of ephemeral semi-desert vegetation north and south of the equator respectively, while the red and migratory locusts inhabit valleys and plains subject to flooding (Uvarov 1977). In the permanent habitat, the conspicuous colouration and diurnal activity of the gregarious phase is abandoned for camouflaging colours and a cryptic habit. This defence against predators is complemented by spacing-out behaviour (Lea 1968, Ellis 1959), of possible advantage to prevent visual predators developing a search image.

As far as is known, *A. planci* does not have such flexible adaptations as locusts, whatever the superficial analogy between their life strategies in alternating between permanent and temporary habitats. However, knowledge of *A. planci*'s ecology may be still at the stage equivalent to that of locusts before the discovery of phases. For instance, growth rates of *A. planci* could increase with population density, as studies have shown for the temperate-water asteroid *Asterias rubens* (Moore & Campbell 1985), a starfish with a similar propensity to temporarily invade a high-risk super-abundant prey resource. Another factor of potentially great significance is the occurrence of two sizes of egg produced by *A. planci* in different studies: one of 0.1 mm diameter, the other 0.19 mm (Yamaguchi 1973). This size difference represents a volume ratio of 7:1, and thus could significantly affect fecundity as well as, presumably, larval development rates. What is its significance to *A. planci*'s ecology and physiology? I suggest that these may be stimulating and fruitful areas for further investigation.

Acknowledgements. I thank the anonymous referee who suggested improvements to the manuscript; and the Australian Institute of Marine Sciences for providing funds to enable me to attend its *Acanthaster* Modelling Workshop.

REFERENCES

Alvarez-Borrego, S. (1983), Gulf of California. In: Ecosystems of the world, 26: Estuaries and enclosed seas (Ketchum, B.H., ed.), Elsevier, pp. 427-449.

Bakus G.J. (1981), Chemical defense mechanisms on the Great Barrier Reef, Australia. Science, 211: 497-9.

Bakus, G.J. & Green, G. (1974), Toxicity in sponges and holothurians: a geographic pattern. Science, 185: 951-3.

Barham, E.G., Gowdy, R.W., & Wolfson, F.H. (1973), *Acanthaster* (Echinodermata, Asteroidea) in the Gulf of California, Fish. Bull., 71(4): 927-942.

Beach, D.H., Hanscomb, N.J., & Ormond, R.F.G. (1975), Spawning pheromone in crown-of-thorns starfish. Nature, 254: 135-6.

Birkeland, C. (1982), Terrestrial runoff as a cause of outbreaks of *Acanthaster planci* (Echinodermata: Asteroidea). Mar. Biol., 69:175-185.

Birkeland, C. (1984), Influence of topography of nearby land masses in combination with local water movement patterns on the nature of nearshore marine communities. In: Productivity and processes in island marine ecosystems, UNESCO Report No. 27, 16-31.

Branham, J.M., Reed, S.A., Bailey, J.H., & Caperon, J. (1971), Coral-eating sea stars *Acanthaster planci* in Hawaii. Science, 172: 1155-1157.

Cameron, A.M. (1977), *Acanthaster* and coral reefs: population outbreaks of a rare and specialised carnivore in a complex high- diversity system. In:Proc.3rd Int. Coral Reef Symp., Vol.1, Rosensteil School of Marine & Atmospheric Sciences, Uni. of Miami, 193-199.

Cassie, R.M. (1954), Some uses of probability paper for the graphical analysis of size frequency distributions. Austr. J. Mar. Freshwat. Res., 5: 513-22.

Cheke, R.A. (1978), Theoretical rates of increase of gregarious and solitarious populations of the desert locust. Oecologia, 35: 161-171.

Chesher, R.H. (1969), *Acanthaster planci* impact on Pacific Coral reefs. Final Rep. Res. Westinghouse Electric Corp. to U.S. Dept. Interior, 151 pp.

Cody, M.L. & Diamond, J.M. (eds.)(1975), Ecology and evolution of communities. Harvard (Belknap) Press.

Conand, C. (1985), Distribution, reproductive cycle and morphometric relationships of *Acanthaster planci* in New Caledonia. In: Echinodermata (Keegan, B.F. and O'Connor, B.D.S., eds.), Balkema, Rotterdam, 499-506.

Connell, J.H. (1973), Population ecology of reef-building corals. In: Biology and geology of coral reefs (O.A. Jones and R. Endean, eds.), Vol. 2, Biology 1, Academic Press, N.Y. and London, pp. 205-245.

Connell, J.H. (1975), Some mechanisms producing structure in natural communities: a model and evidence from field experiments. In: Ecology and evolution of communities (M.L. Cody and J.M. Diamond, eds.), Harvard (Belknap), pp.460-490.

Connell, J.H. (1985), The consequences of variation in initial settlement vs. post-settlement mortality in rocky inter-tidal communities. J. Exp. Mar. Biol. Ecol., 93: 11-45.

Crossland, C. (1911), Reports on the marine biology of the Sudanese Red Sea. A physical description of Khor Dunganab. J. Linn. Soc. (Zool.), 31: 265-286.

Crossland, C. (1907), Reports on the marine biology of the Sudanese Red Sea: IV, The recent history of the mid-west shores of the Red Sea. J. Linn. Soc. (Zool.) 31: 14-30.

Crossland, C. (1919), Dangers of pearl diving. Sudan Notes and Records, 11: 234-236.

Dana T. & Wolfson A. (1970), Eastern Pacific crown-of-thorns starfish populations in the lower Gulf of California. Trans. San Diego Soc. Nat. Hist., 16(4), 83-90.

Dana, T.F., Newman, W.A. & Fager, E.W. (1972), *Acanthaster* aggregations: interpreted as primarily responses to natural phenomena. Pacific Sci., 26: 355-372.

Doherty, P.J. & Davidson, J. (1989), Monitoring the distribution and abundance of juvenile *Acanthaster planci* in the central Great Barrier Reef. Proc. 6th Int. Coral Reef Symp. 2, 131-136.

Doherty, P.J. & Williams, D. McB. (1988), The replenishment of coral reef fish populations. Ocenogr. Mar. Biol. Ann. Rev., 26: 487-551.

Done, T.J. (1982), Patterns in the distribution of coral communities across the central Great Barrier Reef. Coral Reefs, 1: 95-107.

Ebert, T.A. (1983), Recruitment in echinoderms. In: Echinoderm Studies (M. Jangoux and J.M. Lawrence, eds.), Vol.1, Balkema, Rotterdam, 169- 203.

Ellis, P.E. (1959), Learning and social aggregation in locust hoppers. Anim. Behav., 7: 91-106.

Endean, R. (1973), Population explosions of *Acanthaster planci*. In: Biology and Geology of Coral Reefs (O.A. Jones & R. Endean, eds.), Vol. 2, Biology 1. Academic Press, New York, pp. 389-438.

Endean, R. (1974), *Acanthaster planci* on the Great Barrier Reef. Proc. Second Int. Coral Reef Symp., 1. Great Barrier Reef Committee, Brisbane, 563-576.

Endean, R. & Chesher, R.H. (1973), Temporal and spatial distribution of *Acanthaster planci* population explosions in the Indo-West Pacific region. Biol. Conserv., 5(2): 87-95.

Fricke, H.W. (1974), [Possible influence of predators on the behaviour of *Diadema* sea urchins] (in German). Marine Biology, 27: 59-62.

Gaines, S., Brown, S., & Roughgarden, J. (1985), Spatial variation in larval concentrations as a cause of spatial variation in settlement for the barnacle Balanus glandula. Oecologia, 67: 267-272.

Glynn, P. (1973), *Acanthaster*: effect on coral reef growth in Panama. Science 180: 504-6.

Glynn, P.W. (1972), Observations on the ecology of the Caribbean and Pacific coasts of Panama. Bull. Biol. Soc. Wash., No. 2, 13-30.

Glynn, P.W. (1974), The impact of *Acanthaster* on corals and coral reefs in the eastern Pacific. Environ. Cons., 1(4): 295-304.

Glynn, P.W. (1982), Individual recognition and phenotypic variability in *Acanthaster planci* (Echinodermata: Asteroidea). Coral Reefs, 1: 89-94.

Glynn, P.W. (1985), Corallivore population sizes and feeding effects following El Nino (1982-3) associated coral mortality in Panama. Proc. 5th Int. Coral Reef Congr. (Tahiti, 1985), Vol. 4, p. 183-188.

Goreau T.F. (1964), On the predation of coral by the spiny starfish *Acanthaster planci* in the southern Red Sea. Bull. Sea Fish. Res. Stn. Haifa, Israel, 35: 23-36.

Grahame, J. & Branch, G.M. (1985), Reproductive patterns of marine invertebrates. Oceanogr. Mar. Biol. Ann. Rev., 23: 373-398.

Grant, A., Morgan, P.J. & Olive, P.J.W. (1987), Use made in marine ecology of methods for estimating demographic parameters from size-frequency data. Mar. Biol., 95: 201-208.

Grasshoff, K. (1975), The hydrochemistry of landlocked basins and fjords. In: Chemical Oceanography (Riley, J.P. & Skirrow, eds.), Academic Press, 455-597.

Grassle, J.F. (1973), Variety in coral reef communities. In: Biology and Geology of Coral Reefs (Jones, O.A. & Endean, R., eds.), Vol.2, Biology 1, Academic Press, 247-270.

Harding, J.P. (1949), The use of probability paper for the graphical analysis of polymodal frequency distributions, J. Mar. Biol. Assoc. U.K., 28: 141-53.

Head, S.M. (1987), Corals and coral reefs of the Red Sea. In: Red Sea (Key Environments) (A.J. Edwards & S.M. Head, eds.), IUCN and Pergamon, Oxford, pp. 128-151.

Henderson, J.A. (1969), Preliminary observations on rearing and development of *Acanthaster planci* larvae. Queensland Fish. Notes, 3: 69-75.

Henderson, J.A. & Lucas, J.S. (1971), Larval development and metamorphosis of *Acanthaster planci* (Asteroidea). Nature, 232: 655- 657.

Highsmith, R.C. (1980), Geographic patterns of coral bioerosion: a productivity hypothesis. J. Exp. Mar. Biol. Ecol., 46: 177-196.

Holling, C.S. (1959), The components of predation as revealed by a study of small-mammal predation of the European pine sawfly. Can. Entomol., 91: 293-320.

Huffaker, C.B. & Messenger, P.S. (1964), The concept and significance of natural control. In: Biological Control of Insect Pests and Weeds (DeBach, P., ed.), Chapman & Hall, London, pp. 74-117.

Johnson, D.B. & Stoddart, J.A. (1988), Report on surveys of the distribution, abundance and impact of *Acanthaster planci* on reefs within the Dampier Archipelago (Western Australia), April 1987. A.I.M.S., Townsville, 15 pp.

Kenchington, R.A. (1977), Growth and recruitment of *Acanthaster planci* (L.) on the Great Barrier Reef. Biol. Conserv., 11: 103-118.

Kennedy, J.S. (1956), Phase transformation in locust biology. Biol. Rev. 31: 349-370.

Kettle B.T. & Lucas J.S. (1987), Biometric relationships between organ indices, fecundity, oxygen consumption and body size in *Acanthaster planci*. Bull. Mar. Sci., 41(2): 541-551.

Krebs, C.J., Gaines, M.S., Keller, B.L., Myers, J.H. & Tamarin, R.H. (1973), Population cycles in small rodents. Science 179:35-41.

Lea, A. (1968), Natural regulation and artificial control of brown locust numbers. J. Entomol. Soc. S. Africa 31: 97-112.

Lewin, R. (1986), Supply-side ecology. Science 234:25-27.

Lobel, P.S. & Robinson, A.R. (1986), Transport and entrapment of fish larvae by ocean mesoscale eddies and currents in Hawaiian waters. Deep-Sea Res., 33(4): 483-500.

Loosanoff, V.I. (1964), Variations in time and intensity of setting of the starfish, *Asterias forbesi*, in Long Island Sound during a twenty- five year period. Biol. Bull., 126: 423-439.

Lucas, J.S. (1973), Reproductive and larval biology of *Acanthaster planci* (L.) in Great Barrier Reef waters. Micronesica, 9: 197-203.

Lucas, J.S. (1982), Quantitative studies of feeding and nutrition during larval development of the coral reef asteroid *Acanthaster planci* (L.). J. Exp. Mar. Biol. Ecol., 65: 173-193.

Lucas, J.S. (1984), Growth, maturation and effects of diet in *Acanthaster planci* (L.) (Asteroidea) and hybrids reared in the laboratory. J. Exp. Mar. Biol. Ecol., 79: 129-147.

Lucas, J.S., Nash, W.J., & Nishida, M. (1985), Aspects of the evolution of *Acanthaster planci* (L.) (Echinodermata, Asteroidea). Proc. 5th Int. Coral Reef Congress (Tahiti, 1985), Vol.5, Antenne Museum-EPHE, Moorea, Fr. Polynesia, pp. 327-332

McCallum, H.I. (1987), Predator regulation of *Acanthaster planci*. J. Theor. Biol., 127(2): 207-220.

McKnight, D.G. (1978), *Acanthaster planci* (L.) at the Kermadec Islands. N.Z. Oceanogr. Inst. Records, 4(3): 17-19.

Moore, R.J. (1978), Is *Acanthaster planci* an r-strategist? Nature, 271: 56-57.

Moore, R.J. (1979), Locust phase colouration. Nature, 281: 632.

Moore, R.J. (1985), A study of an outbreak area of the crown-of-thorns starfish (Report of the Queen Mary College 1984 Red Sea Expedition). School of Biological Sciences, Queen Mary College (Uni. of London), 91 pp.

Moore, R.J. (in press), The nature, diversity and conservation of Red Sea coral reefs in relation to water productivity. Proc. 1st Symp. on Potential of Wildlife Conservation in Saudi Arabia (Riyadh, Feb. 1987).

Moore, R.J. & Campbell, A.C. (1985), An investigation into the ecological and behavioural basis for periodic aggregations of *Asterias rubens*. In: Echinodermata (Keegan, B.F. and O'Connor, D.S., eds.), Proc.5th Int. Echinoderm Conf. (Galway, 1984), Balkema, Rotterdam, p.596.

Moran, P.J. (1986), The *Acanthaster* phenomenon. Oceanogr. Mar. Biol. Ann. Rev., 24: 379-480.

Moran, P.J., Bradbury, R.H. & Reichelt, R.E. (1988), Distribution of recent outbreaks of the crown-of-thorns starfish (*Acanthaster planci*) along the GBR: 1985-1986. Coral Reefs, 7: 125-137.

Ne'eman, I. Fishelson, L., & Kashman, Y. (1974), Sarcophine - a new toxin from the soft coral Sarcophyton glaucum (Alcyonaria). Toxicon, 12: 593-8.

Nichols, J.D., Conley, W., Batt, B. & Tipton, A.R. (1976), Temporally dynamic reproductive strategies and the concept of r- and K-selection. Amer. Nat., 110(976): 995-1005.

Nishida, M. & Lucas, J.S. (1988), Genetic differences between geographic populations of the crown-of-thorns starfish throughout the Pacific region. Mar. Biol., 98(3): 359-368.

Olson, R.R. (1987), In situ culturing as a test of the larval starvation hypothesis for the crown-of-thorns starfish *Acanthaster planci*. Limnol. Oceanogr., 32(4): 895-904.

Ormond, R.F.G. & Campbell, A.C. (1971), Observations on *Acanthaster planci* and other coral reef echinoderms in the Sudanese Red Sea. Symp. Zool. Soc. Lond., 28: 433-454.

Ormond, R.F.G., Campbell, A.C., Head, S., Moore, R.J., Rainbow, P.S. & Sanders, A. (1973). Formation and breakdown of aggregations of *Acanthaster planci* in the Red Sea. Nature, 246: 167-9.

Ormond, R.F.G. & Campbell, A.C. (1974), Formation and breakdown of *Acanthaster* aggregations in the Red Sea. Proc. 2nd Int. Coral Reef Symp., 1. Great Barrier Reef C'ttee, Brisbane, pp. 595-619.

Paine, R.T. (1974), Intertidal community structure. Experimental studies on the relationship between a dominant competitor and its principal predator. Oecologia 15: 93-120.

Pearson, R.G. (1975), Coral reefs, unpredictable climatic factors and *Acanthaster*. Crown-of-thorns starfish seminar proceedings (Austral. Govt. Publ. Serv., Canberra), pp. 127-9.

Pearson, R. & Endean, R. (1969), A preliminary study of the coral predator *Acanthaster planci* on the Great Barrier Reef. Fish. Notes, Queensland Dept. Harb. Marine, Brisbane, 3: 27-55.

Pianka, E.R. (1970), On r- and K-selection. Am. Nat., 104: 592-7.

Potts, D.C. (1981), Crown-of-thorns starfish - man-induced pest or natural phenomenon? In: The Ecology of Pests (R.L. Kitching & R.E. Jones, eds.), CSIRO, Melbourne, pp. 55-86.

Sale, P.F. (1980), The ecology of fishes on coral reefs. Oceanogr. Mar. Biol. Ann. Rev. 18: 367-421.

Sammarco, P.W. & Andrews, J.C. (1988), Localized dispersal and recruitment in Great Barrier Reef corals: the Helix experiment. Science 239:1422-1424.

Scheltema, R.S. (1986), Long-distance dispersal by planktonic larvae of shoalwater benthic invertebrates among central Pacific islands. Bull. Mar. Sci., 39(2): 241-256.

Scheltema, R.S. (1986), On dispersal and planktonic larvae of benthic invertebrates: an eclectic overview and summary of problems. Bull. Mar. Sci., 39(2): 290-322.

Simpson, C.J. & Grey, K.A. (1989), Survey of crown-of-thorns starfish and coral communities in the Dampier Archipelago, Western Australia. Environmental Protection Authority, Perth, W.A., Tech. Ser. No. 25, 24 pp.

Steinbeck, J. & Ricketts, E.F. (1941), The Log of the Sea of Cortez. Viking Press, N.Y., 598 pp.

Taylor L.R. & Taylor R.A.J. (1977), Aggregation, migration and population mechanics. Nature, 265: 415-421.

Thorson, G. (1950), Reproductive and larval ecology of marine bottom invertebrates. Biol. Revs. 25: 1-45.

Tinbergen, L. (1960), The natural control of insects in pine woods. I. Factors influencing the intensity of predation in song birds. Arch. Neerl. Zool., 13: 265-343.

Tinbergen, N., Impekoven, M. & Franck, D. (1967), An experiment on spacing out as defence against predators. Behaviour, 28: 307-321.

Underwood, A.J. & Fairweather, P.G. (1989), Supply-side ecology and benthic marine assemblages. TREE, 4(1): 16-20.

Uvarov, B.P. (1977), Grasshoppers and Locusts. Centre for Overseas Pest Research, London.

Vermeij, G.J. (1978), Biogeography and Adaptation. Harvard University Press, Cambridge, Mass. and London, 332 pp.

Vine, P.J. (1970), Densities of *Acanthaster planci* in the Pacific Ocean. Nature (Lond.), 228: 341-2.

Vine, P.J. & Vine, M.P. (1980), Ecology of Sudanese coral reefs with particular reference to reef morphology and distribution of fishes. Proc. Symp. 'The coastal and marine environment of the Red Sea, Gulf of Aden and tropical western Indian Ocean' (Khartoum, Jan.1980), 1. ALECSO, Jeddah, 87-140.

Watt, K.E.F. (1965), Community stability and the strategy of biological control. Can. Entomol., 97: 887-895.

Whittaker, J.B. (1971), Population changes in Neophilaenus lineatus (L.) (Homoptera: Cercopidae) in different parts of its range. J. Anim. Ecol., 40: 425-443.

Wilbur, H.M., Tinkle, D.W. & Collins, J.P. (1974), Environmental certainty, trophic level and resource availability in life history evolution. Amer. Natur., 108: 805-817.

Williams, D., Wolanski, E. & Andrews, J. (1984), Transport mechanisms and the potential movement of planktonic larvae in the central region of the Great Barrier Reef. Coral Reefs, 3: 229-236.

Wilson, B.R. & Marsh, L.M. (1975), Seasonal behaviour of a 'normal' population of *Acanthaster* in Western Australia. Crown-of-thorns starfish Seminar Proc., Brisbane, 6 Sept. 1974, Australian Govt. Publ. Serv., Canberra, pp. 167-179.

Wilson, B.R. & Marsh, L.M. (1974), *Acanthaster* studies on a Western Australian coral reef. Proc. 2nd Int. Coral Reef Symp. 1. GBR C'ttee, Brisbane, Oct. 1974, pp. 621-630.

Wilson, B.R., Marsh, L.M., & Hutchins, B. (1974), A puffer fish predator of the crown-of-thorns in Australia. Search, 5 (11-12): 601- 602.

Wilson, E.O. (1975), Sociobiology. Harvard, Belknap Press.

Yamaguchi, M. (1973), Early life histories of coral reef asteroids. In: Biology and Geology of Coral Reefs, Vol. 2, Biology 1. Academic Press, N.Y. and London, pp.369-387.

Yamaguchi, M. (1974), Growth of juvenile *Acanthaster planci* (L.) in the laboratory. Pacific Sci., 28(2): 123-138.

Yamaguchi, M. (1975), Coral reef asteroids of Guam. Biotropica, 7: 12-23.

Yamaguchi, M. (1977), Larval behaviour and geographic distribution of coral reef asteroids in the Indo-West Pacific. Micronesica, 13: 283- 296.

Yamaguchi, M. (1986), *Acanthaster planci* infestation of reefs and coral assemblages in Japan: a retrospective analysis of control efforts. Coral Reefs, 5: 23-30.

Zann L., Brodie J., Berryman C. & Nagasima M. (1987), Recruitment, ecology, growth and behaviour of juvenile *Acanthaster planci*. Bull. Mar. Sci., 41(2): 561-575.

WHAT CONTROLS OUTBREAKS?

R.H. BRADBURY[1]

Australian Institute of Marine Science, Townsville, Queensland 4810

AND

P.L. ANTONELLI

Mathematics Department, University of Alberta,
Edmonton, Alberta T6G 2G1

Abstract. We build a catastrophe theory model of outbreaks of the crown-of-thorns starfish to provide a general framework with which to compare the many hypotheses about the phenomenon. We argue that a minimal model needs two state variables to represent the starfish and coral for which the simplest elementary catastrophe is the elliptic umbilic. This requires three control variables which have meaning to both starfish and coral. We find these variables in three of Andrewartha & Birch's (1954) five fundamental factors controlling the distribution of all animals: the weather, other organisms of different kinds and a place to live. The model predicts, *inter alia*, that alternative stable states exist and depend on a place to live, and that in 'good' coral reefs, three qualitatively distinct outbreak paths exist. Of the three paths, one clearly represents the adult aggregation hypothesis, another the recruitment initiated predation hypothesis, while the third is a 'missing' hypothesis driven partly by a changing competitive regime in the benthos. We conclude that single factor hypotheses such as the larval recruitment hypothesis, the terrestrial runoff hypothesis and the predator removal hypothesis are oversimplified representations of the phenomenon, and that the recruitment initiated predation hypothesis describes the most likely outbreak path on the Great Barrier Reef.

INTRODUCTION

The population dynamics of crown-of-thorns starfish outbreaks are poorly understood, and we may be certain of very little, perhaps only that the outbreaks are a distinctive ecological phenomenon. The dynamics of the various outbreaks that have been studied in detail are similar enough for it to be reasonable to assume that some common underlying

---

1 Present address: National Resource Information Centre, Canberra, ACT.

organizing processes are at work (Antonelli & Kazarinoff, 1984; Bradbury *et al.*, 1985). Yet the picture is only sketched in, as it were, so that work on establishing just what these 'organizing processes' may be has resulted in a plethora of hypotheses. Hypotheses on physical control of the phenomenon vie with those stressing biological control which themselves may emphasize either endogenous processes or trophic interactions (Moran, 1986; Reichelt *et al.*, 1990).

There are two problems with this situation: a serial problem and a parallel one. The serial problem is that many of the hypotheses are untestable in their present form, that is untestable, *seriatum*, against appropriate null hypotheses. Reichelt *et al.* (1990) tackled this difficulty with a simulation approach, building specific models which captured the essential features of specific hypotheses, and then observing their behaviour. The parallel problem is more subtle: there is no common framework for simultaneously comparing the hypotheses in order to tease out their strengths and weaknesses. We think that such a framework should be of great utility given our present understanding of the phenomenon and should assist in the development of new hypotheses which build on the strengths of current ones.

We seek an analytical framework which is general enough to embrace all the hypotheses, but yet specific enough to capture the essence of the phenomenon. That is, a framework more general than, say, Lotka-Volterra systems with their assumption of the primacy of interspecies interactions in the generation of the observed dynamics, but more specific than so-called 'general systems theory' with its assumption that everything is connected to everything else. Such a framework should allow models that are recognizably ecological and also allow the examination of the family of hypotheses together.

Catastrophe theory provides an obvious framework for this exercise. Crown-of-thorns outbreaks are characterized by abrupt changes in the abundance of both coral and starfish, and catastrophe theory is an extension of the topological theory of dynamical systems for modelling discontinuous changes in natural phenomena (Thom, 1975). The recurrence of the outbreak phenomenon suggests a structurally stable dynamics with abrupt discontinuities, which together meet the formal requirements of catastrophe theory nicely. Catastrophe theory itself is couched in terms of state and control variables which themselves need only be broadly defined, so that it meets our requirements of sufficient but not excessive generality. It is also a topological, qualitative theory and is therefore robust to quantitative differences between particular outbreak episodes.

CHOOSING THE VARIABLES

The essential feature of the outbreak phenomenon is an interaction between the predator - the starfish - and its prey - the coral. Neither is ontologically prior to the other (Levins & Lewontin, 1982): both are needed *a priori* for the interaction and hence for the outbreak. Thus, minimally, any model must have two state variables. If this is so, then it is a fundamental theorem of catastrophe theory that, minimally, any model must have three control variables (Thom, 1975). Therefore we seek three control variables which have meaning to both predator and prey, and yet are general enough to embrace all the hypotheses.

Because the phenomenon is based on an ecological interaction, we seek the control variables amongst those heroic syntheses of ecological principles which attempt to describe the totality of ecological phenomena. Elton (1927), Lindeman (1942), Andrewartha & Birch (1954), Macarthur (1972), May (1976) and Hutchinson (1978) have all provided notable schema. Of these, Andrewartha and Birch's *The Distribution and Abundance of Animals* provides a control-oriented theory which most closely meets our requirements. In this famous work, they argued that all animals are controlled by five things: the weather, other animals of the same kind, other organisms of different kinds, food, and a place in which to live.

In the our case, not all of Andrewartha and Birch's 'factors' are *a priori* of equal importance in determining the dynamics. By using our knowledge of the natural history of the phenomenon (Moran, 1986), we can tailor their general scheme to our present needs by reducing the factors as follows:

The weather. This variable has meaning to both prey and predator. We use it to represent those density-independent aspects of the animal's physical environment which enhance or reduce its chances of survival. Thus it tokenizes for hazards, chance and variability, and embraces particularly the planktonic stages of the life histories of both prey and predator.

Other animals of the same kind. Both starfish and corals show density dependent effects in their population dynamics, and, indeed, an aggregation term has been used with significant success in a series of models based on Volterra-Hamilton dynamics (Antonelli & Kazarinoff, 1984; Antonelli *et al.*, 1989). However, this variable generates an essentially 'second order' effect, and we consider it a less suitable candidate for inclusion as a control variable in a minimal model than the one below.

Other organisms of different kinds. For starfish, these are *their* predators, which figure significantly in many models (McCallum, this volume; Ormond *et al.*, this volume; Parslow, this volume; Reichelt *et al.*, 1990), and for corals, they are their competitors for space, such as soft corals and algae, since recent research has shown that these interactions may be of great significance in the overall dynamics of the system (Bradbury & Mundy, 1989).

Food. For starfish, this variable is already in the model as a state variable - the prey; and for corals, the variable, place to live, covers its needs. This variable is therefore not considered further.

A place in which to live. This is critical for the survival of young starfish and corals after settlement and embraces the juvenile stages of the life histories of both prey and predator.

BUILDING THE MODEL

In order to build our model, we need some terminology to describe the various structures associated with a catastrophe. For this we follow Poston & Stewart (1978).

Catastrophe theory describes the behaviour of systems where control variables interact with state variables in a phase space. In some parts of the phase space, smooth changes in the control variables produce smooth changes in the state variables. In other parts, they produce abrupt and discontinuous changes - the catastrophes. Thus the theory is particularly useful where the object of the modelling exercise is to understand the generation of the discontinuities rather than to 'paper over' them in order to understand the more continuous parts of the system.

The theory allows us to map the locations of the parts of the phase space where these different behaviours occur. It does this by defining a potential V, which is a function of the state and control variables. In our case, this potential or pressure may be thought of as a system property: the intrinsic tendency, tuned through evolution, of the starfish and coral to form an interacting system. For any V, we can define an equilibrium surface. This is a manifold (a smooth surface) in the phase space made up of all the equilibria (stable or otherwise) of the system.

From this we find the bifurcation set, which is the projection down from the phase space to the control space of the set of all points where changes in the form of V occur. The bifurcation set separates the control space into regions that produce catastrophes and regions that do not.

These sets are not arbitrary. Thom (1975) has shown that, for systems with 4 or fewer control variables (the so-called elementary catastrophes), there are only 7 distinct sets. Of these, the elliptic umbilic is the simplest to contain 2 state variables and 3 control variables. The potential for the elliptic umbilic is given by

$$V_{(x,y)} = x^3 - 3xy^2 + w(x^2 + y^2) - ux - vy$$

and the equilibrium surface exists in a 5-dimensioned space, while the bifurcation set, the projection of the catastrophe manifold onto the control space, exists in a 3-dimensioned space as shown in Figure 1.

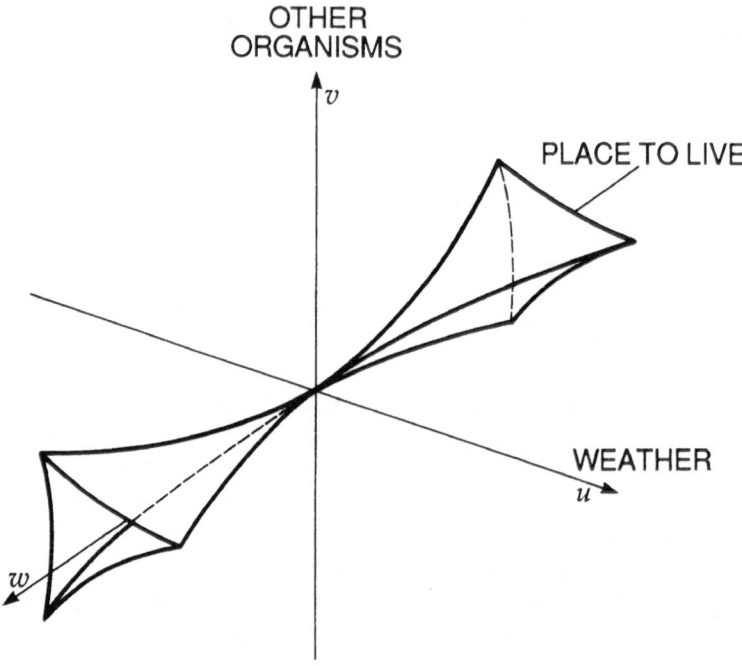

Figure 1. Bifurcation set of elliptic umbilic model of *Acanthaster* phenomenon. See text for details.

Assigning the state and control variables

In what follows, we will argue from simplicity and postulate the validity of the elliptic umbilic (Thom, 1976). We will build our model heuristically by examining features of the natural history of the phenomenon. In particular, we seek empirically observed relationships between pairs of state and control variables with qualitatively distinct behaviours which can be interpreted in catastrophe theory formalism. We will attempt to assign each of the variables of our model, one to one, to each of the variables of the potential function.

The first thing to note is that the two state variables, x and y, have different roles in $V_{(x,y)}$. The potential equation contains an expression of the form x^3 - ax which is the form of the simplest catastrophe, the fold. This is because the elliptic umbilic catastrophe has simpler catastrophes embedded in it. On the other hand, y has a much more complex and subtle role in the generation of V. Corals have more subtle and complex relationships with their control variables than starfish do. Corals exhibit a wide range of ecological interactions including predation, competition and symbiosis. In comparison, starfish are relatively much simpler, being involved mainly in prey-predator interactions. From this we conclude that x represents *starfish* and y represents *coral* in the model.

Now consider the three control variables, u, v and w, in the generation of V. The control variables u and v generally degrade V, while w always enhances it. The control *other organisms*, as predator of starfish and competitors of corals always degrades, the control *the weather* is variable, while the control *a place to live* always enhances. Clearly w best represents *a place to live*.

The two remaining terms, u and v, degrade V in different ways. u interacts only with x, v only with y. There is neither a uy nor a vx term. These would disappear for hyperbola-type relationships with the general form, pq = constant, since the constant would disappear in the potential equation. Simple hyperbolic relationships are unlikely to be found between either of the two state variables and *the weather*, since the weather is a complex, scale-dependent process. Nor are they likely to be found between the state variable *coral* and the control *other organisms* since this relationship is a very complex one. It can involve networks of 'competitive dominance' (Bradbury & Young, 1983; Buss & Jackson, 1979) and may be mediated through allelochemicals (Antonelli *et al.* , in press). However they are likely to be found in a simple predator-prey relationship, since such a simple relationship is essentially an inverse one. We argue that this is particularly the case for the relationship between starfish and their predators when these predators are viewed as controls (e.g. Ormond *et al.*, this volume). Thus we should expect no vx term in the model. Therefore the ux term must be, by elimination, *the weather x starfish*. From this we induce that u best represents *the weather* and v *other organisms*.

We are now able to label the axes of the bifurcation set in Figure 1. It describes the qualitative structure of the control variables within which the predator-prey system is expressed. The surface of the bifurcation set shows, for relative values of the controls variables, where and how the system undergoes abrupt change. Theory tells us that the system is structurally stable only for values of the control variables inside the bifurcation set, and that only one half of the set is stable at a time. That is, the other half is essentially virtual in the sense that an instance of the system can exist only in one or other halves of the set. It also tells us that simple fold catastrophes, which suggest abrupt cessation or breakdown of the system, occur everywhere on the three smooth faces of the bifurcation set. More complex cusp catastrophes, which suggest abrupt and discontinuous changes in the state of the system - the classic outbreak - occur along each of the three ridges.

Table 1. Summary of model structure

Name of elementary catastrophe: elliptic umbilic
Potential function: $V_{(x,y)} = x^3 - 3xy^2 + w(x^2 + y^2) - ux - vy$
Number of state variables (= dimension): 2
Number of control variables (= codimension): 3

Variable in $V_{(x,y)}$	Type	Represents in model
x	state	*starfish*
y	state	*coral*
u	control	*the weather*
v	control	*other organisms*
w	control	*a place to live*

PREDICTIONS OF THE MODEL

Fold catastrophes are more common than cusps in the bifurcation set, since fold catastrophes are generated when the control variables push the system through the faces of the set, and cusps are generated when the system is pushed through the ridges. This suggests *a priori* that the system is more likely to collapse than undergo an abrupt change. The empirical evidence is confusing on this point, since 'the nature of outbreaks and their effects on coral communities are highly variable' (Moran *et al.* (1988) p 128). Moran *et al.* show in their Table 3 that very few reefs with active outbreaks have had low impact on the coral community. Thus while the utter collapse of the predator-prey system, as observed on Green Island or John Brewer Reef, may be an extreme example of a fold catastrophe, folds may nevertheless be more common than cusps on the GBR. This analysis suggests that our classification of outbreak phenomena is wanting. It may be more logical to consider such severe outbreaks as canonical fold catastrophes for this system. Then less severe outbreaks, such as those currently active on Davies Reef or in the Swain Reefs (Johnson *et al.*, 1989), can be considered as cusp catastrophes which are manifested as abrupt changes in, rather than collapse of, the system. This interpretation of the model gives a qualitative as well as a quantitative basis for discriminating between these different sorts of outbreaks, in contrast to the more common quantitatively based classification (Moran *et al.*, 1988.). In addition, the structure of the model and theory suggest that we may consider the cusp type of outbreak as a special, attenuated case of the more severe fold, occurring under special conditions. We will therefore confine the rest of this discussion to the behaviour of the model in generating severe outbreaks or folds.

Only one half of the bifurcation set is structurally stable at a time, and which half exists depends absolutely on the value of w *a place to live.* This suggests that alternative stable states are possible for this system, and that what determines which state exists on a particular reef is the availability of suitable habitat for juvenile starfish and corals. Here the empirical support for the prediction is clearer. The metastable reefs described by Bradbury *et al.* (1985) were significantly poorer 'places to live' than nearby reefs. In a far more detailed study of such habitats in the Red Sea, Moore (this volume) has described the dynamics of such alternative stable states for this system, and their significance for understanding the dynamics on richer reefs.

Within the stable regime associated with richer reefs (which is the regime typically associated with the *Acanthaster* phenomenon), as *a place to live* gets worse, the system moves towards narrower and narrower regions of the bifurcation set. Thus the range of structurally stable values for *the weather* and *other organisms* declines until no values for the control variables can sustain expression of the system. In this sense, catastrophes of either type are more likely from any cause as *a place to live* gets worse. This interpretation suggests that otherwise rich reefs with relatively poor habitat, particularly for juvenile starfish and corals, are more likely to suffer outbreaks. The empirical evidence is poor here since we know little about what makes for a good place to live. We know, for example, that the midshelf reefs of the GBR have been more severely affected by outbreaks than either the onshore or shelf-edge reefs (Moran *et al.*, 1988). However, it is not yet clear what, if anything, is distinctive about these reefs. Done (1982) has argued that the diversity of the coral community is higher on the midshelf than on the nearshore or shelf break reefs, but Wilkinson & Cheshire (1989) point out that this diversity is dominated by the remarkable radiation of one genus, *Acropora*, there. They also note that midshelf reefs are not distinctive in coral growth or recruitment rates.

Conversely, within the other stable regime, as *a place to live* gets worse, the range of structurally stable values for the other control variables, *the weather* and *other organisms* increases until many values can sustain the system. Thus outbreaks in the chronic metastable system are unlikely if the habitat is very poor. It should be remembered however that *a place to live* cannot decline indefinitely on real reefs, since they would very quickly cease to be reefs, and that real metastable reefs would be likely to be of two broad types: relatively stable, quiescent systems in quite poor habitats (perhaps like those observed by Bradbury *et al.*, 1985) or somewhat less stable systems in somewhat richer habitats, with their instability manifested as occasional outbreaks which then may propagate to neighbouring richer systems as observed by Moore (this volume).

Within the 'normal' half of the bifurcation set (the half where *a place to live* is high and the reefs are relatively rich), for any given value of *a place to live*, three qualitatively different sorts of events can push the system through a smooth surface of the manifold and hence generate severe outbreaks (Figure 2). The first of these is a deterioration in *the weather* alone: this by itself increases the likelihood of a severe outbreak since it increases the chance of the system being pushed through the 'weather face' of the manifold. The other two sorts of events are symmetrical about the weather axis, requiring an improvement in *the weather* to operate in concert with either a decrease or increase in *other organisms*.

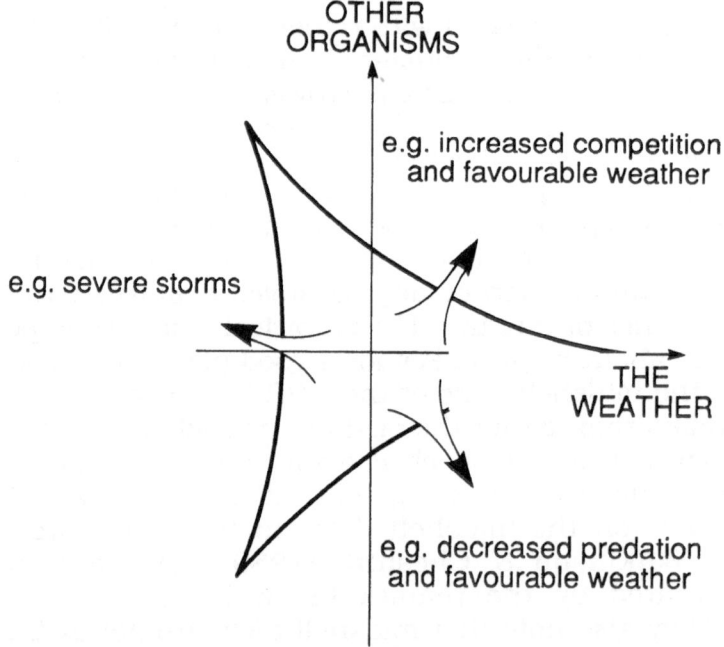

Figure 2. The *weather* x *other organisms* plane of the bifurcation set shown in Figure 1. The arrows show the paths of three qualitatively different types of outbreaks.

These three sorts of events may be interpreted in biological terms as follows. In the first case, a deterioration in *the weather* would directly reduce the chance of survival of either or both of the starfish or coral. This is the outbreak path proposed by the adult aggregation hypothesis of Dana, Newman & Fager (1972). In this hypothesis, severe storms destroy sufficient coral for it to become limiting for the starfish, which are then forced to aggregate in the remaining areas of a reef with high coral cover. In the other two cases, an improvement in *the weather* would improve the survival of either or both coral and starfish, and this would act together with an appropriate change in *other organisms* to produce an outbreak. This suggests, for example, that a 'good' year for the plankton (producing an unusually heavy settlement of starfish larvae) would operate in concert with a reduction in the starfish predators to generate an outbreak. This is the outbreak path proposed by the recruitment initiated predation hypothesis (Ormond *et al.*, this volume).

In its insistence on favourable *weather* operating together with another control, our model stands in contrast to those hypotheses, such as Lucas's (1972) larval survival hypothesis and Birkeland's (1982) terrestrial runoff hypothesis, which stress the primacy of favourable *weather* in the generation of outbreaks. Our model is also in contrast with Endean's (1977) predator removal hypothesis, which seeks to explain outbreaks solely in terms of a change in *other animals*.

However this model goes further than just suggesting that outbreaks involving favourable *weather* also involve a reduction in starfish predators. It also says that increases in *other animals* may generate outbreaks under favourable *weather* conditions just as decreases may, and it insists that *other animals* includes the competitors of corals as well as the predators of starfish. This raises some interesting possibilities, some of which may not have been observed or at least recognized as part of the phenomenon.

Collapse of the predator-prey interaction could well follow a path, if starfish predators so increased or if coral competitors so decreased, where the starfish disappeared from the system altogether, allowing an 'outbreak' of coral. This pattern would fit the model perfectly well, but may not be thought of as part of the *Acanthaster* phenomenon. Similarly an increase in coral competitors might have the same result as a decrease of starfish predators - an outbreak of starfish - under favourable weather conditions. This may be thought of as a 'missing' hypothesis, but one which deserves some attention.

CONCLUSIONS

The model is able to accommodate a wider range of observables about the phenomenon than any individual hypothesis. It allows alternative stable states, controlled by the quality of *a place to live*, and, within the rich reef habitat, suggests that three qualitatively distinct outbreak paths are likely to occur: a deterioration in *the weather* alone, and an improvement in *the weather* together with an increase or decrease in *other animals*. Many of the current hypotheses map uneasily onto this model, particularly those, such as the terrestrial runoff hypothesis, larval survival hypothesis and the predator removal hypothesis, which emphasize only one of the favourable *weather/other animals* pair. Of the other main hypotheses, the adult aggregation hypothesis and the recruitment initiated predation hypothesis describe two of the three distinct outbreak paths, while the third path forms a 'missing' hypothesis.

Even though the model suggests a multiplicity of outbreak paths are possible, the present evidence (Ormond *et al.*, this volume) points to the recruitment initiated predation hypothesis as the most likely explanation for the current series of outbreaks on the Great Barrier Reef.

REFERENCES

Andrewartha, H.G. & Birch, L.C. (1954) *The distribution and abundance of animals.* University of Chicago Press, Chicago.

Antonelli, P.L. & Kazarinoff, N.D. (1984) Starfish predation of a growing coral reef community. *J. theor. Biol.* 107, 667-684.

Antonelli, P.L., Kazarinoff, N.D., Reichelt, R.E., Bradbury, R.H. & Moran, P.J. (1989) A diffusion-reaction-transport model for large-scale waves in crown-of-thorns starfish outbreaks on the Great Barrier Reef. *IMA J. Math. Appl. Med. Biol.* 6, 81-89.

Antonelli, P.L., Sammarco, P.W. & Coll, J.C. (in press) A model of allelochemical interactions between soft and scleractinian corals on the Great Barrier Reef. *J. inf. deduc. Biol.*

Birkeland, C. (1982) Terrestrial runoff as a cause of outbreaks of *Acanthaster planci* (Echinodermata: Asteroidea). *Mar. Biol.* 69, 175-185.

Bradbury, R.H., Hammond, L.S., Moran, P.J., Reichelt, R.E. (1985) Coral reef communities and the crown-of-thorns starfish: evidence for qualitatively stable cycles. *J. theor. Biol.* 113: 69-81.

Bradbury, R.H. & Mundy, C.N. (1989) Large scale shifts in biomass of the Great Barrier Reef ecosystem. In: K. Sherman & L. Alexander (eds.) *Biomass yields and geography of large marine ecosystems.* Westview, Boulder. pp 143-167.

Bradbury, R.H. & Young, P.C. (1983) Coral interactions and community structure: an analysis of spatial pattern. *Mar. Ecol. Prog. Ser.* 11, 265-271.

Buss, L.W. & Jackson, J.B.C. (1979) Competitive networks: nontransitive relationships in cryptic coral reef environments. *Am. Nat.* 113, 223-234.

Dana, T.F., Newman, W.A. & Fager, E.W. (1972) *Acanthaster* aggregations: interpreted as primarily responses to natural phenomena. *Pac. Sci.* 26, 355-372.

Done, T.J. (1982) Patterns in the distribution of coral communities across the central Great Barrier Reef. *Coral Reefs* 1, 95-107.

Elton, C. (1927) *Animal ecology.* Sidgwick & Jackson, London.

Endean, R. (1977) *Acanthaster planci* infestations of reefs of the Great Barrier Reef. *Proc. 3rd Int. Coral Reef Symp.* 1, 185-191.

Hutchinson, G.E. (1978) *An introduction to population ecology.* Yale University Press, New Haven.

Johnson, D.B., Bass, D.K., Miller-Smith, B.A., Moran, P.J., Mundy, C.M., & Speare, P.J. (1989) Outbreaks of the crown-of-thorns starfish (*Acanthaster planci*) on the Great Barrier Reef: Results of surveys 1986-1988. *Proc. 6th Int. Coral Reef Symp.* 2, 165-169.

Lindeman, R.L. (1942) The trophic-dynamic aspect of ecology. *Ecology* 23, 399-418.

Levins, R. & Lewontin, R. (1982) Dialectics and reductionism in ecology. In: E. Saarinen (ed.) *Conceptual issues in ecology.* Reidel, Dordrecht. pp 107-138.

Lucas, J.S. (1972) *Acanthaster planci:* before it eats coral polyps. *Proc. Crown-of-thorns starfish Sem.* AGPS, Canberra, pp 25-36.

Macarthur, R.H. (1972) *Geographical ecology.* Harper & Row, New York.

May, R.M. (1976) *Theoretical ecology: Principles and applications.* Blackwell, London.

McCallum, H.I. (this volume) Effects of predation on *Acanthaster:* age-structured metapopulation models. In: R.H. Bradbury (ed.) The *Acanthaster* phenomenon: a modelling approach. Springer-Verlag, Berlin.

Moore, R. (this volume) Persistent and transient populations of the crown-of-thorns starfish, *Acanthaster planci.* In: R.H. Bradbury (ed.) The *Acanthaster* phenomenon: a modelling approach. Springer-Verlag, Berlin.

Moran. P.J. (1986) The *Acanthaster* phenomenon. *Oceanogr. Mar. Biol. Ann. Rev.* 24, 379-480.

Moran, P.J., Bradbury, R.H. & Reichelt, R.E. (1988) Distribution of recent outbreaks of the crown-of-thorns starfish (*Acanthaster planci*) along the Great Barrier Reef: 1985-1986. *Coral Reefs* 7, 125-137.

Ormond, R.F.G., Bradbury, R.H., Bainbridge, S., Fabricius, K., Keesing, J., De Vantier, L.M., Medlay, P. & Steven, A. (this volume) Test of a model of regulation of crown-of-thorns starfish by fish predators. In: R.H. Bradbury (ed.) The *Acanthaster* phenomenon: a modelling approach. Springer-Verlag, Berlin.

Parslow, J.S. (this volume) Stochastic and spatial effects in predator-prey models of *Acanthaster*-coral interactions. In: R.H. Bradbury (ed.) The *Acanthaster* phenomenon: a modelling approach. Springer-Verlag, Berlin.

Poston, T. & Stewart, I. (1978). *Catastrophe theory and its applications*. Pitman, London.

Reichelt, R.E., Greve, W., Bradbury, R.H., Moran, P.J., 1990. *Acanthaster planci* on the Great Barrier Reef: a starfish-coral site model. *Ecol. Modelling* 49: 153-177.

Thom, R. (1975) Structural stability and morphogenesis. Benjamin Addison-Wesley, New York.

Wilkinson, C.R. & Cheshire, A.C. (1989) Cross-shelf variations in coral reef structure and function - influences of land and ocean. *Proc. 6th Int. Coral Reef. Symp.* 1, 227-233.

TRANSITION MATRIX MODELS,
CROWN-OF-THORNS AND CORALS

T. J. DONE

Australian Institute of Marine Science, Townsville, Queensland 4810

Abstract. Transition matrix models have been used to investigate a number of questions regarding long and short term affects of crown-of-thorns starfish on coral communities and populations. The characteristics of damage to massive *Porites* corals at 21 sites on 5 reefs were expressed as 'disturbance' matrices containing the proportions of corals in each of 5 size classes which were killed, or had their living surface area reduced by various amounts. 'Normal' matrices were developed to describe recruitment, growth and regeneration between outbreaks. The long term (200 y) effects of outbreaks on the field populations were simulated by introducing the 'disturbance' matrix once every 8 - 30 iterations (= years) among 200 iterations of 'normal' matrix x vector multiplications. The matrices were also used to develop indices of the severity of starfish impact and vulnerability to additional outbreaks. Finally, trajectories of total coral cover (all species) over the next 20 y were predicted using 'normal' matrices based on 4 logistic coral growth functions, and population vectors based on field estimates of density and size frequency.

INTRODUCTION

In these days of chaos and cellular automata, what role can there be for the matrix model in crown-of-thorns research? Perhaps its greatest use is to help us ask better questions, such as the following: Is it possible that there have been *Acanthaster planci* outbreaks at 1 - 3 decade intervals throughout the Holocene? How long will it take coral populations to recover their previous densities and size frequency distributions? How vulnerable are coral populations of particular size frequency distributions to outbreaks of particular characteristics? How long before coral cover returns to pre-outbreak levels? These are all questions which have been addressed using matrix models (Done, 1987; 1988; Done *et al.*, 1988).

Transition matrix models were first used to study dynamics of coral populations by Hughes (1984), who circumvented problems of aging corals by monitoring and modelling transitions among size classes. Using records of the annual proportions of corals growing, shrinking, or remaining in their present size class, Hughes defined the dynamics of field populations of several coral species. Through iterative vector x matrix multiplication, the matrix of probabilities was used to extrapolate an initial population size structure (= vector) through time. Fortuitously, in 1980, Hurricane Allen destroyed many of Hughes' Jamaican study areas, an event which Hughes duly quantified in an alternative 'disturbance' matrix. Long term simulations involving many iterations of 'normal' matrices interspersed by random 'disturbance' matrices allowed Hughes to assess the responses of the various coral species to repeated hurricanes.

Hughes' model was timely, as it had obvious application in evaluating disturbances of any sort, including *A. planci* predation on coral populations and communities. Using estimates of population densities and size structures of massive *Porites* corals in a number of areas, and data on the degree of damage caused to the different size classes, I was able to calculate a disturbance matrix for massive *Porites* in each area (Done, 1987; 1988). I then modelled the long term (200 y) effects of outbreaks of various intensities and frequencies on population structure in massive *Porites*.

I also used a matrix model to investigate short term recovery of other corals on a damaged reef (Done *et al.*, 1988). Once new corals had begun to settle on damaged reefs, their densities and size frequencies by species were estimated using belt transects. Increases in colony sizes, population size frequency distributions, and percent cover, were simulated for a period of 20 y.

In both cases, the scales of the field observations were over hectares rather than square metres (cf. Hughes, 1984) and I did not have the temporal observations of marked colonies needed to generate the 'normal' matrix. While initially I felt that this was a disadvantage, I soon concluded it was not, because it encouraged me to explore the effects of a wider range of intuitively reasonable population dynamics than I might otherwise have done. For example: what if survivorship took such and such a form? What if the corals grew a bit faster between outbreaks? What if there were lags in recruitment, or what if there was no recruitment? This paper briefly describes the results of a series of simulations described more fully in Done (1987, 1988) and Done *et al.* (1988).

METHODS

Quantitative surveys of coral abundance, size classes and damage by size class were conducted using belt transects on several reefs in the central Great Barrier Reef within a few years of them being damaged by *A. planci* (Done, 1987, 1988). Massive *Porites* corals, which are important framework and ecosystem components of coral reefs, were categorized into one of 5 diameter classes (0.10 - 0.30 m; 0.31 - 0.60 m; 0.61 - 1.00 m; 1.01 - 2.00 m; > 2 m) and 4 injury classes (dead; <1/3 live; 1/3 - 2/3 live; >2/3 live). Other corals as small as 1 cm diameter were surveyed in separate belt transects (Done *et al.*, 1988), identified to species where possible, and allocated to one of 26 size categories: 1 to 10 cm by 1 cm increments; 11 - 20 cm by 2 cm increments; 21 - 50 cm by 5 cm increments; 51 - 100 by 10 cm increments; > 100 cm.

Outbreak matrix - limits of unequal diameter classes (cm)

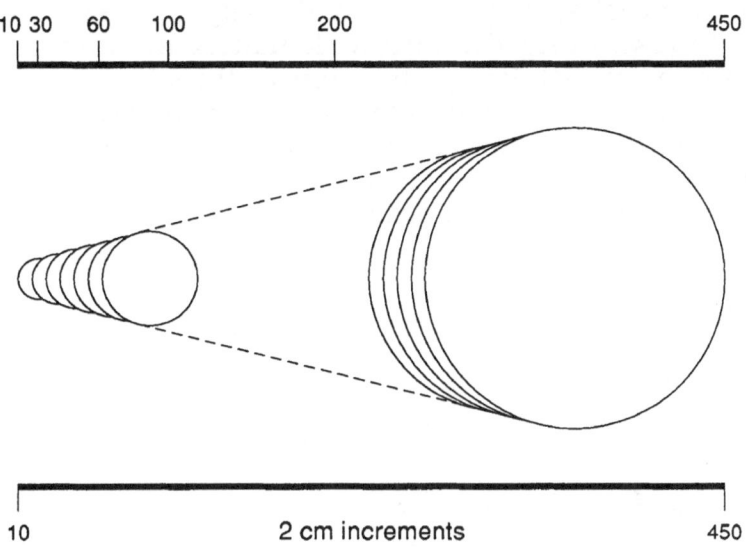

Normal matrix - limits of 221 equal diameter classes (cm)

Figure 1. Illustration of differences in size classes used in 'normal' and 'disturbance' matrices. In 'normal' years, depending on colony growth rates and size-specific survivorship, different proportions will remain in their present 2 cm wide size class or advance to one of the following two size classes. In 'outbreak' years, a proportion of colonies have their live tissue area reduced by starfish predation. This is expressed in a matrix defining the damage as the proportion of colonies in large size classes (boundaries at 10, 30, 60, 100, 200 and 450 cm) whose live tissue area is reduced to the equivalent of that of a smaller size class.

200 y simulations of massive Porites.

The 'disturbance' matrix (one for each area) was calculated assuming colonies were evenly spread throughout their 5 size classes, and were at the mid-point of their damage category (see Done, 1987 for details). Numerous 'normal' matrices were calculated, allowing the effects of different combinations of colony growth rate and size-dependent survivorship to be investigated. These matrices had 221 size classes each 2 cm wide (Figure 1). (It is essential that the size categories in a growth matrix approximate the growth possible in the time period iterated, otherwise serious over-estimates of inter-disturbance performance occur (Vandermeer, 1975; *pers. obs.*)). All corals were assumed to grow at 0.75, 1.0, 1.25 or 1.5 cm y^{-1} (radial growth rate), so the complete set of transition probabilities was contained within the diagonal and the first 2 rows of the upper triangle. Size-dependent survivorship functions were based on the assumptions that a small coral recruiting to the population had either a 0.7 or 0.9 probability P_i of surviving its first year, and a gradually increasing probability of surviving successive years. The rate of increasing survivorship was determined by a threshold size, at which P_i became 1.0.

The effects of outbreaks on population density and structure were investigated by substituting the outbreak matrix at intervals among many iterations of multiplying the non-outbreak matrix by the population vector, initially as estimated in the transects. Prior to each simulated outbreak, the 221 size classes were pooled into the 5 field size classes, and after being depleted by the outbreak, were re-distributed into their 221 size classes until the next outbreak.

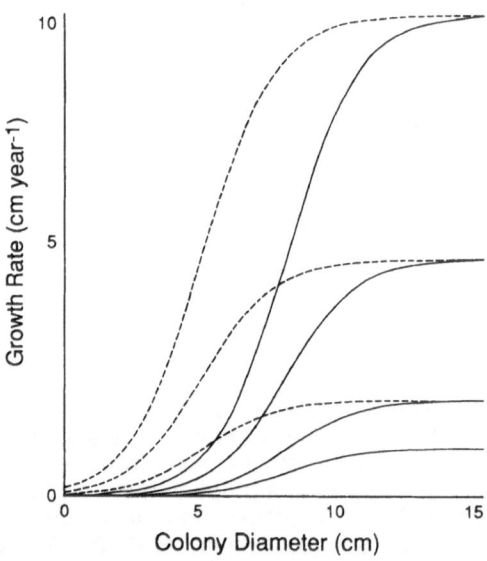

Figure 2. Simulated size-dependent colony growth rate curves used to calculate transition matrices for growing corals. The curves are described by the logistic function:-

$$g_s = K/\{1+\exp[-r(s-s_0)]\}$$

where   g = growth rate in cm y^{-1} at size s,
                    s = maximum lateral dimension of the colony,
                    s_0 = the intercept of the curve with the x axis,
                    r = the slope of the curve (0.08), and
                    K = the terminal colony growth rate (- 1, 2, 5 and 10 cm y^{-1}).

Broken lines indicate 'early start' growth (s_0 = - 7.5 cm); solid lines indicate 'late start' growth (s_0 = - 5.0).

20 y simulations of total coral trajectories.

In these simulations, where the objective was to investigate the total and relative contributions of different categories of coral to the total community as it developed between outbreaks, there was no introduction of a simulated disturbance. Unlike the *Porites* simulations, where growth in corals (which were all > 10 cm diameter) was assumed to be constant with size, I assumed coral growth for smaller corals followed a logistic function (Figure 2). Starting slowly as a small coral spat, fragment or remnant, each coral accelerates and then levels out at one of four terminal growth rates K_l. Each species was allocated to one of 4 categories based on its K_l, for the purposes of simulating their continued growth. The effect of the timing of the acceleration phase was examined explicitly.

RESULTS AND DISCUSSION

Is it possible that there have been *Acanthaster planci* outbreaks at 15 y intervals - the interval between the 1960s and 1980s outbreak - throughout the Holocene? The answer suggested by the simulations to this question and all that follow is, 'Well, it depends....'. I chose a moderately rigorous criterion to investigate the question - the capacity of *Porites* populations to repeatedly restore pre-outbreak densities of colonies 1-2 m diameter before the next outbreak. The simulations were used to define a recruitment/growth-rate phase space within which there were areas in which the criterion was met (Figure 3). Faster growth allowed populations to be restored with a lower annual recruitment, which nonetheless, had to be reliable. Populations having poor inter-outbreak survivorship, perhaps due to disease, other predators, bleaching or overgrowth by other benthic organisms, required both high colony growth rates and high annual recruitment - higher than any subsequently observed in the field (*pers. obs.*). This is all very intuitive, but until recently, not reflected in any of the debate on the effects of the starfish on coral communities.

How long will it take coral populations to recover their previous densities and size frequency distributions? This time the answer depends on what happens to the corals we censused on the reef, and those which recruit subsequently. There was a very strong sensitivity to position of the logistic growth curve along the X axis (colony size) when we simulated the growth of censused corals with an assumption of 100% survival. For example, an early acceleration suggested there would be 100% coral cover at 3 m depth by the year 2004, whereas a delayed acceleration suggested there would be about 50% cover at that time. In the former case, it is clear that one of the major assumptions of the model would have been violated within the 20 y time period of the simulation, viz. that there are no density dependent limits to size in corals. Interference among corals in crowded assemblages would decrease rate of increment in individual colony sizes and total coral cover in the manner suggested for terrestrial plants by Westoby (1988).

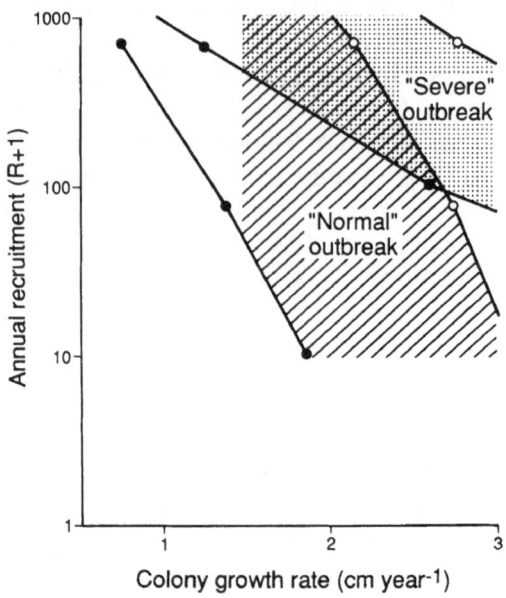

Figure 3 Recruitment/growth rate phase space showing viable domains for a *Porites* population subject to outbreaks of constant characteristics at 15 y intervals. The shaded and hatched areas are bounded by lines showing where size class IV (corals 1 - 2 m) will repeatedly recover prior densities, assuming low (solid symbols) or high background mortality (open symbols) between outbreaks.

How vulnerable are coral populations of particular size frequency distributions to outbreaks of particular characteristics? This question was investigated by compiling the *Porites* size and damage data at a total of 25 sites on 5 reefs (Done, 1988). Each area had unique population size structures and damage characteristics, i.e unique initial population vector and disturbance matrix. 'Vulnerability' was tested by a very artificial simulation in which no further recruitment was introduced. Using the technique of substituting the disturbance matrix among many iterations of the normal matrix, some populations were sent locally extinct within two more outbreaks, for intervals of 8 to 30 y between outbreaks. Others, which had less damage and/or a more favourable initial population vector, could withstand > 20 further outbreaks before local extinction.

The matrix model has thus been used to put easily understood interpretations on complex field data on coral populations modelled in isolation of all other populations. More complicated models may be constructed to incorporate community level processes. For example, benthic community trajectories (diversity, composition) following an outbreak may be predicted by transition matrices in the same way they have been used to predict succession in vegetation on a tree by tree basis (Horn, 1975). Alternatively, the outcomes of interactions among different types of corals and other benthos may be represented as transition matrices, and used in discrete spatial models of benthic recovery (Hogeweg & Hesper, this volume). However to this point, it has been the very simplicity of the models which has made them so valuable in investigating the consequences of crown-of-thorns attacks.

REFERENCES

Done TJ (1987) Simulation of the effects of *Acanthaster planci* on the population structure of massive corals in the genus *Porites*: evidence of population resilience? Coral Reefs 6:75-90.

Done TJ (1988) Simulation of recovery of pre-disturbance size structure in populations of *Porites* spp. damaged by the crown-of- thorns starfish *Acanthaster planci*. Mar Biol 100:51-61.

Done TJ, Osborne K, Navin KF (1988) Recovery of corals post- *Acanthaster*: progress and prospects. Proc 6th Int Coral Reef Symp 2:137-142.

Hogeweg P & Hesper, B. (this volume) Crowns crowding: an individual oriented model of the *Acanthaster* phenomenon. In: R.H. Bradbury (ed.) The *Acanthaster* phenomenon: a modelling approach. Springer-Verlag, Berlin.

Horn HS (1975) Markovian processes of forest succession. In ML Cody and JM Diamond (eds). Ecology and Evolution of Communities. Cambridge, Mass. Harvard University Press pp196-211.

Hughes TP (1984) Population dynamics based on individual size rather than age: a general model with a reef coral example. Am Nat 123:778- 795.

Vandermeer JH (1975) On the construction of the population projection matrix for a population grouped in unequal stages. Biometrics 31:239- 242.

Westoby M (1988) Vegetation in relation to grazing comparing rangelands with shallow hard bottom ecosystems. Proc 6th Int Coral Reef Symp 2:59-64.

THE RELEVANCE OF STOCHASTIC EFFECTS TO THE NOTION OF *ACANTHASTER PLANCI* AS A NEAR-OPTIMAL PREDATOR

R.M. Seymour

Department of Mathematics, University College London,
Gower Street, London WC1E 6BT

Abstract. A stochastic extension of the (deterministic) model of the starfish-coral predator-prey population dynamics introduced by Seymour (1989) is defined and analysed. In particular, it is shown to admit a stationary probability density which is a function of a basic "efficiency of predation" parameter. Properties of this density are used as a basis for discussion of the potential role of (small scale) stochastic effects both in the ecology of interactions of starfish and coral, and with respect to possible selection pressures pertinent to the evolution of *A. planci* into a predator prone to outbreak (i.e. a *near-optimal predator* in the terminology of Seymour, 1989).

INTRODUCTION

Underlying the dispute between proponents of the view that outbreaks of crown of thorns starfish (*Acanthaster planci*) on coral reefs in the Indo-Pacific region are primarily unnatural, man-induced phenomena, and those who take the view that such outbreaks are natural, are two different sets of ecological assumptions (Potts, 1981). This polarization is succinctly captured by Potts:

"Each of these arguments [for the view that outbreaks are unnatural] assumes that a reef community is basically a stable, biologically accommodated system ... where an equilibrium among ecologically specialised species is maintained by complexes of interactions developed during a prolonged evolutionary history in a predictable environment. In such systems, return to the original equilibria is likely to be slow following major (man-induced) disturbances." (Potts, 1981, p. 56).

On the other hand:

> "Most proponents of 'natural phenomenon' viewpoints base
> their arguments on very different ecological assumptions from
> those underlying the 'human interference' hypotheses. In
> particular, many ecological properties of tropical reef
> communities are interpreted as the direct and short term
> results of largely stochastic environmental disturbances
> (physical or biological) rather than as equilibria derived from
> long periods of environmental stability." (Potts, 1981, p. 57).

With particular reference to *A. planci*, it is certainly true that this starfish
is a relatively long-lived carnivore which is highly specialised for feeding
and defence (Cameron & Endean, 1982). These features imply a high
degree of coevolution between the starfish predator and its coral prey (*A.
planci* is essentially a *K-strategist*). On the other hand, Moore (1978) has
noted that many features of the life history of *A. planci* are characteristic
of an *r-strategist*, which suggests that the second (stochastic) view of
population outbreaks is likely to be correct.

In Seymour (1989), I proposed that the traditional r-strategist/K-
strategist continuum might more fruitfully be replaced by what I called a
suboptimal/superoptimal continuum in the context of starfish
population outbreaks. This latter continuum was derived from a
particular model of starfish/coral dynamics which was constructed
around a precisely defined notion of *optimal predation*. The position of a
particular starfish population on this continuum is then determined by
an efficiency of predation parameter, c, which measures the deviation of
the population from optimality in terms of various ecological influences.
The optimal predation condition is $c = 0$, while a population with $c > 0$ is
called suboptimal, and one with $c < 0$ is superoptimal. It was suggested
that 'normal' (non-outbreak) populations of *A. planci* are *near-optimal* on
this continuum (i.e. c small and positive), and that observed population
outbreaks are caused by large scale environmental disturbances which
push the predator-prey dynamics (temporarily) into the superoptimal
range (which is essentially unstable).

The discussion in Seymour (1989) was given in terms of a particular
deterministic model, and featured properties of equilibria of this model. It
is therefore not ideally suited to reflect the second (stochastic) view of reef
communities described by Potts. In the present paper, I consider a
stochastic extension of my (1989) model, which incorporates persistent
random fluctuations in two of the key starfish parameters (net death
rate and primary aggregation potential). A suitably simplified version of
this stochastic model is then analysed with particular reference to the
role of the efficiency parameter c. The principal conclusions to be reached
are that the stochastic effects incorporated in the model may be

interpreted as generating selection pressures on starfish populations, and that the view of *A. planci* as a near-optimal predator on a suboptimal/superoptimal continuum can be seen as a natural consequence of these pressures. This perspective then cuts across the polarity of view as to the nature of relevant ecological interactions, as described by Potts (above).

A brief review of the Seymour (1989) model is given in §1 for the readers' convenience. The stochastic extension is constructed in §2, together with a suitable approximation, which latter is analysed in §3. The consequences of this analysis are then discussed in §4.

1. THE DETERMINISTIC MODEL

The model discussed in Seymour (1989) is given by the 2-dimensional dynamical system,

$$\frac{dN}{dt} = N[\lambda - \alpha N - \rho G]$$

$$\frac{dG}{dt} = G[-\mu + \beta N + \phi(G)] \qquad\qquad 1.1$$

$$\phi(G) = \delta G + e/G^B \text{ , where } B = \frac{(\delta + \rho)\alpha}{(\beta\rho - \alpha\delta)} .$$

Here, G represents the starfish density on a reef (or complex of reefs), and N the density of its (ideal) coral prey. All the scalar coefficients are positive.

A distinguishing feature of this model is the *aggregation term*, $\phi(G)$. The product $G\phi(G)$ is to be interpreted as the *enhanced reproductive efficiency* of the population due to aggregation of starfish (a phenomenon which is thought to increase reproductive success through enhanced fertilization - Moran, 1986). The particular form of $\phi(G)$ given in (1.1) is derived from an 'optimality principle' in Seymour (1989), and (under mild assumptions) is the unique form satisfying such a principle. I shall assume that the *primary aggregation coefficient*, δ, is *large* compared to the *secondary aggregation coefficient*, e. Also, I shall assume that

$$\Delta = (\beta\rho - \alpha\delta) > 0, \qquad\qquad 1.2$$

so that B in (1.1) is positive. In fact, under normal circumstances, it is biologically reasonable to suppose $0 < B < 1$ (Seymour, 1989, §3).

Now define new parameters as follows,

$$A = \frac{(\alpha + \beta)\delta}{\Delta}, \qquad c = \frac{\mu B - \lambda A}{\mu B + \lambda A},$$

$$D = \frac{1}{2AB}(\mu B + \lambda A), \qquad E = \frac{\alpha}{B} = \frac{\beta}{A + 1}, \qquad F = \frac{\delta}{A} = \frac{\rho}{B + 1}. \qquad 1.3$$

Solving equations (1.3) for the parameters in (1.1) gives,

$$\lambda = BD(1 - c), \qquad \mu = AD(1 + c)$$

$$\alpha = BE \qquad , \qquad \beta = (A + 1)E \qquad\qquad 1.4$$

$$\rho = (B + 1)F \ , \qquad \delta = AF.$$

All parameters in (1.3), except possibly c, are positive. D, E and F are scale parameters, while A, B and c are *scale independent*, and hence carry the intrinsic ecological information in the model. In particular, c measures the *efficiency of predation* of the starfish population (with respect to its prey), in a sense made precise in Seymour (1989), §3. The population is *optimal* if c = 0, *suboptimal* if c > 0 and *superoptimal* if c < 0. A suboptimal population underutilises its ecological opportunities, while a superoptimal population overutilises the same.

Putting e = 0 in (1.1), I obtain a unique positive quadrant equilibrium,

$$(\bar{N}_0, \bar{G}_0) = (\frac{\rho\mu - \delta\lambda}{\Delta}, \frac{\beta\lambda - \alpha\mu}{\Delta}) \qquad\qquad 1.5$$

$$= (\frac{A}{(A+B+1)}\{1+(2B+1)c\}. \frac{D}{E}, \frac{B}{(A+B+1)}\{1-(2A+1)c\}. \frac{D}{F})$$

For e small enough, there is an equilibrium $(\bar{N}_e, \bar{G}_e)$ close to (1.5) (Seymour, 1989, §4). This equilibrium is *stable* if c > 0 (suboptimal population) and *unstable* if c < 0 (superoptimal population). Furthermore, there is a Hopf - bifurcation at c = 0 leading to small amplitude *stable limit cycles* in the superoptimal range. From (1.5), c must satisfy

$$\frac{-1}{2B + 1} < c < \frac{1}{2A + 1} \qquad\qquad 1.6$$

It is a fundamental assumption of the discussion of outbreaks given in Seymour (1989), that 'normal' (non-outbreak) starfish populations are *near-optimal*; i.e. suboptimal but with c small. Thus, normal populations exist at stable equilibrium. Outbreaks occur when suitable large scale environmental perturbations (which change the parameters in 1.1) push the population into the superoptimal range for a time, so that the coral/starfish interaction describes (at least part of) a stable limit cycle before conditions gradually return to normal and a stable equilibrium is resumed. A necessary condition for such sustained outbreaks to occur was shown to be that the effect of the perturbation on the starfish parameters (μ, β, δ) must be significantly *larger* (and *faster*) than on the coral parameters (λ, α, ρ). In addition, it was shown that a given level of disturbance has the greatest effect on the starfish population dynamics if it acts through μ rather than through β or δ.

The parameter μ in (1.1) represents the natural death rate of the indigenous population less any recruitment of sexually mature adults from outside the population (either by direct immigration, or the successful establishment of larvae from external sources). In addition, μ includes losses due to non density-dependent predation of adult starfish (assumed to be low level under normal circumstances). In what follows, I shall consider the role of random fluctuations in the two key parameters μ and δ.

2. THE STOCHASTIC MODELS

The sources of random variation in μ are twofold. Firstly, variations in recruitment of new individuals due to immigration, and secondly, variations in predation pressure, particularly on juveniles and adults. (Predators include various species of fish, the painted shrimp, *Hymonocera picta*, and, if one is willing to discount the notion of its being a keystone predator, the giant triton, *Charonia tritonis* - see Potts, 1981, p. 76, or Moran, 1986, p. 414). Other general factors affecting recruitment are: degree of fertilization, temperature, salinity, dispersal and availability of suitable substrate for settlement (Lucas, 1975).

Variations in the extent and duration of aggregations of starfish, and the degree to which these enhance the reproductive success of the population, are also clearly likely to occur. Indeed, aggregations of all sizes, from isolated individuals to clusters of thousands of individuals seem to occur (Potts, 1981, p. 66), and to vary both spatially and temporally (Moran, Bradbury & Reichelt, 1985). Clearly, local topographic factors influence such aggregations; other factors are: the presence of spawning attractants and coral extracts, age of starfish, distribution and abundance of coral, type of coral (Moran, 1986, p. 410). Also, the extent to which aggregation behaviour enhances the reproductive success of the population is likely to be variable, depending on general factors affecting recruitment.

To incorporate the random variation in μ and δ described above into the model discussed in §1, I shall replace μ and δ by $\mu + v\dot{\xi}(t)$ and $\delta + \omega\dot{\eta}(t)$, where v, ω are positive constants, and $\dot{\xi}(t)$ and $\dot{\eta}(t)$ are *independent* white noise terms. Thus, μ and δ are now to be interpreted as mean values with respect to the revelant (normally distributed) fluctuations, with v^2 and ω^2 the associated variances. Effecting this replacement in equations (1.1) and writing (as usual), $d\xi = \dot{\xi}(t)dt$ and $d\eta = \dot{\eta}(t)dt$, 1 obtain the (Ito) stochastic differential equations.

$$dN = N[\lambda - \alpha N - \rho G]dt \qquad\qquad 2.1$$

$$dG = G[-\mu + \beta N + \delta G + e/G^B]dt - vGd\xi + \omega G^2 \, d\eta.$$

Note that the coral parameters λ, α, ρ are kept constant. This is justified by the greater degree of responsiveness of the starfish parameters to fluctuations (see §1 and Seymour, 1989, §5). On the other hand, in Seymour (1989) changes in β were interpreted as due to changes in age distribution in the starfish population. While it would undoubtedly be more complete to include such variations in (2.1), I have not done so because the additional complexity does not justify it, and the conclusions to be obtained in subsequent sections remain substantially unaffected. One may also argue that changes in β occur over a longer time scale than the proposed short time random variations in μ and δ.

Similarly, the assumption of the independence of the noise terms $\dot{\xi}(t)$ and $\dot{\eta}(t)$, is made largely for technical convenience. Clearly, to some extent μ and δ will be influenced by variations in the same external factors. However, in assuming independence, I have tacitly supposed that each variable factor has its most significant impact on only one of the two parameters, with the other influenced to a negligible extent in comparison. Again, this assumption could be weakened somewhat without sacrificing the qualitative conclusions of the sequel, but only at the price of additional technical complexity in the analysis.

The model (2.1) is too complicated to analyze as given, and so I shall consider a suitable 1-dimensional approximation involving the starfish population density only. Such an approximation may be obtained by invoking the fundamental hypothesis (see §1) that *starfish processes occur an order of magnitude faster than coral processes*. Thus, I can assume that fundamental changes may occur over a short time scale to the starfish regime without significant change in the corresponding coral regime. This is expressed by the coral equilibrium condition, $dN/dt = 0$ (i.e. the rate of response of the coral population to changes in the starfish population is small). From the first equation of (2.1), I then obtain the *coral response equation*.

$$N = \frac{\lambda}{\alpha} - \frac{\rho}{\alpha} \cdot G. \qquad\qquad 2.2$$

Substituting in the second equation of (2.1), this gives the 1- dimensional S.D.E.,

$$dG = G[a - bG + e/G^B]dt - vGd\xi + \omega G^2 \, d\eta, \qquad\qquad 2.3$$

where $a = (\beta\lambda - \alpha\mu)/\alpha$ and $b = (\beta\rho - \alpha\delta)/\alpha = \Delta/\alpha$. Note that both a and b are *positive* (by 1.2 and 1.5).

The noise term in (2.3) may be put in standard form by defining a new stochastic process $\tilde{\xi}(t)$, by

$$d\tilde{\xi} = \frac{- d\xi + \sigma Gd\eta}{(1 + \sigma^2 G^2)^{1/2}}; \quad \tilde{\xi}(0) = 0; \quad \sigma = \omega/v. \qquad\qquad 2.4$$

(to be interpreted as an Ito equation). It then follows from the independence of ξ and η, together with P. Levy's characterisation (Elliot, 1982, corollary 12.29, p. 143) that ξ is a standard Brownian motion, so that (2.3) may be written as an Ito S.D.E.,

$$dG = G[a - bG + e/G^B]dt + vG(1 + \sigma^2 G^2)^{1/2} . d\tilde{\xi} \qquad\qquad 2.5$$

This is the equation which I shall take as representing the starfish process, and which will be analysed in §3 below. For those readers who do not wish to pursue the details of the analysis, a summary of the main conclusions is given at the end of §3.

Remark. One could consider coral response functions other than (2.2). In fact, all that is required to obtain the general conclusions of the next section is a response function of the form,

$$N = \frac{\lambda}{\alpha} - Gh(G),$$

where h(G) is a positive, bounded function which is independent of the efficiency parameter c.

3. ANALYSIS OF THE MODEL

The model to be analysed in this section is the scalar (Ito) S.D.E. (2.5) defined on $0 < G < \infty$.

Consider a general scalar S.D.E. defined on $0 < x < \infty$, of the form,

$$dx = f(x)dt + g(x)d\theta, \qquad\qquad 3.1$$

where θ(t) is a standard, scalar valued Brownian motion, and g(x) > 0 for all x. Both f(x) and g(x) are assumed smooth. Recall that the boundary point x = 0 is *attracting* if the function,

$$\psi(x) = \exp\{ - \int^x \frac{2f(u)}{g(u)^2} \, du\} \qquad 3.2$$

is integrable in a neighbourhood of x = 0, and that x = 0 is *unattainable* otherwise (e.g. Gard, 1988, p. 146). Similarly for the boundary x = ∞.

For the model (2.5), the function (3.2) has the form,

$$\psi(G) = \Big(\frac{G^2}{1 + \sigma^2 G^2} \Big)^{-u} . \exp\{(2b/\sigma v^2)\tan^{-1}(\sigma G) - (2e/v^2)F(G)\}, \qquad 3.3$$

where $u = a/v^2$ and

$$F(G) = \int \frac{1}{G^{B+1} (1 + \sigma^2 G^2)} \, dG. \qquad 3.4$$

Clearly F(G) is a strictly increasing function with F'(0) = ∞ and F'(∞) = 0. By setting x = 1/G and performing suitable integrations in series in neighbourhoods of x=0 and x=∞, one also easily shows that,

$$\begin{aligned} F(G) &\sim -1/G^{B+2} & \text{as } G \to \infty \\ \text{and } F(G) &\sim -1/G^B & \text{as } G \to 0 \end{aligned} \Biggr\} \qquad 3.5$$

It follows, in particular, that F(G) is *negative* for G ε (0, ∞). [Note that the arbitrary constant implicit in the definition of F(G) may be ignored, since strictly speaking ψ(G) is only defined up to multiplication by an arbitrary, positive constant.]

In view of the second property in (3.5), I conclude that ψ(G) is never integrable in a neighbourhood of G = 0, *provided e > 0. Thus, G = 0 is an unattainable boundary* in this case.

Now contrast this with the case e = 0. In a neighbourhood of G = 0, I have ψ(G) ~ G^{-2u}, which is integrable if and only if 2u = $2a/v^2$ < 1. *Thus, providing the fluctuations are large enough* ($v^2 > 2a$), *G = 0 is an attracting boundary when e = 0.*

Turning to the boundary $G = \infty$, I have (in view of 3.5),

$$\psi(G) \to \sigma^{2u} .\exp\{\pi b/\sigma v^2\} \text{ as } G \to \infty, \qquad\qquad 3.6$$

from which it follows that $\psi(G)$ is not integrable in a neighbourhood of $G = \infty$, so that $G = \infty$ *is unattainable*. This conclusion is certainly reasonable from a biological point of view.

Returning to the general equation (3.1), recall (Gard, 1988, pps. 148-149) that there is a *stationary probability density* for the process defined by this equation, if the function

$$q(x) = [g(x)^2 \ \psi(x)]^{-1} \qquad\qquad 3.7$$

is integrable on $(0, \infty,)$, and that this density is given by,

$$p(x) = q(x)/K,$$

where $K = \int_0^\infty q(x)dx.$ \qquad\qquad 3.8

For the model (2.5), the function (3.7) is

$$q(G) = \frac{G^{2(u-1)}}{v^2(1 + \sigma^2 G^2)^{(u+1)}} \cdot \exp\{-(2b/\sigma v^2)\tan^{-1} (\sigma G) + (2e/v^2)F(G)\}, \quad 3.9$$

which has the form,

$$q(G) = G^{-4}\left(\frac{G^2}{1 + \sigma^2 G^2} \right)^{(u+1)} .h(G). \qquad\qquad 3.10$$

Using (3.4), one shows that $h(G)$ has a unique maximum at $G = [e/B]^{1/B+1}$. Thus, $h(G)$ is a bounded positive function on $(0, \infty)$. Also, from (3.5), I have

$$\left. \begin{array}{l} h(G) \to (1/v^2).\exp\{-\pi b/\sigma v^2\} \text{ as } G \to \infty \\ \text{and } h(G) \to 0 \text{ as } G \to 0. \end{array} \right\} \qquad 3.11$$

In view of the above remarks, it follows that $G^n q(G) \sim G^{-4+n}$ in a neighbourhood of $G = \infty$, and is therefore integrable over such a neighbourhood provided $n < 3$. On the other hand, it follows from (3.5) that $G^n q(G) \to 0$. *Thus $G^n q(G)$ is integrable on $(0, \infty)$ for $n = 0, 1, 2$, so that a stationary probability density, together with its first and second moments, exists.*

I now consider the stationary density obtained above as a function of the fundamental efficiency of predation parameter c (see §1). From (1.4) and the definitions of a and b in (2.3), I have

$$a = D\{1 - (2A + 1)c\},\qquad\qquad 3.12$$

and b *is independent of* c. It follows that

$$\frac{du}{dc} = -k,\qquad\qquad 3.13$$

where $k = D(2A + 1)/v^2$ is a positive constant (recall from 3.3 that $u = a/v^2$).

Let K be the normalising constant for the density q(G), and let p(G) = q(G)/K be the corresponding probability density, as in (3.8). In the remainder of this section I shall prove the following two results.

Given $l > 0$, *there is a value* $G_0(l)>0$ *such that*

$$p(G) > p_0(G)e^{lc} \ for \ 0 < G < G_0 \ (l).\qquad\qquad 3.14$$

Given $l > 0$, *sufficiently small, there is a value* $G_\infty(l) > 0$ *such that*

$$p(G) < p_\infty (G)e^{-lc} \ for \ G_\infty \ (l) < G < \infty.\qquad\qquad 3.15$$

Here, $p_0(G)$ and $p_\infty (G)$ are independent of c, and (3.14) and (3.15) hold for c in the interval (1.6) and all $\sigma \geq 0$.
For a function f(G), let <f(G)> denote its *expected value*,

$$<f(G)> = \int_0^\infty f(G)p(G)dG.\qquad\qquad 3.16$$

Now define,

$$m(\sigma,c) = \frac{1}{K} \frac{\partial K}{\partial u}.\qquad\qquad 3.17$$

Then

$$m(\sigma,c) = \frac{1}{K}\frac{\partial}{\partial u}\int_0^\infty q(G)dG = \frac{1}{K}\int_0^\infty \log (\frac{G^2}{1+\sigma^2 G^2}) \ q(G)dG,$$

since h(G) in (3.10) is independent of u. [Note that the right hand integral exists since the logarithm is bounded as $G \to \infty$, and also the integrand $\to 0$ as $G \to 0$ by 3.5]. Thus,

$$m(\sigma,c) = <\log (\frac{G^2}{1+\sigma^2 G^2})>.\qquad\qquad 3.18$$

Now observe that $\log(G^2/1 + \sigma^2 G^2)$ is a strictly increasing function of G with,

$$-\infty < \log(\frac{G^2}{1+\sigma^2 G^2}) < \log(1/\sigma^2)$$

for $0 < G < \infty$. It follows from (3.18) that

$$-\infty < m(\sigma,c) < \log(1/\sigma^2) \qquad 3.19$$

and hence that

$$0 \le \sigma^2 \, e^{m(\sigma,c)} < 1. \qquad 3.20$$

Next consider

$$\frac{\partial p(G)}{\partial c} = - k \cdot \frac{\partial}{\partial u}(q(G)/K) = -k\{ - \frac{1}{K^2} \cdot \frac{\partial K}{\partial u} \cdot q(G) + \frac{1}{K} \frac{\partial q(G)}{\partial u}\}$$

$$= k\{ m(\sigma,c) - \log(\frac{G^2}{1+\sigma^2 G^2})\}.p(G). \qquad 3.21$$

It follows from (3.21) that $\partial p(G)/\partial c > lp(G)$ if and only if

$$m(\sigma,c) > \log(\frac{e^{l/k}G^2}{1+\sigma^2 G^2})$$

which reduces to

$$G^2\{e^{l/k} - \sigma^2 e^{m(\sigma,c)}\} < e^{m(\sigma,c)} \qquad 3.22$$

But, (3.20) ensures that $\{e^{l/k} - \sigma^2 e^{m(\sigma,c)}\} > 0$ if $l > 0$, whence,

$$G^2 < \frac{e^{m(\sigma,c)}}{\{e^{l/k}-\sigma^2 e^{m(\sigma,c)}\}} = G_0 \, (l,c)^2 .$$

Since (3.20) holds for all c in the *closure* of the interval (1.6), we may take $G_0(l) > 0$ to be the minimum of the $G_0(l,c)$'s for this range of c. Thus, $\partial p(G)/\partial c > lp(G)$ for each $0 < G < G_0(l)$ and each c, which, on integration, gives (3.14).

A similar argument shows that $\partial p(G)/\partial c < -lp(G)$ if and only if

$$G^2 \{e^{-l/k} - \sigma^2 e^{m(\sigma,c)}\} > e^{m(\sigma,c)} . \qquad 3.23$$

Further, it follows from the fact that (3.20) holds for all c in the *closure* of the interval (1.6), that $\sigma^2 e^{m(\sigma,c)} \le l_1 < 1$ for all c. Hence, if l is chosen sufficiently small, $[l < -k \log(1/l_1)]$, then $\{e^{-l/k} - \sigma^2 e^{m(\sigma,c)}\} > 0$, and (3.23) defines a $G_\infty(l) > 0$ such that $\partial p(G)/\partial c < -l p(G)$ for $G > G_\infty(l)$ and each c. Integration of this inequality then gives (3.15).

Summary of analysis

Roughly speaking, for a suitable stochastic process X(t) with values in an open interval (x_0, x_1), a boundary point x_i (i = 0 or 1) is *attracting* if X(t) eventually hits x_i with non-zero probability. A boundary point which is not attracting is said to be *unattainable* [For precise definitions, see Gard, 1988, pp. 145-146]. These definitions apply when $x_0 = 0$ and $x_1 = \infty$.

For the starfish process, G(t), determined by equation (2.5), it is shown that $G = \infty$ is an unattainable boundary. This is certainly reasonable from a biological point of view. On the other hand, $G = 0$ is unattainable if $e > 0$ in (2.5), but is attracting if $e = 0$ and the fluctuations are large enough $(v^2 > 2a)$. In view of this, the ecological significance of the secondary aggregation coefficient, e, is clear; *its role is to prevent the extinction of small starfish populations in the presence of large fluctuations in recruitment (or predation)*. However, this conclusion only applies if one grants the absolute validity of the model for starfish populations of all sizes - see the discussion in §4 below.

The existence of a *stationary probability density* for the starfish process is shown (together with the existence of first and second moments). This density, p(G), is also a function of the parameters in the model (2.5); in particular of the fundamental efficiency parameter, c (see §1). It is shown (see 3.14) that for G in a neighbourhood of 0, p(G) *increases at least exponentially as c increases*. Similarly (see 3.15), for G in a neighbourhood of ∞, p(G) *decreases at least exponentially as c increases*.

4. DISCUSSION

The notion of an *optimal* predator, as defined in Seymour (1989), and briefly recalled in §1, represents a highly idealised interaction defined only in terms of the (noise free) deterministic model (1.1). Thus, for the stochastic model (2.1) there is no non-zero stochastic equilibrium, and so a starfish outbreak cannot be viewed as corresponding to the onset of instability at some particular value of the efficiency parameter c. Instead, there is (at least for the simplified model 2.5) a stationary probability density with respect to which the probability of high and low density populations can be measured.

Nevertheless, one may still conceive of points on the suboptimal-superoptimal continuum (given by 1.6) as representing possible coevolved states between the coral prey and its starfish predator, with the suboptimal end (c large) representing underutilisation and the superoptimal end (c small) representing overutilisation of available ecological opportunities. The main conclusions (3.14) and (3.15) of §3 then imply that the probability of finding a small population of starfish increases as the population becomes more suboptimal, while *the probability of finding a large population (i.e. an outbreak) increases as the population becomes more superoptimal*. These findings are clearly consistent with those obtained from the deterministic model.

It is believed that *A. planci* has evolved fairly recently as a specialist coral predator from a more generalist ancestor (Moran, 1986, p. 386). As such, this starfish may not yet have reached an evolutionarily stable state with respect to its coral prey - a state of affairs which may be reflected in the observed population instabilities. On the other hand, the ancient order of scleractinian corals is presumably relatively evolutionarily stable. I shall therefore consider below possible selection pressures on *A. planci* in the light of the model analysed in §3.

A priori, one expects selection pressure to increase the fecundity of starfish. Thus, from (2.1), such pressure should act to decrease μ and to increase β and δ (I ignore e for the present); i.e. to increase longevity and to render fertilization more efficient by increasing the tendency to aggregate, and also increasing the efficiency of gametogenesis. The efficiency parameter c is easily seen to be an increasing function of μ, and a decreasing function of β and δ (Seymour, 1989, §5), so that these selection pressures act to *decrease* c; i.e. to push the starfish towards a more *superoptimal* state.

It is customary nowadays to view the evolution of population regulation mechanisms as a by-product of intraspecific competition (e.g. Krebs, 1985, p. 346). Thus, in the context of a coral reef (or complex of reefs), which necessarily has a limited resource base (as far as *A. planci* is concerned), the above selection pressures will operate up to the point at which mutual interference reduces relative fitness. There is some evidence of these conflicting pressures at work in that a pheromone-like compound produced in the gonads of both male and female starfish is thought to promote aggregation (and possibly synchronous spawning - Moran, 1986, p. 388). On the other hand, in at least one incidence, the mean size of aggregated starfish was found to be smaller than that of sparsely aggregated individuals (Moran, 1986, p. 396), suggesting interference due to excessive aggregation. In addition, Lucas (1984) has reported that starfish are affected by a potentially lethal bacterial disease, and it has been suggested by Moran, Bradbury & Reichelt (1985) that this pathogen is a possible cause for the rapid disappearance of large aggregations of starfish in the field. The effect of competition could certainly increase the susceptibility of individuals to such diseases, which would then be transmitted easily through aggregations.

According to (3.15), the more superoptimal a population is, the more probable it is that the population will become very (arbitrarily) large [i.e. the probability that G lies in a suitable neighbourhood of ∞ *increases* as c *decreases*]. But clearly, for a very large population the coral response equation (2.2) cannot continue to hold. Instead, it follows from (2.1) that dN/dt becomes increasingly negative as G grows (though on a time scale which is long compared to the average duration of the stochastic fluctuations in starfish density). It follows then that the coral population must crash if the starfish population stays above some threshold for sufficiently long - the latter population being driven to such a state by *persistent*, small scale (i.e. v and ω small in 2.1) variations in μ and δ. The starfish population will then also crash (either through mass mortality, or mass emigration or both). This process can be viewed as an extreme consequence of intraspecific competition, and its operation becomes (at least exponentially - see 3.15) more likely with increasing superoptimality of the population. Thus, as a consequence, selection pressure will be exerted to push populations towards a more suboptimal state.

A major conclusion of §3 was that, in the presence of the secondary aggregation effect (i.e. e > 0), the boundary G = 0 is unattainable. This means that small starfish populations are unlikely to become extinct through stochastic fluctuations. It appears then that such small (usually widely dispersed) suboptimal populations may well be evolutionarily stable. However, realistically, this is unlikely. Firstly, one can question the validity of the simple model (1.1) for very small populations, whose dynamics must be controlled by many special and local contingencies. Secondly, even in the context of the model, small populations are likely to be extinguished due to fluctuations not considered here (such as variations in e). Thus, it seems reasonable to suppose that starfish populations whose density is below some positive threshold will (with very large probability) become extinct. However, in view of the conclusion obtained from the model referred to above, this threshold is likely to be very small.

Now according to (3.14), the probability of a population being in a specified (small) neighbourhood of G = 0 increases (at least exponentially) as c increases; i.e. as the population becomes more suboptimal. The increased likelihood of extinction of such small populations will then set up selection pressure *against* suboptimality and towards super-optimality (by decreasing μ and increasing β and δ - as discussed previously). However, this effect is likely to be small.

In the light of the above discussion, the actual position occupied by extant starfish populations on the suboptimal-superoptimal continuum should be viewed as the resultant of two opposing pressures, pushing inwards from both the suboptimal and superoptimal extremes, but with pressure from the superoptimal end being the stronger effect. Thus, the 'normal' population is likely to be in the suboptimal range, though not too far. In Seymour (1989) I argued that normal populations of *A. planci* should be viewed as *near-optimal* in precisely this sense. The selectionist arguments given above provide a deeper theoretical underpinning for this view.

The stochastic effects considered in this paper are to be thought of as persistent, small scale fluctuations in average conditions, and should not be confused with the large scale perturbations of limited duration considered in Seymour (1989) as being the principal cause of observed population outbreaks (see §1). With the proviso that outbreaks be viewed as probabilistic events rather than deterministic ones, this view remains consistent with the model considered here.

Finally, from the point of view of the polarisation of view as to the nature of relevant ecological interactions mentioned in the introduction, the moral of this tale is as follows. Although it may be true that many ecological properties of reef communities are the short term results of stochastic environmental effects, nevertheless, if such effects are persistent, they will have evolutionary consequences, which can be seen in major trends (in particular, the tendency of populations of *A. planci* to outbreak from what appear to be stable equilibrium states of low density, widely dispersed populations into unstable, large and aggregated populations). These trends take the form of propensities to respond in characteristic ways to changes in ecological circumstances.

REFERENCES

Cameron, A.M. & R. Endean: In: Proc. Fourth Int. Coral Reef Symp., 2, 593-596 (1982)

Elliot, R.J.: Stochastic Calculus and Applications. Applications of Math. No. 18. Springer-Verlag (1982)

Gard, T.C.: Introduction to Stochastic Differential Equations. Marcel Dekker, Inc. (1988)

Krebs, C.J.: Ecology, 3rd Edition. Harper and Row (1985)

Lucas, J.S.: In: Proc. Crown of Thorns Starfish Seminar, 6 Sept. 1974, AGPs, Canberra, 103-121 (1975)

Lucas, J.S.: J. Exp. Mar. Biol. Ecol., 79, 129-147 (1984)

Moore, R.J.: Nature, London, 271, 56-57 (1978)

Moran, P.J.: Ocean. Mar. Biol. Ann. Rev., 24, 379-480 (1986)

Moran, P.J., R.H. Bradbury & R.E. Reichelt: In: Proc. Fifth Int. Coral Reef Congress, 5, 321-326 (1985)

Potts, D.C.: In: The Ecology of Pests, Ed. by R.L. Kitching and R.L. Jones, CSIRO, Melbourne, 55-86 (1981)

Seymour, R.M.: Ecol. Modelling, 46, 239-260 (1989)

STOCHASTIC ANALYSIS AND APPLICATIONS, 6(4), 349-363 (1988)

NONLINEAR PREDICTION OF CROWN-OF-THORNS OUTBREAKS ON THE GREAT BARRIER REEF

P. ANTONELLI[*]

Mathematics Department, University of Alberta,
Edmonton, Alberta   T6G 2G1   Canada

R. BRADBURY, R. BUCK, R. REICHELT

Australian Institute of Marine Science, CAPE Ferguson, P.M.B. No. 3
Townsville, M.C., Q. 4810, Australia

R. ELLIOTT[**]

Dept. of Statistics and Applied Probability, University of Alberta,
Edmonton, Alberta   T6G 2G1   Canada

ABSTRACT

The Zakai form of nonlinear prediction theory is used to estimate year-to-year state changes in crown-of-thorns starfish populations of the Great Barrier Reef. Taking the previously defined coral-state diffusion as observation process and the starfish-state diffusion as signal process, a least squares polynomial regression curve recently derived for simultaneously collected starfish/coral state data is used to set up the prediction preliminaries. Numerical results are not inconsistant with starfish outbreak values of the last 5 years and yield an expected high value for 1987. However, because of large error in the data it is doubtful that a more refined mesh for the Mihlstein approximations used to evaluate the Îto stochastic integrals involved in Zakai theory would improve the accuracy of predictions.

---

[*]Research of P. Antonelli. Partially supported by NSERC A-7667.
[**]Research of R. Elliott. Partially supported by NSERC A-7964 and U.S. Army Research Office under contract DAAL03-87-K-0102.

349

1. INTRODUCTION

The crown-of-thorns starfish, Acanthaster planci, is an outbreaking species which preys on the hard corals of tropical Indo-Pacific coral reefs [13]. It is the major source of disturbance for these complex ecosystems. Over the last twenty years starfish outbreaks have been reported from many parts of the Indo-Pacific but limitations of both empirical data and analytic methods preclude a treatment at this oceanic scale. Here, we report an analysis of the phenomenon at the macro-scale of a whole reef province - the Great Barrier Reef (GBR). Sets of reefs independent of their location change from one proportion of reef-state type to another in a way treated as continuous and representable as a Markov diffusion process.

The data set was derived from an all-source database held by the Great Barrier Reef Marine Park Authority. Ecologists with field experience of the phenomenon, examined, assessed and extracted 1125 reliable records of starfish and/or corals observed on a reef during a calendar year. The set involved 495 reefs between the years 1966 and 1985 ([2]). We transformed the mixture of data types to 2 categories of starfish abundance (low: 0-15 starfish; high: > 15 starfish) and 2 categories of coral abundance (low: 0-10% cover; high:$> 10\%$ cover) observed on each reef. Although this represents a radical compression of information, the categorization efficiently captures the distinction between normal and outbreaking population densities of starfish, and that between starfish – disturbed and normal coral communities [13].

We then searched this data set for pairs of successive years where an adequate number of reefs were sampled in both years of the pair. We found 9 such pairs for corals and 9 for starfish where the number of reefs sampled in both years of a pair was 10 or more. This was irrespective of their location on the GBR.

Let Ω denote the set of ca. 2500 reefs of the GBR and Σ any subset consisting of 10 or more reefs perhaps in different locations. Let x_t denote the percentage (proportion) of reefs in Σ which are in the <u>low coral state</u> at time t, measured in years. The value of this proportion at $t + \triangle t$ is

$$x_{t+\triangle t} = P_{11}\, x_t + P_{21}\, (1 - x_t) \qquad (1.1)$$

where P_{11}, P_{21} denote the percentage per unit time starting off in the low state and staying there and the proportion per unit time leaving the high coral state for the low state, respectively. The interval $\triangle t$ is to be treated as an infinitesimal and data indicates that a period of a fortnight or less will suffice for this. The actual values of P_{11}, P_{21}, P_{12}, P_{22} come from the data set discussed above. We now apply calculus to (1) to obtain

$$dx_t = \frac{1}{\varepsilon^2} f(P_{ij}) dt \qquad (1.2)$$

$$f(P_{ij}) = [P_{21} - (P_{12} + P_{21})x] \qquad (1.3)$$

where ε^2 is a necessary time-scaling parameter to be determined later. The <u>starfish-state</u> variable is denoted $Y_s \in [0,1]$ and satisfies

$$dy = \frac{1}{\eta^2} f(G_{ij}) dS. \qquad (1.4)$$

Thus, y_s denotes the percentage or proportion of reefs in Σ in the <u>low starfish-state</u> and the G_{11}, G_{12}, G_{21}, G_{22} have the analogous meanings as the P_{ij}. The η^2 factor is another time-scale parameter to be determined. We wished to examine the hypothesis that outbreak processes are simply statistical fluctuations away from equilibrium, caused by environmental noise. In order to do this we chose to use the simplest noise functional appropriate to $[0,1]$ state space ([1], [7]). Thus, using standard Brownian motion W_t we have the stochastic differential equation:

$$dx_t = \frac{1}{\epsilon^2} f(P_{ij}) dt + \frac{1}{\epsilon} \sqrt{x_t(1 - x_t)} dW_t \qquad (1.5)$$

defined on $[0, 1]$. The corresponding backward equation for the transition density of the related Markov process is:

$$\frac{1}{2\epsilon^2} x(1 - x) \frac{\partial^2 P}{\partial x^2} - \frac{1}{\epsilon^2} f(P_{ij}) \frac{\partial P}{\partial x} = -\frac{\partial P}{\partial \tau_t}. \qquad (1.6)$$

There are analogous equations to equations (5) and (6) for the starfish, adding:

$$\frac{1}{\eta} \sqrt{y_s(1 - y_s)} dB_s$$

to equation (4). Here, B_s is standard Brownian motion independent of W_t, the standard Brownian motion for corals in equation (5). Thus:

$$\frac{1}{2\eta^2} y(1 - y) \frac{\partial^2 \rho}{\partial y^2} - \frac{1}{\eta^2} f(G_{ij}) \frac{\partial \rho}{\partial y} = -\frac{\partial \rho}{\partial \tau_s}. \qquad (1.7)$$

There are two important assumptions we make to relate our multiple time-scale model of near-equilibrium starfish and coral processes. The first is that the "starfish-process is an order of magnitude faster than the corresponding coral-process". This is consistently observed in the data ([13], [8]). Thus, we postulate:

$$\triangle S = \frac{\eta^2}{\epsilon^2} \triangle t \simeq 10 \triangle t \qquad (1.8)$$

as the defining time-scale relationship. Note that ϵ^2, η^2 appear in both the drift and noise parts of equation (6) and the corresponding equation (7) for starfish. Thus, both the mean speed as determined by the drift and the "speed measure" (near equilibrium) are related by equation (8). Also, consistantly observed by us is that the starfish state can transit from very low to outbreak values in a single year.

We therefore postulate that the transition density ρ^{star} $(0, 0.2, 1, 0.8)$, *which depends on* ϵ^2, *shall be as close to 1 as possible.* The maximum likelihood method can be applied to the short-time asymptotic solution of (7) resulting in specification of ϵ^2 (and hence η^2). Such solutions are accurate for brownian motion on $[0, 1]$ and the result, $\epsilon^2 = 0.17$, follows from (a) in Theorem I, below. Here we have made use of the work of Keller and Voranka, [12].

In the previous paper, [2] , we announced two theorems:

Theorem I: Main Theorem From Data

(a) The starfish-process equation (7) is indistinguishable from Brownian motion away from the boundaries of the state space $[0, 1]$. For Brownian motion $G_{12} = G_{21} = 0.25$ in equation (1.7). This property is independent of the parameter η^2.

(b) $\epsilon^2 = 0.17$ and ρ^{star} $(0, 0.2, 1, 0.8) = \rho^{star}$ $(0, 0.8, 1, 0.2) = 0.95$ where ρ^{star} $(0, y, t, \bar{y})$ is the transition denisty for the process to be at $\bar{y}$ at time $t > 0$ having started at time zero at y. On the other hand, P^{coral} $(0, 0.8, 1, 0.2) = 0.02$.

(c) Mean first exit times out of small intervals for the starfish-process are the same throughout $[0.2, 0.8]$.

(d) For all data sets $\Sigma \in \Omega$, G_{12} and G_{21} are positive in equation (1.2). The pure low state for corals is inaccessible while the pure high state is accessible. The stochastic equilibrium density ρ_s^{coral} exists.

(e) For all data sets $\Sigma \in \Omega$, G_{12} and G_{21} are positive in equation (1.7). Both pure states are accessible for starfish and the equilibrium density ρ_s^{star} exists.

(f) For both starfish and coral processes the pure states are non-absorbing. The mean of ρ_s^{coral} is x_e and $x_e > 1/2$. Likewise, the mean of ρ_s^{star} is y_e,

$$y_e = \frac{G_{21}}{G_{21} + G_{12}}. \tag{1.9}$$

Theorem II: Main Theorem from Model Structure

If the pure low state for corals is inaccessible, then $x_e > 1/3$. (If $P_{12} < 1/2$ then $x_e > 1/2$.) The data always gave $P_{12} < 1/2$. Typical bootstrap ([10]) values of P_{ij} (denoted by $\bar{P}_{ij}$) are:

$$\bar{P}_{21} \simeq \bar{P}_{11} \simeq .76, \quad \bar{P}_{12} \simeq \bar{P}_{22} \simeq 0.2. \tag{1.10}$$

The P and G values derived for each year produced relatively similar curves in the probability density functions of the stationary densities. The mean values for these were derived using the bootstrap method. The starting set of 9 values of P (or G) was used to derive a distribution of means for this set. Using the mean of this distribution then enables us to discuss a 'typical' starfish result or a 'typical' coral result derived from the set of P's and G's.

The model allows for the first time some measure of prediction of the dynamics of the GBR as a whole, that is prediction at the macro-scale as distinct from prediction of the behaviour of individual reefs. For example, the GBR has recently experienced a major outbreak episode. We may predict the likelihood that the system will recover from such an episode by estimating the probability that the ensemble will shift from a predominance of low-coral state reefs. From Theorem I(b), we see that P^{coral} (0,0.8,1,0.2) = 0.02. That is, there is a probability of 0.02 that the ensemble will change from 0.8 low coral state reefs to 0.8 high coral state reefs within one year. On the other hand, ρ^{star} (0,0.8,1,0.2) = 0.95. This means that recovery of the system is a drawn out process over many years and it is highly likely that

it will be beset by starfish outbreaks during that time. These predictions
need to be appreciated against the fundamental dichotomy in the model's
behaviour: the starfish dynamics are dominated by random excursions, while
the coral dynamics are embedded in a drift field which forces the system
always towards the low coral state.

We believe that the model may have even greater predictive capacity
to be revealed through non-linear filtering and prediction techniques applied
on the starfish as signal and the corals as observations process, see, for example, ([3], [4]). Our approach is based on an empirically derived functional
relationship between starfish and coral states, connecting equations (2.6) and
(2.7). Starfish and coral data was collected simultaneously and put into 15
to 20 pairs of states (x_s, y_s). Four specific subsets were considered depending
on whether or not data from the early 70's was included and whether the
deviant "townsville point" was included, or not. A BMDP5R polynomial
regression software package was used to assess these 4 sets. The least error
fit of the four data sets turned out to include the 70's data, but not the
"townsville point". The resulting regression curve was linear:

$$x_s = -1.46y_s + 1.50 \qquad (1.11)$$

with 13% and 18% error in the x-intercept and slope, respectively.

We consider the starfish process as least knowable of the two and so
took it to be the signal, and reserved the role of observations process for the
coral. Specifically, the signal is

$$y_t = y_0 + \frac{10}{\varepsilon} \int_0^t (a'y_s + b')ds + \frac{10}{\varepsilon} \int_0^t \sqrt{y_s(1 - y_s)}dB_s \qquad (1.12)$$

where $a' = \bar{G}_{12} + \bar{G}_{21} = .5$, $b' = -\bar{G}_{21} = -.25$, $\varepsilon^2 = .17$. Whereas, the
observation process is

$$x_t = x_0 + \frac{1}{\varepsilon} \int_0^t (amy_s + aq + b)ds + \frac{1}{\varepsilon} \int_0^t \sqrt{x_s(1 - x_s)}dW_s \qquad (1.13)$$

where $a = \bar{P}_{12} + \bar{P}_{21} = .96$, $b = -\bar{P}_{21} = -.76$, $m = -1.46$ and $q = 1.50$; W_s and B_s are independent Brownian motions. The <u>prediction problem</u> is to obtain the least squares expected value $E[y_t | \mathcal{F}_s]$ where $\mathcal{F}_s$ is the observation σ-field of x_t. Following Antonelli and Elliott, [3], we use the Zakai form of non-linear prediction theory to solve our problem formally and then we proceed with suitable approximation schemes which utilize the data. In the next section we give a brief introduction to Zakai theory.

2. ZAKAI NONLINEAR PREDICTION

Consider a probability space (Ω, G, P) and for $0 \leq t \leq T$ an increasing family of sub σ-fields $G_t \subset G$ such that $G_{t+} = \cap_{t > s} G_s = G_t$, where each G_t contains all null sets of G.

We suppose the *signal process* $\{y_t\}$ is a real $\{G_t\}$ semi-martingale of the form

$$y_t = y_0 + \int_0^t \beta_u du + N_t. \qquad (2.1)$$

Here, $E[y_0^2] < \infty$, the process β is adapted to $\{G_t\}$, $E[\int_0^t \beta_u^2 du] < \infty$ and N_t is a square integrable $\{G_t\}$-martingale.

The *observation process* is an m-dimensional process of the form

$$x_t = \int_0^t h(y_u) du + \delta(x_t) W_t \qquad (2.2)$$

where $\{W_t\}$ is a standard m-dimensional G-Brownian motion, $h : \mathbb{R} \to \mathbb{R}^m$ is measurable, $E[\int_0^t |h|^2(y_u) du] < \infty$, $\delta(x_t)$ is a nonsingular matrix function with norm bounded away from zero, and the predictable <u>quadratic variations</u> <u>process</u> $\langle N, W_i \rangle_t = \int_0^t \lambda_u^i du$, $1 \leq i \leq m$. The <u>observation</u> σ-field, $\mathcal{F}_t$ at time t is obtained from the right continuous completion of the family of σ-fields $\sigma\{x_u : u \leq t\}$. If ϕ is a random variable with finite expectation, its projection on the observation σ-fields will be denoted by $\hat{\phi} \equiv \pi_t(\phi) = E[\phi | \mathcal{F}_t]$. Then x_t

can be written

$$x_t = \int_0^t \pi_u\big(h(y_u)\big)\,du + \nu_t \qquad (2.3)$$

where ν_t is the "innovations process" and is an $\{\mathcal{F}_t\}$-Brownian motion (see [11]). In our case here,

$$d\nu_s \equiv \frac{\epsilon\,dx_s}{\sqrt{x_s(1-x_s)}} - \frac{am\hat{y}_s + aq + b}{\sqrt{x_s(1-x_s)}}\,ds$$
$$= dW_s + \frac{am(y_s - \hat{y}_s)}{\sqrt{x_s(1-x_s)}}, \qquad (2.4)$$

where $h(y_u) = amy_u + aq + b$, is the information channel connecting signal and observation (see (1.11)).

The <u>prediction problem</u> discusses the derivation of an equation for $\pi_s(y_t) := E[y_t|\mathcal{F}_s]$, where $0 \le s \le t \le T$. For fixed t consider the $\{\mathcal{F}_s\}$ martingale $X_s := E[y_t|G_s]$, $s \le t$, and suppose there is an $\{G_s\}$-predictable process $\{\alpha_u\}$ such that the <u>quadratic variations process</u> is of the form

$$\langle X, W\rangle_s = \int_0^s \alpha_u\,du. \qquad (2.5)$$

Then it is shown in Elliott, 1984[1] that

$$\pi_s(y_t) = \pi_0(y_t) + \int_0^s \left(\pi_u(\alpha_u) + \delta^{-1}\left(\pi_u(X_u h(y_u)) - \pi_u(y_t)\pi_u(h(y_u))\right)\right)'\cdot d\nu_u. \qquad (2.6)$$

This equation exhibits an unfortunate quadratic term in the conditional expectation π. The Zakai theory remedies this. For any G_t-measurable random variable ϕ with finite expectation write the <u>unnormalized prediction</u> as

---

[1]Martingale Methods in Stochastic Control, a Technical Report of Carleton University Statistics and Probability Laboratory

ANTONELLI ET AL.

$$\sigma_{t,s}(\phi) := E_0[\Lambda_t \phi | \mathcal{F}_s],$$

where E_0 denotes expectation relative to P_0 and this new probability measure on (Ω, G) is defined by

$$\frac{dP_0}{dP} = \Lambda_t^{-1},$$

where

$$\Lambda_t = \exp\{\int_0^t (\delta^{-1}(x_u)h(y_u))' \delta^{-1}(x_u)dx_u - \frac{1}{2}\int_0^t |\delta^{-1}(x_u)h(y_u)|^2 du\}. \quad (2.7)$$

In fact,

$$\Lambda_t = 1 + \int_0^t \Lambda_u (\delta^{-1}(x_u)h(y_u))' \delta^{-1}(x_u)dx_u \qquad (2.8)$$

and by Girsanov's Theorem (see [11]) B_t is a standard m-dimensional Brownian motion relative to P_0, where

$$dB_u = \delta^{-1}(x_u)dx_u. \qquad (2.9)$$

Note that $\sigma_{t,s}(1) \equiv \hat{\Lambda}_s$ and $\sigma_{t,s}(\phi) = \hat{\Lambda}_s \cdot \pi_s(\phi)$. The <u>Zakai prediction equation</u> is now

$$\sigma_{t,s}(y_t) = \sigma_{t,0}(y_t) + \int_0^s \left(\delta^{-1}(x_u)\left(\sigma_{t,u}(\alpha_u) + \sigma_{t,u}(y_t h(y_u))\right) \right)' \delta^{-1}(x_u)dx_u.$$
$$(2.10)$$

This is proved in [3]. For our case $\alpha_u \equiv 0$ so equation (2.6) is

$$\pi_s(y_t) = \pi_0(y_t) + \int_0^t \frac{\varepsilon}{\sqrt{x_u(1-x_u)}} \pi_u \big(\pi_u(y_t)[amy_u + aq + b]$$
$$- \pi_u(y_t)[am\pi_u(y_u) + aq + b]\big)d\nu_u \qquad (2.11)$$

where $\delta = \sqrt{x(1-x)}/\varepsilon$ and

$$\pi_u(h(y_u)) = am\pi_u(y_u) + aq + b$$
$$\equiv am\hat{y}_u + aq + b.$$

Likewise, for our case equation (2.7) is

$$\Lambda_t = \exp\{\int_0^t \delta^{-2}(x_u)h(y_u)dx_u - \frac{1}{2}\int_0^t |\delta^{-1}(x_u)h(y_u)|^2 du\} \qquad (2.12)$$

where $\delta^{-1} = \varepsilon/\sqrt{x(1-x)}$ and $\delta^{-1}dx_u$ is a standard 1-dimensional Brownian motion as in (2.9). It follows that

$$\hat{\Lambda}_s = 1 + \int_0^s \hat{\Lambda}_u(am\hat{y}_u + aq + b)\delta^{-2}dx_u \qquad (2.13)$$

and that

$$\sigma_{t,s}(y_t) = \sigma_{t,0}(y_t) + am\int_0^t \delta^{-2}(x_u)\sigma_{t,u}(X_u y_u)dx_u. \qquad (2.14)$$

Also, we shall have need of the important formula

$$\pi_s(y_t) = \frac{\sigma_s(y_t)}{\hat{\Lambda}_s}. \qquad (2.15)$$

3. APPROXIMATIONS AND DATA

Formula (2.15) is the basis of our computational method. Interpretting s to mean "this year", t to mean "next year" and 0, "last year" we obtain

$$X_0 = E[y_t|G_0] = y_0 = \frac{x_0 - q}{m}$$

$\sigma_0(y_t) = \hat{\Lambda}_0\pi_0(y_t) = \pi_0(y_t)$ and for $t \geq 0$, $\pi_0(y_t) = \sigma_0(y_t) = E[y_t|X_0] = y_0$. Finally,

$$\pi_0(y_0^2) = \sigma_0(y_0^2) = E[y_0^2|X_0] = \left(\frac{x_0 - q}{m}\right)^2.$$

From (2.14) and (2.15) we can then obtain a first order "Euler Approximation" for $\pi_s(y_t)$.

It is

$$\pi_s(y_t) \simeq \frac{x_0(1-x_0)\left(\frac{x_0-q}{m}\right) + am\varepsilon^2\left(\frac{x_0-q}{m}\right)^2(x_s-x_0)}{x_0(1-x_0) + \varepsilon^2[ax_0+b](x_s-x_0)} \tag{3.1}$$

This Euler method treats the stochastic integrals above as if they were ordinary Riemann integrals. However, we shall need a more sensitive method called a Mihlstein type approximation.

Suppose B_t is a 1-dimensional Brownian motion relative to P_0 in (Ω, G) and that

$$dx_t = f(x_t)dt + \sigma(x_t)dB_t$$
$$B_0 = 0 \tag{3.2}$$

has strong solutions (see [11]). Taylor's theorem yields, to terms of order $o(t)$,

$$x_t \simeq x_0 + f(x_0)t + \int_0^t \left(\sigma(x_0) + \sigma'(x_0)dx_s\right)dB_s,$$

so that

$$x_t \simeq x_0 + f(x_0)t + \sigma(x_0)B_t + \int_0^t \sigma'(x_0)\sigma(x_0)B_s dB_s,$$

where by laws of stochastic differentials $\triangle B_s dB_s$ is of first order in ds and $\triangle s \cdot dB_s$ being of higher order is neglected. Evaluation of the remaing Îto integral yields the Mihlstein type approximation

$$x_t \cong x_0 + [f(x_0) - \frac{\sigma'(x_0)\sigma(x_0)}{2}]t + \sigma(x_0)\triangle B_t$$
$$+ \frac{\sigma'(x_0)\sigma(x_0)}{2}(B_t^2 - B_0^2) \tag{3.3}$$

	Euler	Strat.	Mihl.	l.s.	Actual Cot
'83	.45	.32	.56	.55	.60
'84	.99	1.50	1.13	.73	.65
'85	.50	.50	.50	.48	.64
'86	.56	.44	.44	.50	.82
'87	.72	.82	.82	–	–

FIG. 1

Starfish State Predictions on the G.B.R.

NOTE: l.s. *denotes the least square value obtained from (1.11).*

(see [14]). The $B_0 = 0$ condition is not necessary as a euclidean motion preserving B_t will make this happen automatically.

If we neglect the $(B_t^2 - B_0^2)$-term of (3.3) we obtain a <u>Stratonovich type approximation</u> (see Wong and Zakai, 1969). We shall compare the Euler, Stratonovich and Mihlstein approximations for the starfish prediction problem.

With B_s as in (2.9) we need to compute the integrals (Îto) below using (3.3) above.

Evaluation of

$$\left.\begin{array}{c} \int_0^t \dfrac{ax_0 + b}{\sqrt{x_0(1 - x_0)}} \, dB_s \\[2ex] \dfrac{1}{m^2} \int_0^t \dfrac{(x_0 - q)^2}{\sqrt{x_0(1 - x_0)}} \, dB_s \end{array}\right\} \tag{3.4}$$

leads to

$$\pi_s(y_t) \cong$$

$$\frac{x_0(1-x_0)\left(\frac{x_0-q}{m}\right) + \frac{\alpha \epsilon^2}{m^3}(x_0-q)^2 \left[m^2(x_s-x_0) + (x_s^2-x_0^2-1)\left[(x_0-q) - \frac{(1-2x_0)(x_0-q)^2}{4x_0(1-x_0)}\right]\right]}{x_0(1-x_0) + \epsilon^2(ax_0+b)\left[(x_s-x_0) + (x_s^2-x_0^2-1)\left[\frac{a}{2} - \frac{(1-2x_0)(ax_0+b)}{4x_0(1-x_0)}\right]\right]}$$

$$(3.5)$$

The Euler Approximation (3.1) is obtained from (3.5) by neglecting the $(x_s^2 - x_0^2 - 1)$-terms in both numerator and denominator. The Stratonovich Approximation is obtained from (3.5) by setting $x_s^2 - x_0^2$ formally to zero in both numerator and denominator. Using all the data associated with the best fit regression curve (1.11), predictions of the starfish state y_t obtained are listed in the table of Figure I.

ACKNOWLEDGEMENTS

The authors would like to thank Vivian Spak, at the Mathematics Department of the U. of A., for her excellent AMS-Tex typesetting.

REFERENCES

[1] Antonelli, P.L. (1978). An asymptotic formula for the transition density of random genetic drift. J. Appl. Prob., 15, 185-187.

[2] Antonelli, P., Bradbury, R. and Reichelt, R. (1988). Multiple time-scale diffusion models of starfish and coral state changes over the whole Great Barrier Reef. In Press J. Inf. and Deduct. Biol..

[3] Antonelli, P.L. and Elliott, R.J. (1986). The Zakai forms of the prediction and smoothing equations. EEE Trans. on Information Theory, 32, 816-817.

[4] Antonelli, P.L., Elliott, R.J. and Seymour, R.M. (1987). Nonlinear filtering and Riemannian scalar curvature, R. Adv. Appl. Math., 8, 237-252.

[5]   Antonelli, P.L. and Kazarinoff, N.D. (1984). Starfish predation of a
       growing coral reef community. J. Theor. Biol., 107, 667-684.

[6]   Antonelli, P.L. and Kazarinoff, N.D. (1986). Comments on starfish/coral
       cycles after R. Bradbury et al. J. Theor. Biol., 119, 501-502.

[7]   Antonelli, P.L. and Strobeck, C. (1977). The geometry of random
       drift, I. Stochastic distance and diffusion. Adv. Appl. Prob., 9,
       238-249.

[8]   Bradbury, R.H., Hammond, L.S., Moran, P.J. and Reichelt, R.E.
       (1985). Coral reef communities and crown-of-thorns starfish: evi-
       dence for qualitatively stable cycles. J. Theor. Biol., 113, 69-80.

[9]   Bradbury, R.H. and Mundy, C.N. (in press). In: Biomass and Geog-
       raphy of Large Marine, Ecosystems. (Sherman, K. and Alexander,
       L., eds.), Westview, Boulder, CO .

[10]   Efron, B. (1979). Computers and the theory of statistics: thinking
       the unthinkable. SIAM Rev.,21, 460-480.

[11]   Elliott, R.J. (1982). Stochastic Calculus and Applications, New
       York, Springer-Verlag.

[12]   Keller, J. and Voronka, R. (1975). Asymptotic analysis of stochastic
       models in population genetics. Math. Biosci., 25, 331-362.

[13]   Moran, P.J. (1986). The Acanthaster phenomenon. Oceanogr. Mar.
       Biol. Ann. Rev., 24, 379-480.

[14]   Pardoux, E. and Talay, D. (1985). Discretization and simulation of
       stochastic differential equations. Acta Applicandae, 3, 23-47.

[15]   Reichelt, R.E. and Bradbury, R.H. (1984). Spatial patterns in coral
       reef benthos: Multiscale analysis of sites from three oceans. Mar.
       Ecol. Prog. Ser., 17, 251-257.

[16]   Wong, E. and Zakai, M. (1969). Riemann-Stieltjes approximations
       of stochastic integrals. Z. Wahr. Verw. Geb., 12, 87-97.

THE *ACANTHASTER* PHENOMENON: A MODELLING APPROACH

RAPPORTEURS' REPORT

PETER ANTONELLI

Department of Mathematics, University of Edmonton, Edmonton, Alberta T6G 2G1

ROGER BRADBURY[1]

Australian Institute of Marine Science, Townsville, Queensland 4810

LAURIE HAMMOND

Victorian Institute of Marine Sciences, Melbourne, Victoria 3002

RUPERT ORMOND

Tropical Marine Research Unit, Department of Biology, University of York, York, England YO1 5DD

AND

RUSSELL REICHELT[2]

Australian Institute of Marine Science, Townsville, Queensland 4810

Abstract. Stable limit cycles at the scale of individual reefs and hydrodynamic connections between reefs for larval transport are together sufficient to account qualitatively for the large scale dynamics of the phenomenon. Several different factors may generate the stable limit cycles, and these need not be mutually exclusive, but rather reinforcing. An important factor in the limit cycle dynamics in other areas is fish predator pressure on the starfish: increased pressure can suppress limit cycle behaviour. There is some evidence that the same factor may have been coercive on the GBR. Fish predator pressure may have been reduced through the intensification of fishing in the 1960s, leading to the two waves of outbreaks experienced on the GBR since then. There is a possibility that successive waves of outbreaks could degrade the system.

1 Present address: National Resource Information Centre, Canberra, ACT.
2 Present address: Bureau of Rural Resources, Canberra, ACT.

INTRODUCTION

The models presented and discussed at the workshop were of several different complementary types: analytical, explanatory and predictive. They also considered the phenomenon at several complementary scales: the whole GBR, single reef and within reef. The rapporteurs agreed that there is a strong consensus among the different modellers and among their models about the dynamics of the phenomenon. This synthesis is emerging despite the differences in modelling approach and scale.

THE PRESENT STATE OF THE PHENOMENON

Macroscale dynamics

At the scale of the whole GBR, Reichelt's model of the empirical starfish distribution data indicates that during both 1962-1975 and 1979-1988, a wave of outbreaks has propagated southwards from the Cooktown sector in a coherent way (probably through larval transport) with a velocity comparable to the residual southward flowing current. The waves have gradually dissipated in the southern part of the system.

The results of this analysis were broadly supported by Dale, Bradbury & Reichelt's model which used grammar inference procedures to search for patterns in irregular sequences of qualitative historical reports of the status of coral and starfish on individual reefs. This procedure was able to recover a starfish-coral predator-prey cycle at the scale of the whole GBR from this poor data set, and show the underlying structure of the cycle in space and time.

The large scale hydrodynamic simulations of James, Dight & Bode support this interpretation by providing a strong physical basis for the 'Reichelt wave': the individual reefs in the GBR are strongly, but not randomly, connected through currents. In the region north of Green Island, the current conditions show more random connections with only relatively weak southerly trends, while to the south of Green Island, connections with reefs to the south are stronger. This is consistent with the idea of an initiation area for the wave in the Cooktown sector, followed by southerly propagation.

The diffusion-reaction-transport model of Antonelli, Kazarinoff, Reichelt, Bradbury & Moran provides a mathematical and numerical basis for linking the hydrodynamic transport processes to the ecological dynamics of the wave. This model demonstrates clearly that, if individual reefs exhibit a stable limit cycle of starfish and coral from whatever underlying cause, then, given appropriate diffusion and drift processes, the whole system will exhibit a moving wave with a period equal to the period of the cycle. This is precisely the form of wave which the analysis of the empirical data also indicates.

Taken together, these large scale models indicate that once the conditions are favourable for the establishment of stable limit cycles (and hence outbreaks) on individual reefs, the hydrodynamics provide the drift and diffusion processes (presumably through transport of larvae) which are necessary for the establishment and synchronisation of the moving wave. Neither hydrodynamics nor limit cycle controls by themselves are sufficient to initiate the large scale dynamics of the phenomenon. The particular factors which may control the dynamics of the limit cycles on individual reefs were the subject of detailed modelling and analysis of mesoscale dynamics as discussed below.

Mesoscale dynamics

At the scale of individual reefs, the hydrodynamic simulations of Black & Gay and Gay, Andrews & Black provide more detailed and comprehensive descriptions of the nature and extent of connectivity between reefs by larval transport: some reefs are more likely to be self-seeding or retentive than connected with other reefs because of a number of hydrodynamic mechanisms including eddies. These simulations also show that the local shape and orientation of a reef should strongly affect the location of the initial larval settlement on that reef - a prediction confirmed by the empirical data.

The idea that some small regions may be self-seeding or retentive was supported by Moore's interpretation of the long term persistence of a small population of starfish in a poor low-coral reef in a semi-enclosed bay in the Red Sea. Occasional 'spill-over' from this bay into neighbouring areas of high coral may provide a model for the initiation of the large scale dynamics of the phenomenon in the Cooktown sector.

Several different biological models at the scale of individual reefs predict the development of stable limit cycles in the starfish-coral dynamics. They also show that several different features of starfish and coral biology will generate such cycles. It needs to be emphasised that such features are not mutually exclusive, but that in all probability they act synergistically to reinforce the cycling. Indeed, the catastrophe theory model of Bradbury & Antonelli suggests that all the factors postulated as controls may be subsumed in a more general model.

The early Antonelli & Kazarinoff (1984) model is still the most sophisticated mathematically and numerically and generates limit cycles from asymmetries in starfish feeding preferences, their fecundity and from their tendency to aggregate. The different models of McCallum,

Ormond *et al.* and Parslow are essentially similar to each other, all being developed from standard logistic equations of the Lotka-Volterra type that underlie modern theoretical population ecology. Each generates limit cycles from the difference between the rate at which high densities of starfish consume coral (relatively fast) and the rate at which coral can grow (relatively slow): thus starfish populations collapse after exhausting their food supply, and there is a lag before coral and then starfish populations can recover.

Moore argued that the limit cycles of low amplitude observed on low-coral reefs in the Red Sea are primarily a result of the starfish-coral interaction. The cycles become amplified on rich reefs because the higher coral cover allows the starfish to aggregate, and fertilisation is more successful.

Microscale dynamics

The detailed behaviour (i.e. movement) of outbreaks within reefs was modelled by both Green and Hogeweg & Hesper . Each of these models used cellular automata and both showed that, given the known behaviour of the starfish, outbreak paths are likely to be highly structured rather than random with starfish tending to accumulate after time into aggregations or fronts. The structuring processes are driven by properties of both predator (starfish) and prey (coral). Empirical data on distribution of starfish on outbreaking reefs support these models.

The coral growth model of Antonelli introduced the concept of vigour to study the evolution of coral species diversity as a defence against starfish predation. The conclusion of this model - that corals shape their distributions to protect themselves against starfish predation - is in strong agreement with the conclusions of Hogeweg & Hesper's model.

THE PAST DYNAMICS OF THE PHENOMENON

Stable limit cycles at the scale of individual reefs and connections between reefs through larval transport are together sufficient to account for the gross features of the present dynamics. For the population dynamics of starfish and coral to have been dramatically different in the past (for example, for there not to have been outbreaks throughout the system), there would need to have been a change in at least one of these two critical processes: either larval survival and transport (which now seems unlikely) or the generation of limit cycles within populations. The only feature identified in the workshop as capable of suppressing limit cycles is increased predator pressure on starfish, since all other parameters are life history characteristics.

Each of the Lotka-Volterra models of McCallum, Ormond *et al.* and Parslow demonstrate that predation on juveniles and adult starfish can, in theory, control starfish population levels and suppress limit cycles. This holds also for the diffusion-reaction-transport model of Antonelli *et al.* where the effect of predation is to reduce aggregation levels, and through this to suppress the limit cycle behaviour. The rates of predation need not be very high, nor the predators very common. Ormond *et al.*, for example, argued that the likely behaviour and dynamics of the Giant Triton shellfish would not be sufficient to control starfish populations, but a predator that was more abundant and showed more sophisticated switching behaviour (Type III functional response) could exercise effective control. It was also shown earlier by Antonelli & Kazarinoff (1988) that simple predator dynamics could not control starfish populations at the mesoscale.

Ormond *et al.* presented data that demonstrated that in the Red Sea, where no large outbreaks of starfish have occurred, the density of fish predators (particularly lethrinids - emperors) is above their model's calculated threshold density required to control the starfish, whereas in the central and northern regions of the GBR, fish predator density is below the required threshold. Moreover, some GBR reefs which have escaped heavy starfish outbreaks have higher densities of fish predators.

These findings suggest further consideration of the possibility that outbreaks of starfish in several areas of the Indo-West Pacific first reported in the 1960s may be linked to the intensification of reef fisheries that took place at that time. The observed fluctuations in GBR commercial fish landings since 1962, although an imprecise record of fishing activity, are compatible with this interpretation.

THE FUTURE DYNAMICS OF THE SYSTEM

No models point to the possibility of the present starfish-coral dynamics disappearing of their own accord. Instead the models, taken together, point to a strong stability in the phenomenon, and the likelihood of it continuing.

Several models do however generate a possibility that the phenomenon could eventually degrade the system. The models of Done (this volume), Done, Osborne & Navin (1989) and Endean, Cameron & De Vantier (1989) show that existing coral size/age structure can be degraded by cycles of outbreaks at about the frequency presently being experienced. A more recent study by Seymour & Bradbury (*in ms*) fits the historical outbreak data to a simple epidemiological model to provide evidence for a long term degradation of the GBR ecosystem.

Hogeweg argued at the workshop that, after repeated 'Reichelt waves', the system may begin to oscillate synchronously rather than as a moving wave, leaving the system at some times in a globally poor condition with consequent risks to re-establishment. This is also a prediction of the diffusion- reaction-transport model of Antonelli *et al.* above. Recent research using cellular automata models (Van der Laan & Bradbury, 1990; Bradbury, Van der Laan & Macdonald, 1990) and Volterra-Hamilton models (Antonelli & Lin, in press) have confirmed the likelihood of such global pulses in certain situations.

In contrast, earlier work by Bradbury, Hammond, Moran & Reichelt (1985) had identified chronically degraded 'metastable' reefs as one possible outcome of the phenomenon, and this was supported by the models of McCallum, Ormond *et al.* and Parslow which showed that low starfish density, very low coral cover reefs could be stable.

As Antonelli's model shows, the critical balance between accretion and bioerosion could be shifted by, say, algal overgrowth or intensified parrotfish grazing such that recovery is not possible.

Essentially the models agree that the system could become degraded by the phenomenon if other factors, essentially external to the present dynamics, are modified or become involved. These could enter into the dynamics through feedback, such as algae inhibiting coral recovery. The critical aspect here would be the presence of a steady increase or decrease in some other factor through successive passes of the wave.

THE PROBLEM OF UNCERTAINTY

Some important parts of the phenomenon were addressed only indirectly by the modelling work discussed at the workshop. For example:

- If the phenomenon is a classic limit cycle, why then are the outbreaks so variable?

Any answers to this question may lie in processes such as regulation of population size by variation in recruitment, or the influence of weather conditions on the foraging behaviour of the starfish. Such problems have not been examined closely here because there are so few data available that any modelling would have to be done in a completely theoretical fashion.

- Why is coral cover not always driven to its low state when the starfish populations on a reef may be relatively large?

The non-linear prediction model of Antonelli *et al.* was the only model which touched on this aspect of the phenomenon. It established as a theorem that, while the pure high state for corals is accessible, the pure low state is inaccessible. This allows, for the first time, some measure of prediction of the dynamics of the GBR as a whole.

In fact, a positive aspect of the workshop was that it served to highlight the areas where the data are missing or weak. Thus while it is true that the paradigm emerging from the workshop results is supported by the available data, it is also true that, for many parts of the picture, the data are sufficiently weak that hypotheses can be neither supported strongly nor refuted.

We see a picture of the phenomenon developing which is similar to the 'acid rain' phenomenon, where large scale processes are understood by use of models that are difficult to confirm (or, put another way, difficult to pose in a way that would allow unequivocal falsification). Scientists and policy makers must adopt strategies that permit them to work under these conditions of uncertainty.

CONCLUSIONS AND RECOMMENDATIONS

A new paradigm for the *Acanthaster* phenomenon is emerging. The results of this workshop have shown that too much emphasis has been placed in the past on examining, one at a time, each of the several hypotheses that have been proposed to explain the phenomenon. The workshop has shown that many of the hypotheses are not incompatible with each other, and, indeed, a synthesis of several of them to create a new paradigm is now possible.

That synthesis, which we call the recruitment initiated predation hypothesis, argues that stable limit cycles in the abundance of starfish and coral at the scale of individual reefs and hydrodynamic connections between reefs for transport of starfish larvae are together sufficient to account qualitatively for the large scale dynamics of the phenomenon. Of the several different factors which may generate limit cycles, an important factor is fish predator pressure: increased predator pressure can suppress limit cycle behaviour.

The hypothesis suggests that the origin of the two observed waves of outbreaks through the GBR system could lie in the reduction of fish predator pressure through the intensification of fishing on the GBR since the 1960s.

The hypothesis also suggests that future waves of outbreaks could degrade the system.

The results of the workshop also show that the phenomenon is complex. Like the 'acid rain' phenomenon in Europe, it operates over many scales, and is resistant to simple single-factor explanations. As with the 'acid rain' phenomenon, natural history observations, laboratory and field experiments, modelling, data analysis and theory have each a distinctive and complementary role to play in building understanding.

We therefore recommend that:

- **The new paradigm, the recruitment initiated predation hypothesis, needs an integrated research effort to test it.** A rigorous research program should now be undertaken to explore all aspects of the new paradigm with the objective of establishing its validity.

- **But it must be remembered that the problem is complex.** The program should recognise the complementary roles of observational and experimental studies, of laboratory and field studies, and of induction from data and deduction from models and theory in attacking complex multi-factor problems such as the *Acanthaster* phenomenon.

- **And that the data are inherently fuzzy.** The program should recognise that the uncertainty attaching to the problem is partly due to paucity of data (which sometimes can be redressed) and is partly due to the inherent fuzziness of some data (which often cannot be redressed), and it should cope with this uncertainty by establishing, as part of the problem definition, the classes of data that constitute evidence for the different aspects of the paradigm.

REFERENCES

Antonelli, P.L. (this volume) Applied Volterra-Hamilton systems of Finsler type: increased species diversity as a non-chemical defense for coral against the crown-of-thorns. In: R.H. Bradbury (ed.) The *Acanthaster* phenomenon: a modelling approach. Springer- Verlag, Berlin.

Antonelli, P.L., Bradbury, R.H., Buck, R. G., Elliott, R. J. & Reichelt, R. E. (this volume) Nonlinear prediction of crown-of-thorns outbreaks on the Great Barrier Reef. In: R.H. Bradbury (ed.) The *Acanthaster* phenomenon: a modelling approach. Springer-Verlag, Berlin.

Antonelli, P.L. & Kazarinoff, N.D. (1984) Starfish predation of a growing coral reef community. J. theor. Biol. 107, 667-684.

Antonelli, P.L. & Kazarinoff, N.D. (1988) Modelling density- dependent aggregation and reproduction in certain terrestrial and marine ecosystems: a comparative study. Ecol. Modell. 41, 219- 227.

Antonelli, P.L., Kazarinoff, N.D., Reichelt, R.E., Bradbury, R.H. & Moran, P.J. (this volume) A diffusion-reaction-transport model for large-scale waves in crown-of-thorns starfish outbreaks on the Great Barrier Reef. In: R.H. Bradbury (ed.) The *Acanthaster* phenomenon: a modelling approach. Springer-Verlag, Berlin.

Antonelli, P.L. & Lin, X. (in press) Bifurcation analysis on a coral starfish model. Math. Comput. Modell.

Black, K.P. & Gay, S.L. (this volume) Reef-scale numerical hydrodynamic modelling developed to investigate crown-of-thorns starfish outbreaks. In: R.H. Bradbury (ed.) The *Acanthaster* phenomenon: a modelling approach. Springer-Verlag, Berlin.

Bradbury, R.H. & Antonelli, P.L. (this volume) What controls outbreaks? In: R.H. Bradbury (ed.) The *Acanthaster* phenomenon: a modelling approach. Springer-Verlag, Berlin.

Bradbury, R.H., Hammond, L.S., Moran, P.J. & Reichelt, R.E. (1985) Coral reef communities and the crown-of-thorns starfish: evidence for qualitatively stable cycles. J. theor. Biol. 113, 69-80.

Bradbury, R.H., Van der Laan, J.D. & Macdonald, B. (1990) Modelling the effects of predation and dispersal on the generation of waves of starfish outbreaks. Math. Comput. Modell.

Dale, M.B., Bradbury, R.H. & Reichelt, R.E. (this volume) Reef syntax: an exploratory data analysis of the *Acanthaster* phenomenon using strings and grammars. In: R.H. Bradbury (ed.) The *Acanthaster* phenomenon: a modelling approach. Springer- Verlag, Berlin.

Done, T.J. (this volume) Transition matrix models, crown-of- thorns and corals. In: R.H. Bradbury (ed.) The *Acanthaster* phenomenon: a modelling approach. Springer-Verlag, Berlin.

Done, T.J., Osborne, K. & Navin, K.F. (1989) Recovery of corals post-*Acanthaster*: progress and prospects. Proc. VIth Int. Coral Reef Symp. 2, 137-142.

Endean, R., Cameron, A.M. & De Vantier, L.M. (1989) *Acanthaster planci* predation on massive corals: the myth of rapid recovery of devastated reefs. Proc. VIth Int. Coral Reef Symp. 2, 143-148.

Gay, S.L., Andrews, J.C. & Black, K.P. (this volume) Dispersal of neutrally-buoyant material near John Brewer Reef. In: R.H. Bradbury (ed.) The *Acanthaster* phenomenon: a modelling approach. Springer- Verlag, Berlin.

Green, D.G. (this volume) Cellular automata models of crown-of- thorns outbreaks. In: R.H. Bradbury (ed.) The *Acanthaster* phenomenon: a modelling approach. Springer-Verlag, Berlin.

Hogeweg, P. & Hesper, B. (this volume) Crowns crowding: an individual oriented model of the *Acanthaster* phenomenon. In: R.H. Bradbury (ed.) The *Acanthaster* phenomenon: a modelling approach. Springer-Verlag, Berlin.

James, M.K., Dight, I.J. & Bode, L. (this volume) Great Barrier Reef hydrodynamics, reef connectivity and *Acanthaster* population dynamics. In: R.H. Bradbury (ed.) The *Acanthaster* phenomenon: a modelling approach. Springer-Verlag, Berlin.

McCallum, H.I. (this volume) Effects of predation on *Acanthaster*: age-structured metapopulation models. In: R.H. Bradbury (ed.) The *Acanthaster* phenomenon: a modelling approach. Springer-Verlag, Berlin.

Moore, R. (this volume) Persistent and transient populations of the crown-of-thorns starfish, *Acanthaster planci*. In: R.H. Bradbury (ed.) The *Acanthaster* phenomenon: a modelling approach. Springer-Verlag, Berlin.

Ormond, R.F.G., Bradbury, R.H., Bainbridge, S., Fabricius, K., Keesing, J., De Vantier, L.M., Medlay, P. & Steven, A. (this volume) Test of a model of regulation of crown-of-thorns starfish by fish predators. In: R.H. Bradbury (ed.) The *Acanthaster* phenomenon: a modelling approach. Springer-Verlag, Berlin.

Parslow, J.S. (this volume) Stochastic and spatial effects in predator-prey models of *Acanthaster*-coral interactions. In: R.H. Bradbury (ed.) The *Acanthaster* phenomenon: a modelling approach. Springer-Verlag, Berlin.

Reichelt, R.E. (this volume) Dispersal and control models of *Acanthaster planci* populations on the Great Barrier Reef. In: R.H. Bradbury (ed.) The *Acanthaster* phenomenon: a modelling approach. Springer-Verlag, Berlin.

Seymour, R.M. & Bradbury, R.H. (in ms) Is *Acanthaster* degrading the Great Barrier Reef?

Van der Laan, J.D. & Bradbury, R.H. (in press) Futures for the Great Barrier Reef ecosystem. In: Avula, X.J.R. (ed.) Proc. 7th. Int. Conf. Math. Comput. Mod.

Journal of
Mathematical Biology

Editorial Board:

W. Alt, Bonn
K. P. Hadeler, Tübingen
U. an der Heiden, Witten/ Herdecke
S. A. Levin, Ithaca (Managing Editors)
H. T. Banks, Los Angeles
O. Diekmann, Amsterdam
J. Gani, Santa Barbara
F. C. Hoppensteadt, East Lansing
D. Ludwig, Vancouver
J. D. Murray, Oxford
T. Nagylaki, Chicago
S. Sawyer, St. Louis
L. A. Segel, Rehovot

From a recent issue:

F. Cervantes-Pérez, M. A. Arbib:
Stability and parameter dependency analysis of a facilitation tectal column (FTC) model

L. Edelstein-Keshet, G. B. Ermentrout:
Models for contact-mediated pattern formation: cells that form parallel arrays

M. Notohara:
The coalescent and the genealogical process in geographically structured population

L. F. Murphy, S. J. Smith:
Optimal harvesting of an age-structured population

R. Sridhara:
Inference on system parameters in a model for interacting species

Covered by
Current Contents and
Zentralblatt für Mathematik

Springer-Verlag
Berlin
Heidelberg
New York
London
Paris
Tokyo
Hong Kong
Barcelona

*Subscription information
and other details are available
from the publisher at one
of the given addresses.*

□ Heidelberger Platz 3, W-1000 Berlin 33 / F.R. Germany □ 175 Fifth Ave., New York, NY 10010, USA
□ 8 Alexandra Rd., London SW19 7JZ, England □ 26, rue des Carmes, F-75005 Paris
□ 37-3, Hongo 3-chome, Bunkyo-ku, Tokyo 113, Japan □ Citicorp Centre, Room 1603, 18 Whitfield Road,
Causeway Bay, Hong Kong □ Avinguda Diagonal, 468-4° C, E-08006 Barcelona

Volume 18

S. A. Levin, Cornell University, Ithaca, NY; **T. G. Hallam, L. J. Gross,** University of Tennessee, Knoxville, TN (Eds.)

Applied Mathematical Ecology

1989. XIV, 491 pp. 114 figs. Hardcover DM 98,– ISBN 3-540-19465-7

This book builds on the basic framework developed in the earlier volume – "Mathematical Ecology", edited by T. G. Hallam and S. A. Levin, Springer 1986, which lays out the essentials of the subject. In the present book, the applications of mathematical ecology in ecotoxicology, in resource management, and epidemiology are illustrated in detail. The most important features are the case studies, and the interrelatedness of theory and application. There is no comparable text in the literature so far. The reader of the two-volume set will gain an appreciation of the broad scope of mathematical ecology.

Volume 19

J. D. Murray, Oxford University

Mathematical Biology

1989. XIV, 767 pp. 292 figs. Hardcover DM 98,– ISBN 3-540-19460-6

This textbook gives an in-depth account of the practical use of mathematical modelling in several important and diverse areas in the biomedical sciences. The emphasis is on what is required to solve the real biological problem. The subject matter is drawn, for example, from population biology, reaction kinetics, biological oscillators and switches, Belousov-Zhabotinskii reaction, neural models, spread of epidemics.
The aim of the book is to provide a thorough training in practical mathematical biology and to show how exciting and novel mathematical challenges arise from a genuine interdisciplinary involvement with the biosciences. It also aims to show how mathematics can contribute to biology and how physical scientists must get involved.

Volume 20

J. E. Cohen, Rockefeller University, New York, NY; **F. Briand,** Gland; **C. M. Newman,** University of Arizona, Tucson, AZ

Community Food Webs

Data and Theory

1990. XII, 308 pp. 46 figs. Hardcover DM 148,– ISBN 3-540-51129-6

Contents: I. General Introduction: Food Webs and Community Structure. – **II. Empirical Regularities:** Untangling an Entangled Bank. – **A. General Regularities:** Ratio of Prey to Predators in Community Food Webs. – Community Food Webs have Scale-Invariant Structure. – Trophic Links of Community Food Webs. – Food Webs and the Dimensionality of Trophic Niche Space. – **B. Differential Regularities:** Environmental Control of Food Web Structure. – Environmental Correlates of Food Chain Length. – **III. A Stochastic Theory of Community Food Webs:** Theory: Circles of Complexity, Spherical Horses. – Models and Aggregated Data. – Individual Webs. – Predicted and Observed Lengths of Food Chains. – Theory of Food Chain Lengths in Large Webs. – Intervality and Triangulation in the Trophic Niche Overlap Graph. – **IV. Data on 113 Community Food Webs.** – Bibliography. – Subject Index. – Acknowledgements.

Springer-Verlag
Berlin
Heidelberg
New York
London
Paris
Tokyo
Hong Kong
Barcelona

Lecture Notes in Biomathematics